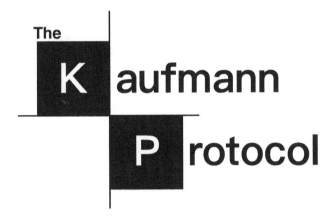

## AGING SOLUTIONS

WRITTEN BY
## DR. SANDRA KAUFMANN

ILLUSTRATIONS BY
**ROSS GOLDSTEIN**

LAYOUT DESIGN BY
**ENRIQUE BOSCH**

COVER PHOTO CREDIT
**LYNN PARKS**

*Legal Notice and Disclaimer*

*The information provided in this book is based upon peer-reviewed and published medical studies. Dr. Kaufmann's analyses and conclusions based on these studies are, however, for informational purposes only and are not meant to replace the guidance and advice of your licensed physician or other healthcare practitioners. Statements in this book have not been evaluated by the Food and Drug Administration for, inter alia, drug interactions, warnings, or alerts.*

*The information provided is meant for general use only and is not intended to diagnose, cure, treat, or prevent any diseases or conditions, or to provide medical advice. Any decision to use the supplements and medicines discussed should be considered in partnership with your licensed physician or other healthcare practitioner.*

*The Kaufmann Protocol*
*Copyright © 2022 Sandra Kaufmann, MD*

*All rights reserved*

*This tome is dedicated to all of those folks who madly scribbled in the margins of the first book.*

AGING SOLUTIONS

| | |
|---|---|
| CHAPTER 1 INTRODUCTION | 7 |
| CHAPTER 2 BONE, BRAIN, GUTS & SKIN | 16 |
| CHAPTER 3 ALPHA-KETOGLUTARATE | 32 |
| CHAPTER 4 ALOE VERA | 46 |
| CHAPTER 5 ANDROGRAPHOLIDE | 56 |
| CHAPTER 6 BERBERINE | 64 |
| CHAPTER 7 BLACK SEED OIL | 75 |
| CHAPTER 8 CENTELLA ASIATICA | 86 |
| CHAPTER 9 CHLOROGENIC ACID | 97 |
| CHAPTER 10 CISTANCHE DESERTICOLA | 106 |
| CHAPTER 11 COLLAGEN | 114 |
| CHAPTER 12 COENZYME Q10 | 129 |
| CHAPTER 13 DELPHINIDIN | 144 |
| CHAPTER 14 ECKLONIA CAVA | 152 |

AGING SOLUTIONS

| | |
|---|---|
| CHAPTER 15 ELLAGIC ACID | 162 |
| CHAPTER 16 FISETIN | 171 |
| CHAPTER 17 GANODERMA LUCIDUM | 179 |
| CHAPTER 18 HYALURONIC ACID | 189 |
| CHAPTER 19 KAEMPFERIA PARVIFLORA | 202 |
| CHAPTER 20 LACTOFERRIN | 211 |
| CHAPTER 21 LEUCINE | 224 |
| CHAPTER 22 MAGNESIUM THREONATE | 233 |
| CHAPTER 23 NARINGENIN | 242 |
| CHAPTER 24 POLYPODIUM LEUCOTOMOS | 250 |
| CHAPTER 25 PYRROLOQUINOLINE QUINONE | 257 |
| CHAPTER 26 SALIDROSIDE | 263 |
| CHAPTER 27 SHILAJIT | 271 |

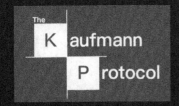

AGING SOLUTIONS

| | |
|---|---|
| CHAPTER 28 SPERMIDINE | 279 |
| CHAPTER 29 URSOLIC ACID | 295 |
| CHAPTER 30 VITAMIN C | 302 |
| CHAPTER 31 THE LAST CHAPTER | 317 |
| CITATIONS | 323 |
| TABLE 1: KAUFMANN RATINGS | 424 |
| TABLE 2: BENEFITS | 425 |
| DOSES | 426 |

**CHAPTER 1**

# INTRODUCTION

*"I'm not afraid of death, but I'm in no hurry to die. I have so much I want to do first."*

*~ Stephen Hawking*

I have come to realize over the past several years that aging means completely different things to different people. Thoughts range from agonizing over wrinkles to the fear of Alzheimer's. To some, aging is the first strand of gray hair or loose skin under the jawline. It might be poor healing after an injury or the presence of a few extra pounds that won't go away. Or it's my personal favorite, the inability to shake a hangover.

Just as the idea of aging varies tremendously, so does the age at which people start to see it in themselves. Up until a decade ago, only the elderly were concerned with the ramifications of aging, cognizant that their time would soon be up. In today's world, however, people seem to be more aware of their mortality and are certainly more proactive in improving their futures. It is not uncommon these days for people in their thirties to initiate life-enhancing strategies.

Attitudes towards aging are also changing dramatically. Most people used to accept the decline with a half-hearted smile, noting that acquiescing to the inevitable was the only option. The folks that attempted to defy this were unfortunately left empty-handed as living longer and better was simply a utopian ideal.

Fortunately, it is a new world; the field of longevity has exploded. Incredibly brilliant scientists and researchers have dissected and examined every nook and cranny of the aging cycle. As a result, there are now a plethora of ingenious solutions. These range from simple things such as caloric restriction diets and alterations in exercise routines to complex treatments such as gene therapy. There are almost too many choices at this point, accompanied by a lot of confusion as to where to start and what to do. The other unfortunate complication is that there are still shysters peddling snake oil, giving the rest of the longevity community a bad rap.

As a side note, the original, actual snake oil from Chinese water snakes had a high concentration of omega-3-fatty acids which is anti-inflammatory. In the states, this was replaced with rattle snake oil, which, unfortunately, does nothing.

Regardless, we now turn to the Kaufmann Protocol. As you might know from book one, the protocol presents an organized, scientific approach to aging that helps us better comprehend why we age. This is coupled with guidelines and information so that individuals can best create an optimal, personalized longevity protocol.

To begin this discussion, understanding exactly why our cells age is vital. This was explained in great detail in the first tome, [some might say painful detail] and aside from new information, it will not be rehashed here. I will refer to these categories and concepts throughout this book, however, so I think a quick review might be advantageous.

## TENET 1
### INFORMATION SYSTEMS OR DNA ALTERATIONS
DNA is the information depot within the cell. Issues in this category include things that negatively affect DNA, such as epigenetic modification, telomeric integrity, and the need for DNA and RNA protection.

## TENET 2
### ENERGY SYSTEMS OR MITOCHONDRIAL HEALTH
Mitochondria are intracellular organelles that serve as our only true energy source, converting sugars to adenosine triphosphate or ATP, the energy currency of the body. Over time, these organelles fail secondary to excessive exposure to free radicals, a decline in the production of endogenous antioxidants, or the limited availability of raw materials.

## TENET 3
### CELLULAR PATHWAYS
Cellular pathways consist of specific genes and their corresponding proteins and enzymes that act in concert to precipitate cellular change. There are

innumerable pathways that directly affect aging, but by far the most important are AMP Kinase, the Sirtuins, and mTOR. Over time, these pathways fail and must be reactivated or deactivated by specific molecules to stabilize cellular homeostasis.

## TENET 4
QUALITY CONTROL

As cells experience stress, both DNA and proteins become damaged and cannot do what they need to do. Thus the body has mechanisms to detect and repair any such issues. Unfortunately, over time, these DNA and protein repair mechanisms slow down or stop altogether. This leads to extensive damage throughout the cell and precipitates tissue and organ failure. In addition to individual molecules, organelles also become dysfunctional, and depending on the specifics, the organelle must be either repaired or recycled. The recycling of these organelles is referred to as autophagy, where the cell disassembles the damaged structure and then reuses whatever is deemed salvageable.

## TENET 5
IMMUNE SYSTEM

The cells that compose the immune system constitute your security system. Over time, as with all else, this system falters, precipitating a state of chronic and systemic inflammation. In addition, the failing immune system leads to an increase in infection rate, a decline in the response to vaccines, and an increased prevalence of leukemias and lymphomas.

## TENET 6
INDIVIDUAL CELL REQUIREMENTS

This category addresses the requirements of individual cells, encompassing very different cell types such as stem cells, brain cells, bone cells, and even senescent cells. As these cells differ in energy and nutrient requirements, lifespan, and niche specifications, their needs will be quite variable.

## TENET 7
WASTE MANAGEMENT

At baseline, every cell requires some form of sugar to function. Unfortunately, these necessary molecules come with detrimental side effects. Glucose and the related sugars bind to lipids, DNA, and proteins throughout the entire body, forming Advanced Glycation Endproducts or AGEs. These complex molecules not only destroy tissue but are highly inflammatory, precipitating even more damage. Lipofuscin falls into this category as well. A byproduct of autophagy, this is a cellular waste product that cannot get recycled. As a result, it accumulates in long-lived cells, taking up space and preventing normal cellular function.

Realizing that I have neatly separated the tenets, they really do not exist in isolation. For example, we can identify a substance as being an epigenetic regulator from Tenet one, but at the same time, this epigenetic alteration specifically controls autophagy, important in Tenet four. We can talk about sirtuin activation in the pathway category, knowing full well that it affects the production of endogenous antioxidants in the mitochondria, Tenet two. Therefore, I like to visualize the tenets as a Venn diagram, where despite treating them as separate categories on one level, we really know that they overlap quite a bit.

Moving along...

The second thing the protocol does is to examine and score the various longevity or molecular agents for every category or tenet. If this is new to you or you have forgotten, every agent has been assigned a seven-digit number, representing how well that agent does in each of the seven tenets.

In brief, an agent is designated either a 0, 1, 2, or 3 in each category.

• Zero is exactly that, zero. If an agent does nothing at all for your DNA in Tenet 1, for example, it gets a zero.

• A score of 1 means there is evidence of efficacy, but minimal evidence and usually only in theory or in a test-tube type condition.

• A score of 2 translates into meaningful results in rodents or any animal that isn't human.

• A score of 3 means there is human evidence that an agent is efficacious in the real world

Therefore, you might see the following: 2.1.3.2.0.2.1.

In this example, the agent does well in Tenet 3 or the cellular aging pathways category, but it does nothing for the immune system in category five.

I realize that there are limitations to this system. I can only evaluate an agent if there is published and appropriate scientific information. This in turn is dependent on the availability of research money, the curiosity of the researcher, and the dissemination of any discoveries. Sometimes it just boils down to the right person asking the right scientific question.

Thus, the rating system might be considered a bit conservative. I'm confident that there are rated agents that should have higher scores, but at present, the evidence is scant or just not available. Therefore, over time, ratings will change - but generally, they will only go up. [It's hard to get rid of evidence that already exists. Unless you work for the government]

I rated fifteen of the best known and readily available agents in the first book and suggested a few protocols based on numerical algorithms and medical conditions. [I would have done more if I thought anyone was actually going to read it]

This book aims to continue the process of examining molecules and offering more choices and more information.

Another thing that I have learned since the first book is that people are very interested in particular organ systems or medical conditions. Therefore, I have added iconography at the beginning of each chapter to make it easier to understand which body systems each agent benefits. See the last page of this chapter for the full list of icons.

As with any endeavor, there will always be caveats.

I probably have not included your favorite adjuvant.

I discarded agents with minimal supporting evidence, have a poor risk/benefit ratio, or are outrageously expensive.

Dose ranges are deliberately vast and nonspecific. Everyone is different, and rarely is there a one-size-fits-all standard recommendation.

Lastly, although I only discuss consumable molecular agents, there are many, many other things one can do to promote longevity. Exercise and diet are the most obvious interventions, and I consider them step one on my ladder of longevity. However, there are plenty of other reputable sources discussing these things, so I'm not going to.

The oral agents we talk about here I consider rung two. These are easy to acquire and consume and are relatively low risk.

Once a daily protocol of oral agents is adopted, we can then move onto the higher rungs, which would include things that one might do less frequently. Peptides, exosomes, stem cell infusions, and even gene therapy fall into this category. There is information about all of these treatments elsewhere, so I am not going to review them. Still, I do need to emphasize that different therapies can undoubtedly be used synergistically. I like to consider the daily protocol as the baseline or basal therapy and the additional treatments as bolus, intermittent interventions. [This is a bit like washing your face every day but getting a real facial every month or so. Could you really do without either?]

Chapter two will review four of the most essential and complex organ systems or at least the ones that people seem most interested in. After that, we plunge into the next batch of amazing and wondrous life-changing molecules.

BONE

 BRAIN

 FAT

 G.I. TRACT

 HAIR

HEART

 JOINTS

 LUNGS

 VIRILITY

 ACCLIMATION

 MUSCLE

 SKIN

 IMMUNITY

# CHAPTER 2

# BONE, BRAIN, GUTS & SKIN

*"Be careful about reading health books. You may die of a misprint."*

*~ Mark Twain*

All organ systems are affected by age, however, some tissues fair better than others over time. Not too many people complain about their aging spleen, for example. There is also extreme variability between individuals, as determined by genetics, lifestyle choices, and luck. That being said, and after listening to patients for many years, the organ systems that attract the most attention are the big four: skeletal, the brain and central nervous system, the gastrointestinal tract, and skin. Apprehension over osteoporosis and the increasing risk of fractures is high on the list of older women. Neurocognitive decline is the main focus for anyone with a family history of dementia. Gastrointestinal failure is more covert, with symptoms ranging from food intolerance to colon cancer. And finally, skin reflects the most apparent external manifestations of cell failure and age-related decline.

Therefore, I think it's beneficial to spend time reviewing key aspects of these systems. Understanding what is supposed to happen can help us understand what fails to occur as we age. More importantly, it guides the direction of proactive, longevity therapies.

## BONE

Bones are fascinating structures and are far more active than we think. They are in a constant state of remodeling, equivalent to living in your house as it undergoes never-ending reconstruction. The process is a lot of work, expense, and inconvenience, but it comes with the benefit of always having a sound foundation and the latest in kitchen backsplashes. Similarly, bone remodeling is metabolically expensive but cost-effective if you want a solid skeleton. It's also paramount when we do something rather stupid or unexpected and break a bone. With enough time, bones repair themselves. Children tend to repair fractures more quickly than adults, but over time, all bones do mend.

Unfortunately, as we get older, the remodeling system doesn't function as well, and we have more deconstruction than construction. This eventually leads to osteoporosis, a disease characterized by low bone mass and micro-architectural deterioration. Presently, about ten million Americans over fifty have osteoporosis, and another thirty-four million are at risk. Skeletal turnover itself is essentially two major processes. First, the old and damaged bone is removed by osteoclasts, followed by new bone formation by osteoblasts. [The B in osteoblasts can be thought of as Builds bone]

In the destruction phase, osteoclasts, multi-nucleated cells derived from the monocyte macrophage cell line, resorb bone by releasing substances that digest proteins. Specifically, hydrogen ions dissolve hydroxyapatite mineral crystals, and hydrolytic enzymes dissolve the organic bone matrix. When these cells complete the resorption process, they undergo apoptosis and disappear.

Following the destruction phase, undifferentiated mesenchymal stem cells are recruited to the site. They proliferate, transform into osteoblasts, and go to work. Osteoblasts are cuboidal cells specialized for the active secretion of the extracellular matrix or ECM and have a relatively short lifespan, estimated at about three months in human bone. These cells first produce collagen and then secrete ground substances that fill in the spaces.

The extracellular matrix, rich in Type I collagen, is known as osteoid when first deposited. Calcium phosphate, in the form of hydroxyapatite, accumulates around the collagen in a process called mineralization.

The osteoblasts then get trapped within the ECM they created. These cells either undergo apoptosis and disappear, or morph into osteocytes, the most abundant cell type in bone. Comprising up to 95% of the cell mass, there are ten times as many osteocytes as osteoblasts and even fewer osteoclasts.
Osteocytes are long-lived cells that are able to survive throughout the life of an individual. They are small, with numerous long, dendritic-like cytoplasmic extensions that form a canal system within the bone. Through this extensive network, osteocytes collect and transmit information regarding the mechanical

environment, the presence of microscopic mechanical fatigue damage, and the extracellular calcium concentration. Using this information, they direct osteoblast and osteoclast differentiation and function.

At the most basic level, there are two types of bone. Cancellous bone is a meshwork of spongy tissue or trabeculae in mature adult bone, typically found in the core of vertebral bones, in the spine, and at the ends of long bones. Cortical bone, meanwhile, is the dense outer surface of bone that forms a protective layer around the internal cavity. Also known as compact bone, it constitutes nearly 80% of skeletal mass. Cortical bone is imperative to body structure and weight-bearing because of its high resistance to bending and torsion.

In the first three decades of life, bone turnover maintains a steady-state between resorption and formation. Peak bone mass and size is achieved around the age of twenty in women and a few years later in men. After reaching this peak, however, bone turnover continues, but at a slower rate, with the breakdown exceeding the speed of reconstruction. As a result, cancellous bone mass starts to decline late in the third decade, while cortical bone begins its descent after fifty.

Bone deterioration manifests as reductions in trabecular and cortical bone density, decreased cortical thickness, a marked increase in cortical porosity, and the loss of trabecular connectivity. This is partially due to the declining number of osteoblasts, attributed to a decrease in the mesenchymal stem cell population, the defective proliferation and differentiation of progenitor cells, and an increase in the apoptotic rate.

The failure of bone is also secondary to age-related changes in bone mineral composition, collagen content, collagen cross-linking, and alterations in water compartments. For example, the non-enzymatically-mediated collagen cross-linking due to Advanced Glycation End-products or AGEs makes bones stiffer, more brittle, and prone to fracture. Water in the bone that is not bound to collagen or mineral increases with age, while the more critical bound water declines. There is also a shift in cell composition. Mesenchymal stem cells that are supposed to become osteoblasts become fat cells instead.

Turning to other cellular challenges, oxidative stress, of course, increases with age and is harmful to bone for two reasons. Oxidative stress triggers osteoclasts, and it decreases the lifespans of osteoblasts and osteocytes. Chronic inflammation, meanwhile, precipitates osteoclast differentiation and bone resorption. In fact, the accelerated bone loss at menopause is linked to increases in the production of proinflammatory cytokines, including Tumor Necrosis Factor (TNF-α), IL-1, and IL-6.

In addition, the loss of estrogen accelerates bone loss, a process that may persist for up to ten years following menopause. Growth hormone production from the anterior pituitary also declines over time, decreasing up to 14% per decade in elderly men and women. Endogenous glucocorticoids actually increase with age, as well as the sensitivity of bone cells to these glucocorticoids. Unfortunately, these steroids are potent inhibitors of bone formation by stimulating osteoblast apoptosis and decreasing osteoblast numbers.

Consequently, the skeleton is subject to innumerable aging insults, resulting in increased destruction and decreased production of bone. Thus, in addition to limiting the standard aging insults such as inflammation, oxidative stress, and AGEs, we will also identify agents that stimulate osteoblasts and reduce or normalize osteoclastic activity. [Sorry, this section wasn't that humerus]

# BRAIN

The brain is the most vulnerable organ affected by aging. It utilizes nearly a quarter of the body's glucose and oxygen stores, leading to significant challenges with reactive oxygen species (ROS) and AGEs. The brain is also composed of fatty acids that are victimized by these radicals, undergoing oxidation and peroxidation, resulting in cell membrane destruction. Making things even more challenging, the brain has a meager reserve of free radial neutralization mechanisms leaving it vastly unprotected. As a result, 15 to 20% of adults over sixty-five develop mild cognitive impairment annually.

Most of the standard aging insults affect brain tissue, leading to alterations in the brain's structure, metabolism, and physiology. Oxidative damage is particularly relevant, as mentioned above, and brain regions with higher levels of reactive oxygen species, such as the hippocampus, also have an increased number of senescent neurons. Another prominent issue is quality control. The accumulation of abnormal protein aggregates or mis-folded, nondegradable proteins and the failure to clear them contributes to the pathogenetic etiology of various neurodegenerative diseases.

Low-level, chronic inflammatory processes are also present in the aging brain, as is lipofuscin deposition. This is especially true in large cortical and thalamic neurons, the inferior olivary nuclei, and spinal motor neurons. Advanced Glycation End-products or AGEs are also detrimental to intracranial arteries making them stiffer, with less control of blood flow.

Examining the brain more closely, in broad terms, there are two cell types: neurons and neuroglia. In the neuroglia category, these cells in the brain, spinal cord, and peripheral nervous system do not produce electrical impulses. This category has several important subtypes, including oligodendrocytes, astrocytes, ependymal cells, and microglia.

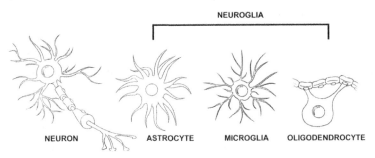

While glia were thought to outnumber neurons by a ratio of 10:1, recent studies suggest an overall ratio of less than 1:1, with substantial variation between different brain tissues.

Oligodendrocytes wrap around the axons of neurons, constituting what is known as the myelin sheath. Like the plastic coating covering an electrical wire, oligodendrocytes insulate the axon and help neurons pass electrical signals at incredible speed and over long distances.

Microglia are the immune cells of the central nervous system. They travel around within the brain and are in constant communication with other glia. If a neuron becomes damaged or infected, these microglia release chemical signals that precipitate an immune response. Over time, unfortunately, these cells produce an overabundance of inflammatory cytokines, creating an exaggerated immune response. They also become less efficient in terms of motility and the response to stimuli.

Astrocytes or astroglia are star-shaped cells, obviously, that surround neurons and support neuron function. They provide biochemical support to endothelial cells that form the blood-brain barrier, bring nutrients to the nervous tissue, maintain the extracellular ion balance, and regulate cerebral blood flow. Aging astrocytes change shape (less star-like and more blob-like), express inflammatory cytokines, and accumulate proteotoxic aggregates. In essence, aging astrocytes resemble and act like senescent cells.

Turning to neurons, there are more than eighty-six billion of them within the cranium. They are primarily post-mitotic cells, meaning they are rarely replaced during an individual's lifetime. Unfortunately, neurons do not age well. They undergo a host of detrimental changes involving synapses, receptors, neurotransmitters, cytological alterations, and electrical transmission.Individual neurons also have reduced dendritic and axonal arborization, meaning intercellular communication is diminished.

Large neurons, having long axons, are more likely to die in response to increased oxidative stress, as are cells in the hippocampus, substantia nigra, amygdala, and frontal cortex. Although the actual loss of neurons is more modest than initial estimates suggested, it affects some areas more than others. The hippocampus and prefrontal cortex, for example, tend to be hardest hit. Even within these areas, some cells are more sensitive, such as the CA1 neurons in the hippocampus. This is unfortunate, as this is where memories are made.

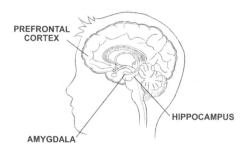

The brain as a whole shrinks a bit as well over time. This atrophy, predominantly cortical thinning or a decrease in gray and white matter volume, is most significant in the prefrontal cortex, temporal lobes, and hippocampus. Overall, the average loss of brain matter is around 0.5% of total brain volume per year.

The frontal lobe, which is the last to mature at around twenty-five years, is also the first to deteriorate. This area plays an essential role in higher-order cognitive processes and a variety of executive functioning processes such as planning, decision-making, problem-solving, and working memory.

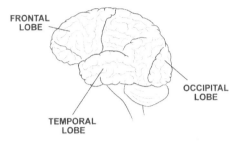

The brain is further divided into gray and white matter. Gray matter is mostly cell bodies, while white matter is the myelinated cell extensions or axons. Unfortunately, white matter suffers significantly over time, from demyelination, atrophy, or lesions to white matter tracts. For example, lesions like leukoaraiosis increase with age. MRI studies have shown their prevalence in over 90% of individuals above the age of sixty-five. The incidence increases not only with age but also with diseases such as diabetes and hypertension, correlating with impaired cognitive function.

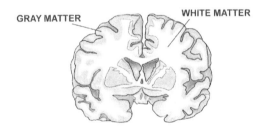

White matter volume gradually increases before the age of forty, peaks around fifty, and then rapidly declines after sixty to seventy years of age. This decline of white matter also shows regional differences, especially in the frontal lobe and the corpus callosum.

Another vital component of the brain is the Blood-Brain Barrier (BBB), a physical barrier that impedes entry from the blood into brain tissue for virtually all molecules, except those that are small and lipophilic, or those that enter the brain through an active transport mechanism.

Composed of pericytes, neurons and astrocyte end feet, the BBB complex surrounds the capillaries that flow through the central nervous system. Due to the unique needs at this barrier, the capillary endothelial cells have an exceptionally high number of mitochondria. With age, of course, this barrier fails. The resulting increase in permeability allows many toxins in, including an influx of inflammatory cytokines, leading to neuronal and glial damage.

## BLOOD-BRAIN BARRIER

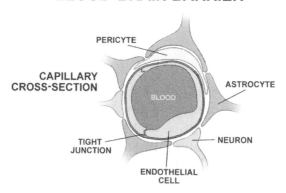

There are innumerable age-related changes in the central nervous system, and certainly too many for this review, but in addition to the other issues, problems of note include neuroendocrine dysfunction and a decline in cerebral spinal fluid or CSF production.

Turning to energy utilization, the brain, only 2% of the body's volume, consumes over 20% of its energy and is particularly vulnerable to changes in metabolism. In fact, the uninterrupted delivery of glucose and oxygen is a well-established requirement for brain homeostasis. Alternative substrates, like ketones, are metabolized by brain tissue under various conditions, but they cannot fully replace glucose or reverse the effects of hypoglycemia.

How much do we use? An adult brain, weighing roughly 1,400 grams, consumes about ninety grams of glucose per day, equivalent to approximately twenty-three teaspoons or cubes of sugar (for simplicity, sucrose is equated with glucose). [For comparison, an average cantaloupe weighs 1,500 grams]

When we think of brain aging, in reality, we are less focused on individual cells and regions, but on how well our brain continues to function. Unfortunately, working memory, long-term memory, and episodic memory all decline with advancing age.

Episodic memory, defined as a type of long-term memory that involves the recollection of a person's own specific experiences, begins to decline during mid-life. On the other hand, semantic memory, i.e., long-term memory that encompasses concepts or facts considered common knowledge, declines much later as individuals become elderly. In addition, mental flexibility and response inhibition show age-dependent decreases as does the ability to focus attention or multi-task.

Perhaps the only good news is that compared with younger adults, older adults react less to adverse situations. They are more able to ignore irrelevant negative stimuli and remember more positive than negative information.

Interestingly, as we age, the female brain has a persistently lower metabolic brain age relative to their chronological age as compared with the male brain.

With an increased awareness of brain aging, it now becomes essential to identify agents that traverse the blood-brain barrier, decrease free radicals and neural inflammation, and address standard cellular challenges. Additionally, these agents need to promote neural plasticity and neuronal development to combat brain aging. [seems like a no-brainer]

## GASTROINTESTINAL TRACT

The gastrointestinal system is crucial to longevity for innumerable reasons, not only for the health of the GI tract but also because it has long-reaching effects all over the body.

In a 2005 U.S. census, at least 50% of people over the age of sixty-five had at least one GI complaint severe enough to precipitate a visit to the physician. I would imagine that significantly more people had symptoms but elected to stay home.

The gastrointestinal system itself can be envisioned as a very long tube with an opening at each end. Food enters at the top, and refuse empties at the bottom. In the middle, we have specialty sections: esophagus, stomach, small and large intestine, and the rectum with attached contributory organs such as the pancreas and liver. Each segment has its complexities and challenges, but suffice it to say that the overall task is to break down complex molecules into smaller pieces, absorb what nutrients are required, and then dispose of the unnecessary items.

Skipping straight to age-related changes, many detrimental alterations occur from top to bottom over time. For starters, taste perception changes, and there is a reduction in the quantity and quality of saliva. Motility decreases, as does enzyme and hormone secretion. In addition, a declining number of nerve cells, dystrophic nerve fibers, and lipofuscin accumulation contribute to enteric neurodegeneration.

In the stomach, there is a decrease in gastric acid secretion called hypochlorhydria, a shift in the microbiota, mucosal protective mechanisms are reduced, and gastric blood flow becomes more limited. There is an overall decrease in pancreatic secretions, especially bicarbonate, lipase, chymotrypsin, and amylase, due to atrophy and fatty degeneration.

Moving distally, the small intestine is about twenty feet long. Secondary to this extensive length, its mucosa is the body's largest surface area that directly contacts the environment and its foreign antigens. As a result, the immune system is particularly critical here, and like all else, it fails over time. This failure is multifactorial, but one standout is the decline in the number and function of dendritic cells, i.e., immune cells that present antigens or foreign proteins to B cells. Under normal conditions, this would precipitate the production of antibodies, especially IgAs, which bind unwanted microbes and antigens, blocking their adherence to the host epithelium and leakage through the gut barrier. The failure of dendritic cells thus leads to a decrease in immunoglobulin A, and increases the risk of infection.

The intestinal mucosa is also a barrier responsible for absorbing vital materials while barring harmful substances. It is composed of specialized epithelial cells joined together by *tight junctions*, where neighboring cell membranes are essentially fused together. In addition, goblet cells produce mucus that coats the inner layer of the intestinal wall, constraining potential invaders.

This loss of this barrier has serious consequences. An influx of bacteria and toxins into the systemic circulation leads to Inflammatory bowel disease, bacteremia, infections, and systemic inflammation. There are even more far-reaching issues, such as increases in fatigue, autoimmune diseases, headache, allergies, dental problems, and depression.

A poorly functioning barrier is also associated with the malabsorption of necessary nutrients. In particular, B-12, calcium carbonate, and ferric iron. Sugar uptake also changes over time as fructose uptake increases in advanced age, whereas glucose uptake declines.

Another much talked about aspect of the GI tract is the microbiota, also referred to as a "second genome." This might seem like a side event, but in fact, it takes center stage because the human superorganism, i.e., our entire body, contains at least the same number of microorganisms (bacteria, fungi, and viruses) as it does its own cells. There are estimated to be trillions of microbes in the human gut due to more than a billion years of mammalian–microbial coevolution, creating an intricate interdependency. [This idea freaks out a lot of folks]

Four phyla, *Firmicutes, Bacteroidetes, Actinobacteria,* and *Proteobacteria,* account for more than 90% of the total microbiota population in an adult. The first two are dominant and represent more than 1,000 different bacterial species. Of note, *Firmicutes* are gram-positive bacteria, while *Bacteroidetes* are gram-negative. There is so much variety in gut microbiota that a person's microbiome is a bit like a fingerprint. Every individual has a specific combination and proportion of different microbial species and subspecies. This is due to a zillion variables, such as diet, medications, geographical region, and the like.

The gut microbiota is also more than just the actual bacteria. The microenvironment consists of genes, proteins, and metabolites that provide energy and nutrients to the host organism. It affects the integrity of the epithelial layer and offers protection from an inflammatory reaction in the GI tract. In addition, the environment influences bone density, lipid storage, and vitamin biosynthesis throughout the entire body.

Over time, there are significant changes in this population; the most important is an overall loss of diversity. At the same time, *Bacteroides, Bifidobacteria, Ruminococcaceae, Lachnospiraceae,* and *Lactobacilli* are proportionally reduced, while the levels of opportunists such as *Enterobacteria, Clostridium perfringens,* and *Clostridium difficile* increase in number.

Another general trend is the escalation in gram-negative bacteria. This is unfortunate as they secrete lipopolysaccharides which act as endotoxins and induce inflammation. The actual ratio of *Firmicutes* and *Bacteroidetes,* the latter being gram-negative, however, is controversial in terms of healthy aging. [Even the gut becomes more negative over time]

Yet another essential function of these bacteria is the production of short-chain fatty acids (SCFAs). These metabolites include acetate, propionate, and butyrate and are produced by the fermentation of fiber-rich diets or indigestible carbohydrates.

Acetate is a substrate for liver lipogenesis and cholesterol synthesis; it helps regulate colonic blood flow and may also prevent liver cancer. Propionate is involved in liver gluconeogenesis, inhibits cholesterol synthesis, and helps regulate adipose deposition. Butyrate, the most studied and beneficial metabolite, is absorbed by colonic epithelial cells, serves as an energy source, and plays an essential role in cell growth and differentiation. As a histone deacetylase inhibitor, it contributes to the prevention of colorectal cancer. Lastly, it is a known anti-inflammatory agent, reducing factors such as IL-6. As could be predicted, the quantity of SCFAs declines with advancing age.

The age-related reduction in gut diversity, as mentioned above, is associated with depression, Alzheimer's, and Parkinson's disease. Known as the brain-gut axis, this is a bidirectional communication; just as the gut affects the brain, brain alterations change the microbiome. The GI bacterial population contributes to the regulation, maturation, and activity of microglia in the brain; it also influences the permeability of the blood-brain barrier. In the other direction, middle cerebral artery occlusion in the brain in mice increases gut permeability.

In patients with Alzheimer's Disease, the intestinal flora has reduced microbial diversity with a reduction in *Firmicutes* and *Bifidobacterium* and an increase in *Bacteroides*. In this population, probiotics improve cognitive function, learning, and memory, and it restores the expression of brain-derived neurotrophic factor. This is thought to act through circulating mediators derived from the microbiota.

As far back as 1903, Nobel Prize winner Elie Metchnikoff suggested that health status could be improved, and senility delayed by replacing the native gut microbes with lactic acid bacteria such as those present in yogurt. This idea has proven to be valid, and presently, there are both prebiotics and probiotics available, which successfully promote a healthy microbiome.

Prebiotics are selectively fermented ingredients that activate specific microbes, such as *Bifidobacterium* and *Lactobacillus*, and are generally comprised of complex carbohydrates such as inulin, galacto-oligosaccharides, fructo-oligosaccharides, manno-oligosaccharides, pecticoligosaccharides, xylo-oligosaccharides, and transgalactosylated-oligosaccharides.

Probiotics are live microorganisms that improve or restore the dysbiosis. Most of the options are based on one or more strains of *Lactobacilli* and *Bifidobacteria*.

From a dietary perspective, the consumption of fiber, fiber-rich diets, or indigestible carbohydrates promotes the production of SCFAs by microbes as well. In fact, a high-fiber diet enriches the abundance of *Bacteroidetes* and decreases the abundance of *Firmicutes*. [I know, its a lot to digest]

## SKIN

On the most basic level, skin can be thought of as a multi-layered birthday cake. [That's why we call it your birthday suit]

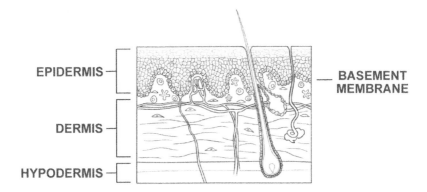

The outermost layer is the epidermis, followed by the dermis below. In between is a thin layer of frosting known as the basement membrane. Beneath all of this is a layer of subcutaneous fat, but that ruins my analogy, so let's return to the metaphor and visualize any protruding hairs as birthday candles.

The epidermis, or the part exposed to the outside world, is itself a layered composite of cells stuck together by sticky and waterproof proteins.

All cells in the epidermis begin at the basement membrane, created from keratinocyte stem cells. As these cells journey towards the skin's surface, they lose their organelles and cytoplasm and become flattened. Glued together with lipid-rich proteins, these terminally differentiated keratinocytes create a protective barrier called the stratum corneum that makes us waterproof. Visually, we can imagine the stratum as a pile of bricks held together with mortar. The bricks, however, are not uniform but become progressively flatter as they approach the surface.

There are many intriguing things about the epidermis, and one of them is the absolute absence of any blood vessels. Therefore, any agents targeting these cells must move by diffusion either from below or from above.

The dermis is a much more active layer that provides strength and elasticity to the skin. The bulk of the cells here are fibroblasts, which create the extracellular matrix, composed of collagen and elastin, as well as proteoglycans and glycosaminoglycans, like hyaluronic acid and dermatan sulfate. This layer is also home to the assortment of necessary accouterments: blood vessels, nerves, sweat glands, and the lymphatics.

The barrier between the two layers, or the frosting in our analogy, is a specialized basement membrane. This undulating stratum keeps the two layers firmly adhered and controls the transport of substances. This barrier, at least in the younger years, vacillates up and down, such that if you looked at the cutout piece of cake, the line would resemble a sine wave. In 3D, it would look more like a piece of foam egg crate. Unfortunately, this stratum flattens with age. Consequently, not only is nutrient transfer limited, but the connection between the epidermis and dermis is weakened, and the skin tears more easily from shear forces.

Aging has adverse effects, and we all know that skin is no exception. Keratinocyte stem cells decline in number; as a result, epidermal cell turnover decreases 50% between the ages of 30 to 70. The keratinocytes also become progressively abnormal, with increasing numbers of senescent cells. Fewer working keratinocytes translate into a failed epidermis with fewer bricks and less mortar.

Other cells malfunction as well. Melanocytes, cells that produce pigment at the base of the epidermis, can be overactive and form lentigines or *liver spots*, or they can be inactive and produce no color, leaving hypo-pigmented spots.

The dermis, meanwhile, suffers from even more remarkable age-related changes; it becomes thinner, declining in thickness 10-50% between the ages of 30 and 80.

Fibroblasts dwindle in number and functionality, leading to a depletion in collagen, elastin, and proteoglycans. As a consequence, skin recoil declines, there is less structural support, and reduced hydration. Increased, abnormal collagen cross-linking from Advanced Glycation Endproducts or AGEs also precipitates increased stiffness and structural weakness.

There is decreased subcutaneous fat below the dermis, which is ironic as there is more fat everywhere else.

In addition to intrinsic skin aging, skin is also the victim of extraneous insults in the form of radiation, smoke, and environmental stressors. Aging precipitated from solar radiation even has its own name, i.e., photoaging, coined in 1986, and contributes up to 80% of extrinsic aging. From a cellular perspective, UV radiation shortens telomeres, melts DNA, depletes endogenous antioxidants, inactivates sirtuins, precipitates collagen breakdown, inhibits collagen synthesis, increases inflammatory processes, and promotes AGE production. In fact, UVA/UVB radiation generates hydrogen peroxide in the skin within fifteen minutes of exposure.

This exposure risk has increased significantly in the last fifty years due to stratospheric ozone depletion. Topical sunscreens have also led to a false sense of security while outdoor leisure activities have skyrocketed.

In addition to the sun, skin is traumatized by innumerable insults, leading to superficial cuts, abrasions, burns, and the like. To repair this damage, the skin has enzymes, matrix metalloproteinases (MMPs), that dissolve damaged areas to make way for new tissue. Unfortunately, these enzymes get triggered too frequently and destroy tissue unnecessarily.

As we now know, both time and environmental stress damages skin with unfortunate consequences. The good news is that there are many things we can do to reverse these changes.

**CHAPTER 3**

# ALPHA-KETOGLUTARATE

(3.2.1.1.2.2.0)

BENEFITS:

BONE    FAT    G.I. TRACT    IMMUNITY    MUSCLE

> *"In terms of antioxidative function, AKG exhibits a crucial role in multiple diseases involved in aging, cancer, cardiovascular diseases, and neurological diseases."*
>
> ~ S. Liu 2018

Alpha-Ketoglutarate, a.k.a. AKG, is a popular supplement driven by bodybuilders seeking more ways to bulk up. But AKG has a plethora of abilities beyond just muscle production; the animal industry utilizes it to improve not only the growth and performance of animals, but also to boost immunity and bone development. In addition, some hospitals use it to improve patient recovery after physical trauma and surgery.

Only recently however, has alpha-ketoglutarate begun to be valued for its potential role in improving general health and promoting longevity. To understand why AKG is so helpful, we need to start at the most basic level. Thus, examining the cellular landscape, we can immediately see that AKG is a very active molecule.

• It's an essential part of the Krebs cycle, which is an intricate pathway in the mitochondria.

• It provides building blocks for proteins.

• It participates in enzymatic control over important epigenetic mechanisms.

Alpha-ketoglutarate goes by several names, depending on the scientific nomenclature. These include 2-ketoglutaric acid, 2-oxoglutamate, 2-oxoglutaric acid, oxoglutaric acid, and 2-oxopentanedioic acid. In reality, this is only important if you feel like perusing the ingredient list on the back of supplement bottles.

Jumping to the obvious question of longevity, AKG has demonstrated extensions in the lifetimes of flies, worms, and mice.[9] More about this later- [Of note, these older mice were rather muscular and sexy]

The List...

## 1) DNA ALTERATIONS: 3

Not exactly a direct epigenetic modifier, alpha-ketoglutarate is more of an indirect participant. It helps to control some of the enzymes that can then directly influence epigenetic chromatin modifications. For example, within the 2-oxoglutarate-dependent dioxygenase (2-OGDD) superfamily, there are two enzyme families which require α-ketoglutarate in order to demethylate various substrates.[16]

Ten–eleven translocation (TET 1–3) hydroxylases, for example, fall into this category and participate in DNA demethylation.[7] [Vitamin C is also important for the TET Enzymes]

In terms of histone epigenetics, alpha-ketoglutarate is a required substrate of the Jumonji C domain, containing lysine demethylases. One key function of these enzymes is to catalyze the removal of histone methyl groups (specifically H3K9me3 and H3K27me3) and activate the expression of important developmental genes during stem cell differentiation.[11]

It turns out that the ability to catalyze DNA and histone demethylation is crucial to stem cell and progenitor cells. These cells contain minimal amounts of alpha-ketoglutarate while having a high demand for catalyzing DNA demethylation and other reactions. Thus, AKG becomes a rate-limiting factor for stem/progenitor cell differentiation. As animals age and the supply gets reduced, the ability of stem cells and progenitor cells begins to fail as well.

*"In summary, through affecting histone and DNA demethylation, aKG is a critical mediator of cell differentiation and tissue homeostasis. While its level declines during aging, dietary supplementation of AKG increases intracellular aKG level, which can rejuvenate stem/progenitor cell function during aging, maintaining the health of adipose tissue."*[11]

## 2) MITOCHONDRIA: 2

Before embarking on this section, first we must take a slight detour to better understand energy production in the mitochondria. In addition to the Electron Transport Chain (ETC), there is a very important metabolic pathway called the Krebs Cycle, or the TCA (tricarboxylic acid) cycle.

This enzymatic process is a cyclic pathway of eight reactions. In essence, it starts with Acetyl-Co-A within the matrix of mitochondria, creates $CO_2$, reduces coenzymes (NADH and $FADH_2$), and forms the backbone of many proteins. The reduced coenzymes feed electrons into the ETC to generate ATP.

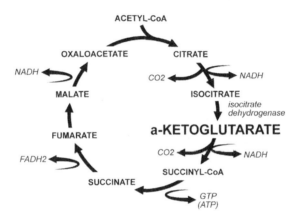

Specific to this conversation, alpha-ketoglutarate sits between isocitrate and succinyl-CoA and is considered one of the key intermediates. More specifically, AKG is generated from isocitrate by oxidative decarboxylation, catalyzed by isocitrate dehydrogenase. AKG is further decarboxylated to succinyl-CoA and $CO_2$ by AKG dehydrogenase.

Interestingly, the amount of AKG produced in mitochondria depends on the oxidative status. If there is more NAD+ than NADH available, the pathway gets pushed toward oxidative decarboxylation of AKG and the formation of succinyl-CoA. In the opposite direction, more NADH and a shortage of NAD+ drives the reductive transamination of alpha-ketoglutarate and, with the participation of glutamate dehydrogenase, produces glutamate.[16]

This production of amino acids, especially glutamate, is important but will be addressed a bit later.

The take-home message here, since I'm pretty sure you just skimmed over that part, is that alpha-ketoglutarate is essential for energy production within the mitochondria. [I probably should have just stated that in the first place]

In an odd, not entirely understood twist of fate, AKG can also block energy production. Recently, it has been discovered that alpha-ketoglutarate is an ATPase inhibitor. In fact, it binds to the beta subunit of the ATP synthase molecule within the ETC, reducing ATP production. This also decreases oxygen consumption and increases autophagy, which has been credited with some of its life-enhancing properties. A complete block is, of course, quite detrimental, and in fact will kill you, but a partial block can be beneficial under certain circumstances.[13]

Moving on, in the antioxidant category, alpha-ketoglutarate also does well. First off, AKG is an antioxidant all by itself.

Second, it enhances anti-oxidative enzymes, especially superoxide dismutase[4] and non-enzymatic agents that decrease oxidative stress, lipid peroxidation[16] and cellular levels of hydrogen peroxide.[4]

Lastly, AKG increases intracellular levels of glutamate, which is a precursor of glutathione.

## 3) PATHWAYS: 1

Following the recognition that alpha-ketoglutarate could inhibit ATP synthase thus precipitating a reduction in energy levels, it was also determined that this activated AMP Kinase.

This was subsequently tied to an increase in SIRT1, at least in flies.[9] At the same time, it was found that this corresponded to an inhibition of mTOR in certain models.[9,13] This would be a remarkable finding, as an inhibited mTOR pathway frequently correlates with an improved lifespan. However, AKG, as we have seen, is rarely straightforward.

For example, in several cell cultures and animal muscle studies, AKG activates the Akt/mTOR pathway, and in the process, builds muscle.[1,12] Similar findings were discovered in human osteoblast cell cultures where alpha-ketoglutarate also increased mTOR. This was reinforced by the finding that rapamycin, a potent mTOR inhibitor, abolished the AKG-promoted osteoblast differentiation.[17] Therefore, it is difficult to say what the effects of AKG on the mTOR pathway really are. It may be cell-specific, tissue-specific, or even niche-specific.

## 4) QUALITY CONTROL: 1

In this category, there is no evidence of any DNA or protein repair system activation, but there is evidence hinting towards an increase in autophagy.[13]

## 5) IMMUNE SYSTEM: 2

Before discussing the effects on the immune system, it's time for another detour. Alpha-ketoglutarate exhibits almost all of its positive effects on the immune system through the production of glutamine. Thus, we need to explore how one can turn into the other.

In terms of reversible pathways, alpha-ketoglutarate is first converted to glutamate, which can then be altered to become glutamine.

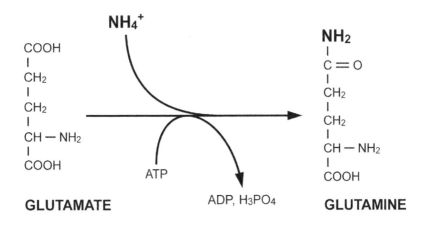

Glutamate can also be converted to Gamma-aminobutyric acid (GABA), which, in this case, is a one-way street.

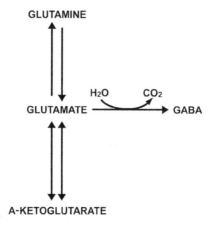

Glutamine, a five-carbon molecule, differs from glutamate, also a five-carbon molecule, in that it has an extra nitrogen group. (This is easy to remember as glutamine has an N in it) Traditionally considered non-essential, glutamine becomes essential during catabolic states and cellular stress. Thus, some folks have deemed it semi-essential.

A healthy adult contains over eighty grams of free glutamine, with greater than 98% of it inside skeletal muscle cells. Why do we need so much? Because glutamine serves many functions in and around the body. It is a precursor for a number of biosynthetic pathways, and serves as an important signaling molecule. Glutamine stimulates anabolic functions such as protein synthesis, cell growth, and differentiation, and inhibits catabolic processes such as protein degradation and apoptosis.

Glutamine is also converted to glutamate, which serves as a substrate for hepatic gluconeogenesis, i.e. making new glucose, a major fuel source for intestinal cells and cells of the immune system. The remaining glutamate is a precursor to essential substances like glutathione, proline, ornithine, and arginine.

In the central nervous system, glutamate is the most prominent excitatory neurotransmitter. In another odd twist, glutamate is also used to synthesize GABA as we saw above, the main inhibitory neurotransmitter in the brain.

So clearly, glutamine and glutamate have far-reaching effects around the body, all linked to AKG.

This then leads to where we were going in the first place.

Alpha-ketoglutarate is a critical nutrient for the immune system, for the most part, because it provides the glutamine necessary for many types of white blood cells.

Glutamine modulates the function of monocytes and neutrophils by increasing phagocytosis and the reactive oxygen species within these cells, making them more capable of destroying pathogens. In addition, lymphocytes utilize glutamine both as an energy source and as a building block for glutathione production.

During severe inflammatory states, glutamine levels become depleted, demonstrating the crucial role they play under stress. Supporting this, several studies have shown that depleted glutamine levels depress the immune system, while supplemental glutamine improves immune function.[16]

Thus, AKG as a glutamine source, has immuno-enhancing properties: it increases the number of immune cells, increases the activity of neutrophils and phagocytosis, and reduces bacterial translocation in the gut.[13] Also, in the form of ornithine alpha-ketoglutarate, it has been shown to partially counteract the involution of the thymus gland, at least in burn-injury rat models.[16]

## 6) INDIVIDUAL CELL REQUIREMENTS: 2

There is reasonable evidence that alpha-ketoglutarate and its derivatives positively influence bone.

Scientists have confirmed the presence of glutamate receptors on osteoblasts and osteoclasts.[13] Additionally, there is evidence that dietary AKG counteracts bone loss in rats with ovariectomy-induced osteopenia.[13]

In fact, in many rodent models, alpha-ketoglutarate treatments have demonstrated a significant increase in bone mineral density, bone mineral content, cross-sectional area, mean relative wall thickness, maximum elastic strength, and overall strength of bones.[13]

Moving up the evolutionary chain, pigs treated with alpha-ketoglutarate demonstrate improved bone mineral density, greater elastic strength and significant increase in overall strength.[13]

In baby sheep, two weeks of treatment improved the trabecular bone mineral density, cortical bone mineral density, and the maximum elastic femur strength.[13]

In human osteoblast cell lines, AKG did not increase the proliferation of osteoblasts; but it up-regulated the expression of transcription factors and proteins involved in osteoblast activation and differentiation. Interestingly, in this study, mTOR was activated and rapamycin abolished the AKG-driven osteoblast differentiation.[17]

And finally, in a study of menopausal women, alpha-ketoglutarate for twenty-four weeks inhibited bone resorption and reduced the effects of osteopenia, especially in the lumbar spine. Researchers concluded that AKG could not only inhibit bone resorption but could also induce reconstruction of bone tissue.[13]

Thus, it appears that whereas we generally do not want to activate mTOR, in this case, it does have a positive effect.

### MUSCLE
Bodybuilders are already aware of this, but now you know it as well: alpha-ketoglutarate helps to build muscle and prevent its breakdown.

Descending back down the evolutionary ladder, given to mice, AKG increased the gastrocnemius muscle weight and fiber diameter. Interestingly, these effects were purely due to AKG, independent of glutamate. The study also demonstrated that this was dependent on the activation of the mTOR pathway.[1]

## WASTE MANAGEMENT: 0
Not a lot to report here.

## OTHER INTERESTING THINGS OF NOTE
AKG has a few other cellular tricks up it sleeve in addition to the seven tenets. One of these is the regulation of several enzymes.

Remember from the epigenetics section that alpha-ketoglutarate is an obligatory co-substrate for 2-oxoglutarate-dependent dioxygenases (2-OGDDs). In humans, as there are more than sixty different 2-OGDDs, the effects are far-reaching. In addition to demethylation, some of these enzymes affect the hypoxic response, fatty acid metabolism, and nucleic acid repair.[6]

# SYSTEMS KNOWN TO BENEFIT
## COLLAGEN PRODUCTION
As it turns out, alpha-ketoglutarate is involved in the production of collagen through several mechanisms.

1) AKG is a cofactor for the enzyme prolyl-4-hydroxylase (as is vitamin C). Located within the endoplasmic reticulum, this catalyzes the formation of 4-hydroxyproline, crucial for collagen triple helix formation.

2) AKG contributes to the pool of proline residues via glutamate. About 25% of dietary AKG is converted to proline in the enterocytes. As proline is a primary substrate for collagen synthesis, this is crucial.[13]

These findings were confirmed in human culture studies, where AKG stimulated the synthesis of procollagen.[16]

## ADIPOSE TISSUE
Another cool effect of AKG is the epigenetic demethylation in fat cells.

As a quick review, white adipose tissue is primarily used in the body for energy storage while brown adipose tissue is utilized for thermogenesis. More specifically, brown fat increases energy expenditure by dissipating chemical energy as heat through an uncoupling of the electron transport chain. More simply, brown fat converts calories into heat, and can reduce obesity.

Once upon a time, it was thought that brown fat disappeared way before adulthood, and in fact, the capacity of brown/beige adipogenesis does decline during the aging process. As well, brown fat converts into white fat as we age. Luckily, we now know that brown fat does not entirely disappear, and we know how to manipulate its existence.

The genetic transcription factor governing the production of brown fat is the protein PRDM16 (PR domain containing 16) from the gene PRDM16. It turns out that AKG is a rate-limiting factor for the demethylation of this gene, promoting the development of brown adipose tissue.[15]

On a practical level, when the formation of cold-induced, brown fat was impeded by a high-fat diet in middle-aged mice, AKG rescued the situation. As a result, these mice gained brown fat and became less obese.[11]

The take-home message: AKG will make your body warmer, but thinner. On a personal note, I can attest to this. I started the supplement, and within a few weeks, without appreciating that this would happen - was feeling a bit warmer than usual. This would have been helpful if I lived in Alaska, but in Miami, I was forced to adjust the air-conditioning.

## GASTROINTESTINAL TRACT

The gut is more complex than most people imagine (but you know this from chapter two); it helps balance our health through an intricate interaction between nutrients, the microbiota, and the host intestinal epithelium.

### ENTEROCYTES

Alpha-ketoglutarate is quite active in the cells of our intestinal walls. It is an energy source[4] and provides amino acids. As a result of supplemental AKG in animal studies, the cellular composition of the intestinal wall improves, increasing the ratio of villus height to crypt depth, consequently increasing absorption capacity.

### INTESTINAL IMMUNITY

Alpha-ketoglutarate also improves the innate immunity of the intestine in mouse models. Specifically, "it promoted ileal expression of mammalian defensins of the alpha subfamily (such as cryptdins-1, cryptdins-4, and cryptdins-5). A-defensins are antimicrobial peptides expressed by goblet cells and have important roles in mucosal defense."[2]

### ANTI-INFLAMMATORY

There is preliminary evidence that AKG has potent inhibitory effects on the NF-$k\beta$ inflammatory pathway in the gut.[3]

### MICROBIOTA COMPOSITION

AKG influences the intestinal microbial composition, in particular by increasing the *Bacteroidetes* to *Firmicutes* ratio.

## HORMONAL EFFECTS

It also turns out that AKG supplements can significantly increase circulating levels of insulin, growth hormone, and insulin-like growth factor-1, which is thought to be a direct result of increased levels of glutamine and glutamate. These amino acids can be transformed into ornithine and then arginine, which both stimulate the secretion of growth hormone and insulin-like growth factor.[13]

## CYANIDE TOXICITY

Interestingly, in case you are a Russian spy on the outs with the KGB, you will be happy to hear that AKG is considered a natural antidote for cyanide poisoning.

AKG binds with cyanide (hydrocyanic acid) to produce cyanohydrin, further preventing cyanide poisoning or cyanide lethality.[4]

## AGING AND LONGEVITY

So, does alpha-ketoglutarate actually elongate life? So far, it seems to in flies, worms and mice. In flies, AKG increased the average and maximum lifespan by 8% and 15%.[9]

In worms, AKG extended the lifespan by about 50%. They also demonstrated delays in age-related phenotypes, such as the decline in rapid, coordinated body movement.[13]

And in mice, Asadi Shahmirzadi et al. reported that late-onset dietary supplementation with calcium alpha-ketoglutarate resulted in increased survival and reduced fraility. They also demonstrated a reduction in systemic inflammatory cytokines.[6]

And in people? We don't know.

We do know that our natural levels of AKG dwindle with age. And as we have now seen, the direct effects of this deficiency may be far-reaching in terms of general health and well-being.

The question then is whether or not oral supplementation actually helps; the answer seems to be yes. [or we wouldn't be talking about it...] We know that much of any supplemented alpha-ketoglutarate is used by enterocytes as it gets absorbed. Still, an appreciable amount is available for use by extra-intestinal tissues as well[12], as roughly 20% of dietary AKG appears in the bloodstream. It is, however, removed very quickly, with a half-life of less than five minutes.

After this discussion, one frequently asked question is why not just take glutamine or glutamate as they are all closely related and interchangeable.

A good idea in theory, it just doesn't work very well. Glutamine, as it turns out, is not stable in aqueous solutions and only works in intravenous form.

Even worse, glutamate is not routinely used because of its relative neurotoxicity and poor permeability across cell membranes.[16]

Ok, so AKG it is. But there are many forms of AKG out there.

There is straight AKG, or it's complexed to accessory components. The most popular options are ornithine, arginine, sodium, or calcium.

So, which to take? The choice depends on personal goals. For example, ornthine AKG consists of two molecules of ornithine and one molecule of AKG. The combination is thought to be more efficient than AKG or ornithine alone in restoring glutamine pools in muscles.

Meanwhile, arginine alpha-ketoglutarate provides arginine in addition to AKG. Arginine is considered a conditionally essential amino acid, meaning that under normal conditions, endogenous synthesis by the human body is sufficient. But it is also the substrate for nitric oxide, which helps relax and open blood vessels; thus, the amino acid can reduce blood pressure. And lastly, arginine is a known AGE inhibitor.

*"Several reports showed that Arg inhibits in vitro non-enzymic glycation and advanced glycation end product formation in serum and tissue proteins from human subjects and animals. The proposed mechanism was that the amino group of Arg can block an early stage of the Maillard reaction to form non-reactive substituted Amadori products, and with its guanidinium group, Arg may react with dicarbonyls (Amadori breakdown products) which results in fewer advanced glycation end products."*[4b]

# DOSE

How much to take? Again this is goal-dependent.

The range is between 300 to 1,000 mg daily, although bodybuilders tend to take a bit more, even up to nine grams a day.

For longevity purposes, somewhere between 500 to 1,000 mg a day should be sufficient.

**CHAPTER 4**

# ALOE VERA

(1.3.0.0.3.2.2)

BENEFITS:

BONE  FAT  SKIN

*"The plant leaves contain numerous vitamins, minerals, enzymes, amino acids, natural sugars and other bioactive compounds with emollient, purgative, antimicrobial, antiinflammatory, antioxidant, aphrodisiac, anti-helmenthic, anti-fungal, antiseptic and cosmetic values for health care."*

~ A.Jadhav, 2020

Aloe comes to the modern age with the most impressive pedigree that I have ever encountered. It is the absolute representation of understated elegance, long-term success, and even better, universal availability.

At least on the written record, aloe's roots date back to ancient Egypt circa 4000 BC, where it was *The Sanctuary Plant of Immortality,* while descriptions of aloe are inscribed in cuneiform on Sumerian tablets from around 2100 BC.[32] Nefertiti, considered the most beautiful queen in history, used it as a part of her daily skin and beauty care. It was also part of Cleopatra's skincare line, with perhaps a little less success; despite winning over both Caesar and Mark Antony, she was not that attractive.

Alexander the Great understood the importance of the plant and, under advisement from Aristotle, captured the island of Socotra in the Indian ocean to raid the aloe groves. In doing so, Alexander acquired sufficient medication to heal the wounds of his battalions. He even had carts with potted plants brought along near the battlefields to treat the less successful of the warriors.

During the reign of Emperor Nero around 50 BC, Dioscorides, the prominent physician of the era, decided that aloe was one of his favorites. He recommended using aloe juice for numerous physical disorders, including the treatment of wounds, gastrointestinal discomforts, gingivitis, aches and pains, skin irritation, sunburn, and even acne.

Christopher Columbus was known to have *Aloe vera* growing in pots on his armada of ships, without which, things may have been very different.

The Maya christened the juice of this desert plant the *Fountain of Youth*.

During the Crusades, the Knights of Templar imbibed a concoction of palm wine, aloe pulp, and hemp, named *The Elixir of Jerusalem*. They believed that this would add years to their health and life.

A favorite of King Solomon's, aloe is frequently mentioned in the bible, and some say it was used to anoint the body of Jesus.

In the modern era, the first original, processed product was less sexy. Aloin, a component of the plant's yellow sap, was produced as a laxative. The aloe gel, meanwhile, having no laxative properties, was introduced in the 60s, and was incorporated into foods, drinks, topical ointments and over-the-counter oral health aids.

The plant itself is a perennial succulent xerophyte (meaning it needs very little water) and not a cactus, as some people think. It is native to southern and eastern Africa along the upper Nile in Sudan and was subsequently introduced to Northern Africa and then into the Mediterranean. From there, it has spread globally. Presently, the most extensive *Aloe vera* plantations can be found in Barbados and in northern California.

Our beloved aloe is of the species *Aloe Barbadensis Miller*, belonging to the family *Xanthorrhoeaceae*. The aloe genus in its entirety is actually 581 species, with *Aloe Vera* the star of the show.

The plant itself is roughly 99% water, with the remaining percent divided between a plethora of substances. Depending on the source, there are between seventy-five to two hundred active components.

The most important, however, seem to be the following:

**FLAVONOIDS:** Naringenin

**ANTHRAQUINONES:** Aloe-emodin, Aloin, Emodin

**CARBOHYDRATES:** Acemannan, Glucomannan, Hyaluronic acid

**ENZYMES:** Alkaline Phosphatase, Amylase, Catalase, Lipase, Superoxide dismutase, Peroxidase

**(7 OUT OF 8 ESSENTIAL) AMINO ACIDS:** Hydroxyproline, Isoleucine, Leucine, Valine, Lysine, Methionine, Phenylalanine, Threonine, Tyrosine

The exact composition of aloe, of course varies by many factors; the soil and growing conditions, the harvest time, climate, the position of leaves on the stem, and the method used for harvesting leaves. Of note, the optimal time to harvest aloe leaves is after three years, when it has the highest content of polysaccharides and flavonoids.[19]

The List...

## 1) DNA: 1
Acemannan, one of aloe's prevalent carbohydrates, is fermented by bacteria in the gut to become butyrate, a natural histone deacetylase inhibitor (HDACi). [2, 56]

## 2) MITOCHONDRIA: 3
There is no doubt *Aloe vera* is chocked full of antioxidants. These include polyphenols, flavonoids, tannins, α-tocopherol (vitamin E), carotenoids, and ascorbic acid to name a few. [22, 35, 38]

Aloe also contains superoxide dismutase and can pass this on orally as well.[19]

As noted above, the three-year-old plants have the highest antioxidant levels, and the rind has more capacity than the inner gel. [21, 22]

It turns out that the additional antioxidant capacity helps us in more ways than we might think. For example, coating fresh raspberries with aloe gel triggers the berry to make more antioxidants, including glutathione peroxidase, glutathione reductase, superoxide dismutase, and ascorbate peroxidase. This serves to increase the shelf life of raspberries, as well as being more advantageous to the diet. It even works for strawberries.[17, 45]

Aloe also triggers increases in endogenous antioxidant production in rats. In rodents inflicted with either cardiac ischemia or diabetes, oral aloe demonstrated the ability to either increase or return to baseline levels of endogenous antioxidants such as superoxide dismutase, catalase, glutathione peroxidase, and glutathione reductase.[16, 40]

## 3) PATHWAYS: 0

There is very little research in this arena, but there is one article worth mentioning. Researchers examined kidney fibrosis and found that aloe-emodin could improve cell survival by suppressing the PI3K/Akt/mTOR pathway in cell cultures. Thus, there is a hint, but little else to go on in this category.[12]

## 5) IMMUNE SYSTEM: 3

Aloe has been put through the rigors of various cell cultures[5, 30, 55] and innumerable abused little rodents[3, 36, 61] with essentially the same outcome. There is no doubt that aloe and its components reduce inflammation at many levels. It reduces TNF-a, NF-κB, $PGE_2$, and several Interleukins (IL-6, IL-8, IL-1β).

In humans, there are very few studies. One of which was a double-blinded, randomized, placebo-controlled trial looking at *Aloe vera* gel as a treatment for mild to moderate ulcerative colitis. After four weeks of oral therapy, patients reported significant clinical improvement.[54]

In terms of immunity, a key study explored the effect of the aqueous extract of aloe on humoral and cell-mediated immunity parameters. It was discovered that the plant, at a dose of 400 mg/kg, significantly enhanced the secondary humoral immune response.

It has also "been shown to activate macrophages and stimulate T cells that are involved in the defense against harmful microorganisms."[56]

## 6) INDIVIDUAL CELL REQUIREMENTS: 2

Several studies demonstrate that aloe improves bone. Specifically, acemannan induces bone formation by stimulating bone marrow stem cell proliferation, its differentiation into osteoblasts, and then extracellular matrix synthesis.[4,23]

In a guinea pig dentition model, aloe increased the number of osteoblasts and the levels of collagen while decreasing osteoclast activity.[29]

## 7) WASTE MANAGEMENT: 2

Historically, aloe has been used for the treatment of diabetes. To evaluate this premise, there have been countless studies in both rodents[33] and people with mixed results.

Human studies have included non-randomized trials, randomized trials, and varying doses of aloe and its components. Whereas these trials have presented conflicting results,[37,46] "the preponderance of evidence suggests a trend towards the benefit of oral *Aloe vera* use in reducing fasting blood glucose concentration and HbA1c."[54]

The second, perhaps more intriguing aspect of aloe is that there is evidence that specific plant components can inhibit the formation of Advanced Glycation Endproducts (AGEs). For example, in test tubes, *Aloe vera* extracts can significantly inhibit albumin glycosylation.[20]

During cooking, "the anthraquinones in A. vera, can effectively inhibit AGE formation and reduce protein crosslinking."[27]

Lastly, "A. vera extract or aloin offered protection against glucose or MG induced glycation of SOD. A. vera extract has earlier been reported to have anti-diabetic effects, and this along with its antiglycating effect as has been observed in this study, makes it an effective product against diabetes and its complications."[60]

## SYSTEMS KNOWN TO BENEFIT

### SKIN

*Aloe vera* is simply incredible for the skin, both orally and topically. At the most basic level, aloe has positive effects on the cellular components. In the dermis, fibroblasts are activated by aloe, increasing the production of collagen and hyaluronic acid.[38]

At least two components of aloe, glucomannan and acemannan, are responsible for stimulating the production of collagen,[18,19,38] and glucomannan is known to influence fibroblast growth factor receptors specifically.[18,19]

In cell cultures of human dermal fibroblasts, aloe gel not only increased the collagen content of the wound but also altered the collagen composition with more type III represented.[44]

In rodents, there are countless studies examining wound healing that demonstrate aloe's positive effects on collagen production, both orally and topically.[47,48] In human studies, Japanese women over forty in a randomized, double-blinded, placebo-controlled trial demonstrated that *Aloe vera* gel powder increased skin collagen and hyaluronic acid production. Researchers noted an increase in skin hydration at eight weeks, with improved wrinkle depth. As a bonus, the percent of body fat after eight weeks was also significantly improved.[49]

Another important component of skin health is keeping the Matrix Metalloproteinases (MMP's) at bay.

MMP's are enzymes that degrade all kinds of extracellular matrix proteins. They are useful when tissues need to be dissolved prior to reconstruction, such as in wound healing. But when MMPs are activated inappropriately, they damage critical components such as collagen, elastin, and hyaluronic acid. For example, when inflamed by UV radiation, cells release immense quantities of MMPs, with consequent tissue destruction. Therefore, inhibitors of these MMPs or reductions in their production can be very useful for tissue preservation.

Thus, in a mouse model where MMP expression was triggered by UVB radiation, oral aloe successfully prevented these increases in MMP-2, -9, and -13 expression.[43]

In human peripheral blood mononuclear cells, aloe showed significant inhibition in the activity of MMP-9.[53]

Moving to humans, thirty healthy female subjects over the age of forty-five took either 1,200 or 3,600 mg per day of *Aloe vera* gel for ninety days. In addition to improvement in facial wrinkles and facial elasticity, MMP-1 mRNA levels were significantly decreased in the higher dose group.[8]

In line with the ancient knowledge of Alexander the Great, both topical and oral aloe are efficacious in wound healing.[37, 51] In addition to standard wounds, it has been used for the treatment of burns[41], frostbite, inflammation, infections and pressure ulcers.[18, 34]

In terms of overall skin health, we are going to turn to two human studies.

In a three-month, randomized, double-blinded, placebo-controlled study, oral aloe was evaluated in sixty-four healthy women, ranging in age from 30 to 59. Results demonstrated significant improvements in skin moisture, transepidermal water loss, skin elasticity, and collagen scores.[50]

Lastly, in a 2020 double-blinded, randomized, placebo-controlled supplementation study lasting three months, scientists reported significant improvements in skin acne and fingernail brittleness and, in addition.... less constipation.[25]

## ADIPOSE TISSUE

In trying to determine if *Aloe vera* was beneficial as a diabetic therapy, it became apparent that it precipitated weight loss.[9, 37, 51] These studies were in overweight to obese diabetics, but in reality, those are the folks who would benefit the most anyway.

Under the microscope, aloe seems to reduce the average size of adipocytes [51], and in rodents, total adipose accumulations are reduced.[59] How this occurs is not clear at this point.[59]

## LIPID PROFILES

A second discovery in the diabetic studies was the realization that aloe could positively affect the lipid profile.

In innumerable rodent studies[33], oral aloe improves many markers. In a model with prematurely aged rodents, 1.0% aloe marked decreased 15% and 20% of triglyceride and cholesterol levels, respectively, and also significantly reduced LDL-cholesterol levels 15%.[10]

This trend was repeated in humans[11] and is thought to be due to the inhibition of pancreatic lipase. By blocking the breakdown in the gastrointestinal tract, there is less systemic absorption of fat into the body, accounting for the improved lipid panels. [51]

## GASTROINTESTINAL TRACT

For better or worse, depending on the state of your bowels, the anthraquinones are quite potent laxatives. This can lead to a bit of unpleasantness, abdominal spasms, and even metabolic acidosis if it gets out of hand. Thus, In 2002, the US Food and Drug Administration declared that aloe extracts as laxatives were not generally recognized as safe.

## POTENTIAL SIDE EFFECTS

The anthraquinones, and more specifically, aloin, have demonstrated some unfortunate side effects, including hepatitis [54], and the risk that aloe-emodin might be a carcinogen.[13] Ingestion of the entire plant is associated with diarrhea, hypokalemia, pseudomelanosis coli, kidney failure, phototoxicity, and hypersensitive reactions.[13] These too are thought to be primarily associated with the anthraquinones.

Thus, the International Aloe Science Council established the industry standard for oral consumption to less than ten parts per million of aloin. [31] [Who knew that Aloe had its own international club?]

Along this vein, the recommendations for drinking aloe juice should not exceed 30 to 40 cc per day.[19]

Topical doses are considered safe if the anthraquinone concentration is below 50 ppm.[31]

The good news is that most processed aloe does not contain sap and therefore has no anthraquinones. As a result, this information becomes a bit less worrisome.

*"The possible benefits of aloe vera use in certain pathologies described in the literature should be weighed against the potential toxic effects of this ingredient, whose activity depends on the purity and characteristics of the product utilized."* [54]

## DOSE

Topically, It's hard to get enough. It's great for the treatment of sunburns, and really, any kind of skin damage.

It's also great in the your bath water. Add either a few mashed up aloe pieces or a few cups of aloe juice to the bathwater, and soak. It's delightful. [Candles, wine and Netflix help too]

I also have a concoction that I slather on my skin twice a day that consists of roughly 60% aloe juice, plus hyaluronic acid, white tea extract, centella, and some peptides. Despite being labelled 'swamp juice' by friends and family, it does wonders on the skin!

In terms of an oral dose, there are numerous suggestions and no absolute answers. 100 to 200 mg is recommended for constipation. Diabetic treatment seems to be more aggressive, up to one to two grams daily. Unfortunately, if you read the labels on commercially available capsules, options range from 25 mg to 10,000 mg, which is truly confusing. The best advice… aim for somewhere between 300 to 600 mg a day.

# CHAPTER 5

# ANDROGRAPHOLIDE

(0.3.1.1.3.1.2)

BENEFITS:

BONE  BRAIN  G.I. TRACT

*"The natural product andrographolide, isolated from the plant Andrographis paniculata, shows a plethora of biological activities, including anti-tumor, anti-bacterial, anti-inflammation, anti-virus, anti-fibrosis, anti-obesity, immunomodulatory and hypoglycemic activities."*

*~ H. Zhang 2021*

Without success, I have hunted for a rationale to explain the name of this agent. It suggests that a man (andro) has written (graph) something. The end of the word makes the most sense; -olide, coming from the Latin olidus, means to smell, particularly an unpleasant smell.

Perhaps this was an attempt for some guy to document the horrible odor of this agent? Who's to know. But I can attest to its notably bitter and atrocious taste.

Onto the facts...
*Andrographis paniculata*, a herbaceous plant belonging to the family *Acanthaceae*, is a traditional folk medicine for the management of inflammation, arthritis, and the treatment of sore throat, flu, and upper respiratory tract infections.

Found throughout India, China, Malaysia and other southeast Asian countries, it tends to live in tropical and subtropical regions.

In India, it's a predominant constituent of at least twenty-six Ayurvedic formulations and has been used to treat tumors since ancient times.

Because of the wide geographical spread, it not surprisingly goes by innumerable aliases. The most common title is *The King of Bitters* [which I find a little distasteful]

In India, it's Kalmegh; in China: Chuan-Xin-Lian; Thailand: Fah Tha Lai; Japan: Senshinren; and in Scandinavian countries, it is known as green chiretta.[8]

Regardless, andrographolide (*14-deoxy-11,12-didehydroandrographolide*) is a labdane diterpenoid, a molecule we have not yet seen, which is mostly found in the leaves.

Isolated in 1951, it is extremely....extremely bitter. [Have I mentioned this already?]

Only in one country is it considered a food, i.e. Cambodia, where the dried root is macerated in alcohol and consumed as a way to stimulate the appetite. Either the alcohol is doing the trick here, or you are forced to eat something right after its ingestion to cover up the horrible taste.

**ANDROGRAPHOLIDE**

In addition, the plant contains several bioactive compounds including other diterpene lactones, diterpene glucoside and flavonoids.

We will see other terpenes in this book, and those too are bitter.

The List...

## 1) DNA ALTERATIONS: 0

There are hints suggestive of epigenetic activity, but nothing too conclusive as of yet. In a study of mouse skeletal muscle regeneration, andrographolide treatment enhanced myotube generation and promoted myotube fusion. The authors "observed that andrographolide treatment significantly promoted histone modification, such as H3K4Me2, H3K4Me3 and H3K36Me2, both in vivo and in vitro." Unfortunately, there are no confirmatory studies.[26]

## 2) MITOCHONDRIA: 3

Andrographolide is a primary free radical scavenger and anti-oxidant, comparable to ascorbic acid.[16,30]

*"The extract of Andrographis paniculata was found to inhibit the formation of oxygen-derived free radicals such as superoxide (32%), hydroxyl radicals (80%), lipid peroxidation (80%), and nitric oxide (42.8%) in an in vitro model."* [16]

In mice, studies demonstrated the same effect, significantly inhibiting superoxide (32.4%) and nitric oxide (65.3%) formation.[16]

Andrographolide also enhances endogenous production of anti-oxidants through the activation of Nrf2.[6,22]

In fact, andrographolide induces nuclear factor-erythroid factor 2-related factor 2 ( aka Nrf2) expression and its translocation to the cell nucleus [25] independently of the cell type studied. This Nrf2 induction is precipitated by the regulation of the Keap-1 inhibitor, via interaction with Cys151. This mechanism is similar to that of sulforaphane, another of my favorite agents in the war against aging.[16]

Through this pathway, the agent increases catalase, superoxide dismutase, glutathione-S-transferase, and glutathione content. [8,16]

Andrographolide is also great for your mitochondria in other ways; for example, in mitochondria isolated from rat brains after treatment, it increased the activity of electron transport complexes I, II, and III.[16]

## 3) PATHWAYS: 1

There is budding evidence of a link to AMP Kinase.

In a mouse model of acute colitis, andrographolide was found to activate the AMP Kinase pathway in LPS-induced macrophages, helping to alleviate symptoms.[9] As well, treatment activated AMP Kinase in rat hippocampal neuronal cultures.[5]

## 4) QUALITY CONTROL: 1
There is some, however slim, evidence that andrographolide can trigger autophagy.[4] (stay tuned for updates)

## 5) IMMUNE SYSTEM: 3
The king of bitters is truly royal in this category, being a very potent anti-inflammatory. [12, 13, 18]

*"Transcription factor NF-κβ is a master regulator of the proinflammatory gene expression program and since it was suppressed by andrographolide, andrographolide is rightly termed a regulator of the master regulator."* [15]

We even know that andrographolide inhibits NF-κβ activation through the modification of a reduced cysteine (62) of the p50 protein. More specifically, andrographolide forms a covalent adduct with the cysteine, blocking the binding of NF-κβ to nuclear proteins and preventing any downstream cytokine expression.[8]

Moving to rodents, a mouse model of acute lung injury demonstrated improved lung outcomes with treatment. Pulmonary inflammation was decreased, as was pulmonary edema. The ultrastructural changes of type II alveolar epithelial cells were minimized, as were the number of neutrophils and macrophages, in addition to the cytokines TNF-α, IL-6 and IL-1β.[32] This trend continues in several rodent models of mice arthritis.[7, 10]

In people, we have a plethora of historical evidence that the agent has been used for millennia to treat all sorts of inflammatory conditions.

A question that frequently pops up in the discussion of anti-inflammatories is which one is best and how many does one need? I wish I had a firm answer here, but there are very few, if any, studies addressing these questions.

I will, however, make a few suggestions based on the following bits of information. Many anti-inflammatories work through different mechanisms, and inhibition of the inflammatory cascade through several, varied targeted points might prove to be more effective than any single therapy. Different agents have different solubilities, for example. Fat soluble versus water soluble molecules are going to get into different tissues and thus, both types should be useful. As well, only particular agents can get through the blood-brain barrier and into the brain, thus in order to treat neural inflammation, this must be considered.

Taking several agents in the same category also means that you can take less of each one individually, and therefore limit any potential side effects.

## 6) INDIVIDUAL CELL REQUIREMENTS: 2

Muscle, bone, and nerve cells all benefit in this category. Information on bone and brain will be presented in the following section. Meanwhile, in terms of muscle, andrographolide promotes mouse skeletal muscle regeneration; specifically, it enhances myotube generation and promotes myotube fusion.[26]

## 7) WASTE MANAGEMENT: 2

There is good evidence in this category that andrographolide can reduce blood glucose levels. This has been shown in non-diabetic rabbits[8] and both diabetic and non-diabetic rats. [8,17]

Of significance, an aqueous extract administered to diabetic rats resulted in a 52.9% reduction in blood glucose levels, while the dry powder of the plant decreased blood glucose levels by 61.8%.[8] This works through several mechanisms. First, it inhibits alpha-glucosidase and alpha-amylase, reducing the absorption of glucose from the intestine.[8,17,20] Next, it increases glucose utilization by increasing GLUT-4, a glucose transporter, thus moving more glucose into cells.[17] And lastly, it has been shown to stimulate insulin release,[17] possibly by promoting the development and maturation of pancreatic β-cells. [31]
[No good human studies here either]

## SYSTEMS KNOWN TO BENEFIT

### BONE

As mentioned above, andrographolide appears to be beneficial to bones. For example, in various models, it suppresses osteoclast differentiation and bone resorption and reduces the expression of osteoclast-specific markers, thought to be through the mechanism of NF-κβ suppression.[29]

As this cause and effect has become a common theme, clearly osteoporosis seems to be associated with chronic inflammation as we age.

## GASTROINTESTINAL TRACT

In the gastrointestinal tract, andrographolide does several instrumental things. It enhances intestinal barrier integrity through the up-regulation of tight junction proteins, which reduces the leak of toxins into the body. It also modifies gut microbiota composition as indicated by an elevated *Bacteroidetes* to *Firmicutes* ratio and enriches the population of *Akkermansia muciniphila*. It also increases the levels of short-chain fatty acids.[20]

## BRAIN

The bitter king deserves an ovation here...

Able to penetrate the blood-brain barrier[14], andrographolide does what we already know it to do: it acts as a free radical scavenger, it up-regulates endogenous anti-oxidants, and it acts as an anti-inflammatory. This is helpful in innumerable situations. For example, in models of Alzheimer's disease, the agent not only reduces Aβ aggregation but it suppresses the neuroinflammatory response.[14]

Beyond the normal expectations, however, andrographolide has proven to be valuable to the brain. In a plethora of models, the agent induces the proliferation and generation of new neurons in the adult hippocampus of mice.[11,14,23]

As a competitive inhibitor of glycogen synthase kinase-3β (GSK-3β), a key enzyme of the Wnt/ β-catenin signaling cascade that regulates neurogenesis in the adult hippocampus, andrographolide acts by increasing neural progenitor cell proliferation and the number of immature neurons in the hippocampus, at least in mice.[23]

It has also been shown in a mouse model of Alzheimer's disease to increase cell proliferation and the density of immature neurons in the dentate gyrus.[23]

Continuing the list of benefits, it ameliorates middle cerebral artery occlusion and reperfusion-induced brain infarct damage and improves neurological deficits in ischemic rodents.[14] It also recovers spatial memory functions that correlate with protecting synaptic plasticity and synaptic proteins in Alzheimer's disease models.[19] And in a different study with a similar Alzheimer's disease model, treatment with andrographolide prevented cognitive decline.[3,4] There is also evidence that andrographolide can help alleviate depression[3,4,24] and improve cognitive performance in various rodent models.[2,24]

Again, there is a paucity of human information.

## DOSE

As per any good thing, there are limitations. Andrographolide has two issues. One, it is insoluble in water, resulting in limited bioavailability after oral administration. However, once in the body, the half-life is quite long at 10.5 hours.[16]

Two, even in capsule form, it tastes terrible. [it is always the last pill on my pile to swallow] The good news, however, is that its toxicity is very low, even at high doses. In terms of dose, the range is quite broad. Studies range from 60 mg to 1,500 mg and can even go as high as 6 grams.[8,14] The most reasonable dose recommendation is probably between 400 to 800 mg per day.

## POTENTIAL SIDE EFFECTS

Potential side effects include the loss of appetite, diarrhea, vomiting, rash, headache, runny nose, allergic reaction, and fatigue. High doses may cause swollen lymph glands and elevations of liver enzymes.

# CHAPTER 6

# BERBERINE

(0.2.3.0.3.0.3)

BENEFITS:

BRAIN　　FAT　　G.I. TRACT　　HEART

*"Berberine shows significant activities such as antimicrobial (bacterial, fungal, protozoans, viral, helminthes), antidiarrhoeal, antitumor and apoptosis, anticarcinogenic, immunomodulatory, antihyperglycaemic, antioxidant, hepatoprotective, cardiovascular and several miscellaneous biological functions associated with human healthcare."*

*~ B. Singh 2020*

The biggest question surrounding berberine is always the same; is it an adequate substitute for Metformin? I promise, we will get to that shortly.

But first, what is berberine? Berberine is a natural isoquinoline alkaloid, with the formal name: *5,6-dihydro-9,10-dimethoxybenzo[g]-1,3-benzodioxolo[5,6-a] quinolizinium.*

It is a large molecule with five rings ($C_{20}H_{18}NO_4$) and a molar weight of 336.36 g/mol.

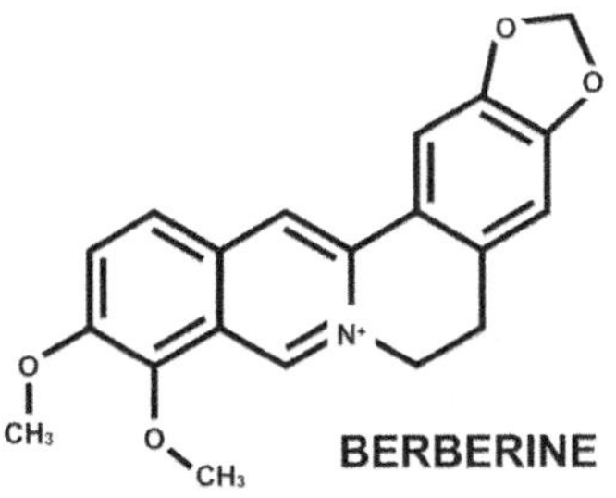

**BERBERINE**

Its most elemental form, an intense yellow powder, is odorless with a characteristically alkaloidal, bitter taste.

The history of berberine is quite long, and as the compound exists in many plant species, it has been popular worldwide.

In the *Berberidaceae* family alone, the genus *Berberis* comprises about 450 to 500 species, representing the primary source of berberine. It is present in other plant species as well, but in much lower concentrations.

Historically, berberine can be traced back to Chinese and Ayurvedic medicine around 3000 BC. However, the first known written use of the barberry fruit *(Berberis vulgaris)* as a blood purifying agent is on clay tablets from the library of the Assyrian emperor Asurbanipal during the 650 BCs.

In traditional Arabic or Persian medicine, healers prescribe *Berberis asiatica* to treat asthma, eye infections, jaundice, skin pigmentation, toothache, general inflammation, and swelling.

Even in southern South America, the leaves and bark of the genus *Berberis* are used in traditional medicine for mountain sickness, infections, and fever. Conveniently, it also serves as an antidote to scorpion stings and snakebites.

Despite being around for thousands of years, berberine wasn't isolated until 1917 from goldenseal or *Hydrastis canadensis*, a North American herb of the buttercup family.

The List...

## 1) DNA ALTERATIONS: 0

Despite evidence that berberine has epigenetic effects on cancer cells, there is little to no information in other, non-cancerous cells.

The only study available utilized reverse-docking techniques and identified berberine as a ligand for lysine-N-methyltransferase. This suggests that berberine may play a role in epigenetics, but there is nothing definite.[25]

## 2) MITOCHONDRIA: 2

In this category, berberine possesses antioxidant capabilities, thought to be equivalent to vitamin C.[16,18]

Under test tube conditions, berberine has direct antioxidant activity in several cell-free systems, such as the DPPH radical scavenging assay.[8]

In cell studies, berberine quenches reactive oxygen species (ROS) and reactive nitrogen species (RNS) by exerting radical scavenging activity against the highly reactive peroxynitrites and hydroxyl radicals. In an in vitro system, berberine scavenged ONOO- and its precursors, nitric oxide and superoxide anion.[28]

Berberine also demonstrates beneficial effects on endogenous antioxidants. In cultured dorsal root ganglion neurons isolated from diabetic rats, berberine augmented the nuclear factor-erythroid factor 2-related factor 2 (Nrf2) mediated endogenous antioxidant defense systems, decreasing neuronal damage and neuroinflammation.[30]

Given to diabetic rats, berberine increased superoxide dismutase activity and decreased levels of malondialdehyde, a marker of lipid peroxidation.[18]

Remaining in the mitochondrial category, berberine blocks complex I of the electron transport chain. It accumulates on the inner membrane of mitochondria, and by inhibiting the ETC, it increases the AMP/ATP ratio and activates AMP Kinase.[8]

Berberine also triggers mitochondrial biogenesis. In mitochondrial dysfunction induced by a high-fat and hyperglycemic diet in skeletal muscle, berberine reverses the dysfunction by increasing biogenesis.[9] This is true for cardiomyocytes as well. Berberine significantly ameliorated mitochondrial function by promoting both mitogenesis and mitophagy. In addition, it improved the status of high glucose-induced cardiomyocyte injury by activating AMP Kinase and stimulating mitochondrial biogenesis.[10]

In cultured hippocampal cells, berberine increased axonal mitochondrial density and length and improved mitochondrial motility.[33]

Lastly, in this category, berberine enhances the uncoupling protein expression

(UCP2), which in turn inhibits oxidative stress. In addition, temporarily uncoupling the electron transport chain makes the mitochondria less efficient. This is detrimental for athletes, but it does help with weight loss.[28]

The only downside in the category is when there is too much berberine. In cultures, excess berberine triggers an increase in oxidative stress, mitochondrial swelling, structural collapse of the mitochondria, and ultimately apoptosis.[29]

The question, however, is how much is too much? [Assume this is a rhetorical question - I don't have an answer]

## 3) PATHWAYS: 3

Before plunging in here, as the briefest of reviews, remember that the cell uses ATP or adenosine triphosphate as its energy source. The energy is trapped within the bonds that hold the three phosphates onto the molecule. Therefore, ATP holds a lot of energy. The loss of one phosphate leads to ADP or adenosine diphosphate, and the loss of two leads to adenosine monophosphate. The enzyme AMP-activated protein kinase, or AMP Kinase, gets turned on when the cell senses elevated levels of AMP and determines that the cell has no energy. The activation of this enzyme promotes catabolic activities and reduces anabolic ones in an attempt to use less energy and harvest it from available sources. This process is well known to be associated with improvements in longevity.

OK, back to business…

There is a plethora of evidence that berberine activates AMP Kinase in many cell types, including endothelium, smooth muscle, cardiomyocytes, cancer cells, pancreatic β-cells, hepatocytes, macrophages, and adipocytes.[8,11,17,20,28]

I could say more, but is it necessary?

There is also evidence that berberine inhibits mTORC1, specifically by suppressing the phosphorylation of S6K at Thr 389 and S6 at Ser 240/244 [17], or phosphorylation of rpS6 on Ser235/236.[28] [I love useless details!]

In MIA-PaCa-2 cells, low dose berberine effects were dependent on AMP Kinase, while higher doses inhibited mTORC1, which was independent of AMP Kinase.[17]

In cardiac fibroblasts, berberine was also shown to down-regulate the mTOR/p70S6K signaling pathway.[2]

In a rat model of cardiac hypertrophy, induced by transverse aorta constriction, treatment with berberine significantly decreased cardiac hypertrophy and fibrosis compared to the control group. This was thought to be due to the inhibition of "mammalian target of rapamycin (mTOR) signaling-related protein expressions, including phospho-mTOR, phospho-4EBP1, and phospho-p70 S6K (Thr389), but not phospho-p70 S6K (Ser371)." [3] [Feel free to gloss over all of these details - but I think the details are pretty cool]

There is also clear evidence that berberine can increase SIRT1, which appears to be completely separate from AMP Kinase. [4, 8, 9, 20]

In a hepatic cell line exposed to hydrogen peroxide, berberine precipitated a significant up-regulation of SIRT1, increasing cell viability and reducing apoptosis.[34]

The berberine effects on mitochondria are partly mediated by its activation of SIRT3. This mitochondrial-based sirtuin is a deacetylase that regulates proteins involved in apoptosis, autophagy, and metabolism. [22, 32]

As compared to Metformin:
*"BBR and MET recruit both direct (as an antioxidant) and indirect mechanisms (Sirt3 content) to deal with arsenic trioxide toxicity. Metformin compared with BBR exhibited a less significant effect on ROS levels and since its direct antioxidant property is minor, depressed the ROS level mainly through the Sirt3 modification."* [12]

## 4) QUALITY CONTROL: 0
There is mixed information in this category. Some studies suggest that berberine can inhibit autophagy [8], while others say it is induced.[6]

The jury is still out.

## 5) IMMUNE SYSTEM: 3
In innumerable models, there is evidence that berberine can act as an anti-inflammatory agent in many tissues around the body. The abridged version: Berberine inhibits IL-6 [8,16,17,36], IL-8 [23,36], IL-1β [8,16], IL-13[36], NF-κβ[8], TNF-a [8,16,17,36], Monocyte chemo-attractant protein 1[16,17], COX-2[16,18], CRP[16], MMP-2 and MMP-9.[16,17,18]

## 7) WASTE MANAGEMENT: 3

The first study examining the anti-diabetic effects in animals was in 1986. Since then, scientists have ascertained that these properties are very similar to those of Metformin. In Type II diabetic patients, various clinical studies have described the safety and effectiveness of berberine (within a dose range of 0.2–10 g/day), reporting a decrease in blood glucose levels by 20–40% with berberine alone. The results are consistent with those of Rosiglitazone and Metformin.[16]

Moreover, a synergistic and hypoglycemic effect was observed in Italian Type II diabetic patients when treated with berberine in combination with sulfonylureas or Metformin.[16]

The improved insulin sensitivity and the decline in blood glucose levels have been attributed to gut-microbiota modulation, pancreatic β-cell regulation, activated AMP Kinase, reduced gluconeogenesis, suppressed mitochondrial functions, and up-regulation of insulin receptors.[8,16]

In addition, berberine inhibits α-amylase and α-glucosidase, enzymes present in the small intestine responsible for the breakdown of complex oligosaccharides and disaccharides into glucose and further monosaccharides suitable for absorption.[8]

*"Modern pharmacological effects of BBR on glucose metabolism are summarized, such as improving insulin resistance, promoting insulin secretion, inhibiting gluconeogenesis in liver, stimulating glycolysis in peripheral tissue cells, modulating gut microbiota, reducing intestinal absorption of glucose, and regulating lipid metabolism."*[19]

Finally, berberine and Metformin were randomly assigned to thirty-six patients for three months in a head-to-head trial. The hypoglycemic effect of both agents were found to be comparable.[23]

As a bonus, berberine, in a study using 1,500 mg/day in Type II diabetics, not only improved the HbA1c comparable to that of Metformin (also 1500 mg/day), but it improved the lipid profile, while Metformin did not.[18]

Moving on to other body systems…

# SYSTEMS KNOWN TO BENEFIT

## BRAIN

The good news here is that berberine can cross the blood-brain barrier [26], and is thought to offer protection in the central nervous system from diseases such as Alzheimer's, cerebral ischemia, depression, schizophrenia, and anxiety. [6]

This occurs through the processes already discussed, as well as a few others. First, it can regulate brain neurotransmitters, especially biogenic amines. Both acute and chronic administration of berberine at low doses results in increased norepinephrine, serotonin, and dopamine in whole-brain, cultured samples. [6] In full-body mice models, both the acute and chronic administration of berberine precipitated an increase in norepinephrine, serotonin, and dopamine levels as well. [6]

It is worth noting, however, that berberine regulates biogenic amines in a concentration dependent manner. At low doses, berberine increases levels. In contrast, high doses of berberine decrease the biogenic amine concentrations. So, again, as noted in the mitochondrial category, too much berberine may be detrimental. [6]

Other effects include both inhibitory as well as stimulating actions. Berberine inhibits acetylcholinesterase and monoamine oxidase, accounting for the impact on neurotransmitters. [28] In the activation category, berberine potentiates nerve growth factor activity, which can modulate synaptic function and plasticity in the CNS.

Tying all of these effects together, treated adult male rats demonstrated improved spatial recognition, improved performance in novel object recognition tasks, and prevented learning and memory dysfunction in passive avoidance tasks. In addition, berberine lowered the hippocampal activity of acetylcholinesterase, malondialdehyde, caspase 3, and reduced DNA fragmentation. At the same time, it augmented the levels of glutathione peroxidase, superoxide dismutase, catalase, and glutathione. It also attenuated inflammation-related indices, as was evident by lower levels of nuclear factor-kappa B, tumor necrosis factor, and interleukin-6. [21]

And lastly, in another mouse study, rodents were less depressed on chronic berberine treatment. This effect was attributed to the inhibition of the NF-κβ pathway in the hippocampus. [6]

## CARDIOVASCULAR DISEASE

There is limited evidence that berberine acts as a vasodilator and lowers blood pressure, at least in rats and guinea pigs.[8]

## LIPID PANEL

There is, however, significant evidence that berberine can improve lipid profiles in humans. In 2004, Kong et al. reported that the oral administration of berberine in patients with hypercholesterolemia for three months could reduce total cholesterol by 29%, triglycerides by 35%, and low-density cholesterol by 25%.[27]

In elderly, hypercholesterolemic patients who were statin-intolerant, berberine ameliorated hypercholesterolemia and plasma LDL-c levels. [23]

In a novel mouse combination study of berberine with resveratrol, mice had more significant lipid reductions on the combined therapy than either mono therapy.

Specifically, in hyperlipidemic mice, the combination of berberine and resveratrol reduced total serum cholesterol by 27.4% and low-density lipoprotein-cholesterol by 31.6%, which was more effective than that of the resveratrol or berberine alone.[35]

## ADIPOSE TISSUE

To start, berberine inhibits the proliferation and differentiation of adipocytes. Then, at least in free-standing adipocytes, treatment with berberine decreases the size and number of lipid droplets. And in mice, treatment reduces the size of adipocytes, acting through a down-regulation of adipogenesis as well as lipogenesis.

## GASTROINTESTINAL TRACT

It is well known that berberine modulates gut microbiota, increasing the relative ratio of bacteria that produce short-chain fatty acids. This is important as there are many known benefits of increasing levels of SCFAs, including protecting the mucosa from pathogen-incited damage, supplying nutrients to the intestinal cells, and mitigating inflammation.[8]

We also know that decreases in SCFA-producing bacteria are associated with diabetes, colorectal cancer, and getting older in an ungraceful manner.

Both berberine and Metformin modulate gut microbiota, with the resulting enrichment of SCFA-producing bacteria.[31]

There are some differences, however. For example, *Akkermensia* was increased in Metformin-treated rats, but not in berberine-treated ones. As well, Metformin showed superior effects on some well-known beneficial bacteria, such as *Prevotella* and *Lactobacillus*.[31]

## LIFESPAN

We have no evidence that berberine prolongs life in people, but it does in *Drosophila melanogaster*. In addition, these fancy flies have improved locomotor activity.

## BIOAVAILABILITY

One of the drawbacks of berberine is its poor bioavailability, starting with absorption in the gut. Berberine tends to self-aggregate, reducing its solubility in the gastrointestinal tract and its ability to permeate the gut wall. [23]

Folks are busy working on ways to increase absorption and bioavailability, such as adding P-glycoprotein inhibitors and improving molecular packaging. At the moment, though, absorption remains limited.

*The P-glycoprotein is located in the epithelial cell membrane and can efflux many drugs (including BBR), thereby limiting their oral bioavailability.*

The good news is that whatever drug remains in the gut can interact positively with the microbiota.[31] The rest enters the systemic vasculature, and consequently gets metabolized quickly in the liver, and is excreted through bile, feces, and urine.[8]

Despite the low plasma concentrations, the tissue concentrations of berberine and its metabolites tend to be higher and get distributed to the liver, kidney, muscle, lung, brain, heart, pancreas, brain and adipose tissue.[8]

## POTENTIAL SIDE EFFECTS

Whereas clinically berberine is thought to be relatively safe, it does come with a few caveats. Standard doses of berberine are usually well-tolerated, and adverse reactions are rare. In contrast, high doses have been associated with low blood pressure, difficulty breathing, flu-like symptoms, cardiac damage, and gastrointestinal discomfort. By far, much like Metformin, the most common side effect is GI-related, reported in 34.5% of patients (on a dose of 500 mg three times daily).[6]

Berberine also interacts with other common medications. For example, it interferes with macrolides, a class of antibiotics that includes erythromycin, roxithromycin, azithromycin, and clarithromycin, potentially leading to dangerous cardiac arrhythmias.

In addition, the risk of cardiac toxicity is increased when berberine is combined with statins.[8]

These issues arise from the effect of berberine on liver enzymes, as high doses can inhibit certain types of CYPs, such as CYP3A11 and CYP3A25. As these metabolize other drugs, their inhibition can lead to overdosing of the medications.[8]

Also recall that high doses of berberine can cause damage to the mitochondria.

## DOSE

Doses generally used in human studies range between 500 mg to 2,000 mg daily, but these studies have been limited to several months or less.[18,23]

But by far, the most common dose is 500 mg two times a day.

The question that remains, however, is should we use berberine or Metformin?

To address this, sixty people with Type II diabetes were divided into groups taking Metformin, berberine, or both. Interestingly, the combination group took home the gold.[23]

Therefore, the answer is not simple. Metformin and berberine are similar but not the same. In terms of diabetic measures, they appear to be equal, but berberine is poorly absorbed with side effects. They are metabolized differently and use different cellular transporter systems. Metformin has a much longer half-life, but berberine has more impact on the lipid profile.

Thus, the real solution may be a combination approach.

# CHAPTER 7
# BLACK SEED OIL
## (1.2.2.0.3.0.3)

BENEFITS:

BRAIN — FAT — HEART — SKIN — VIRILITY

"The pleiotropic pharmacological effects of black cumin, and its main bioactive component thymoquinone (TQ), have been manifested by their ability to attenuate oxidative stress and inflammation, and to promote immunity, cell survival, and energy metabolism, which underlie diverse health benefits, including protection against metabolic, cardiovascular, digestive, hepatic, renal, respiratory, reproductive, and neurological disorders, cancer, and so on."

~ M. Hannan 2021

Black seed oil, or more formally, *Nigella sativa,* is considered one of the most treasured nutrient-rich herbs in all of history and around the world. It has been called many things, ranging from the *Panacea* to the *Seeds of blessing.*

The earliest appearance on the historical record is from residue found in a flask from a Hittite temple, dating about 1650 BC. Of note, the Hittite civilization, ranging roughly from 1700 to 1180 BC, existed in modern Turkey, in the deep mountains of Anatolia or Asia Minor, and coexisted with the empires of Mesopotamia and Egypt. One can only guess why the flask was there; it could have belonged to a priest hoping to curse a cold or as a sacrifice to one of their Gods. [It is these ridiculous and unanswerable thoughts that keep me up at night]

A bottle was also found in the tomb of the Egyptian Pharaoh Tutankhamun, and hopefully the herb did more for him in the afterlife than in his actual life. Known to be dysmorphic or oddlooking, he was noted to have orthopedic irregularities and evidence of several bouts of malaria on autopsy. Regardless, King Tut died at the young age of seventeen in 1325 BC.

The herb has touched almost every civilization. It was widely used by the Romans, the ancient Chinese, the Arabs, and the Celtic civilization. It is cited in the Bible, the Quran, by Hippocrates and even Mohammed, who described the curative powers of the black seed as being able to cure every illness or disease, *except death.*

In the 10th century, it was referenced in the historic work *The Canon of Medicine* by Avicenna, the most renowned Muslim and Persian physician of his time who recommended the seeds for enhancing energy levels and expediting recovery from fatigue and depression.

Even today, Moroccans purchase these seeds in a souk, wrap them in linen, warm the bundle in their hands and inhale the aroma to unblock their nasal passages.

In addition to being a curative medicine, it gets baked into bread and spices up pickles, yogurt, sauces, and salads. In fact, it remains an essential ingredient in many traditional cuisines from Turkey, Egypt, India, and Mexico.

These valuable and utilitarian little black seeds come from an annual flowering plant that puts out delicate flowers in many colors, ranging from yellow, white, pink, pale blue to pale purple. The plant itself originates in the Mediterranean region and western Asian countries, including India, Pakistan, and Afghanistan, but it has clearly spread across the globe.

Research as well has mushroomed. Most of the original scientists were from Egypt and Sudan, but this soon spread to Saudi Arabia, and more recently to Japan, France, England, Canada, and the US.[31]

The historical and international flair of these minute seeds is also reflected in its many names.

| | |
|---|---|
| Arabic: | *Habbah Sawda, Habbah Al-Sauda* or *Habbat el Baraka* translated as *Seeds of blessing* |
| India/ Hindi: | *Kalonji or mangrail* |
| China: | *Hak Jung Chou* |
| Bangladesh: | *Kalo Jeera* |
| Russian: | *Chernushka* |
| Turkish: | *Cörek otu* |
| Persia: | *Cyah-daneh* |
| Amharic: | *Tikur azmud* |
| Assamese: | *Kaljeera* |

Whereas the plant goes by many names, it was officially deemed *Nigella sativa* by Linnaeus in 1753.

As we have noted, the seeds have been used for almost every illness under the sun, except curing death, as noted by Mohammad. Whereas this was probably meant to be uplifting, this is actually rather unfortunate as death is precisely what we are attempting to cure.

Avicenna, circa 1000 AD, reported benefits in terms of improving shortness of breath and for decreasing phlegm. Dioscorides recorded its efficacy for alleviating rough skin and psoriasis. In Arabian countries, the seeds are a traditional remedy for asthma, cough, abdominal pain, colic, general fatigue, rheumatism, and skin diseases.

The complete list of traditional illnesses that are treated with black seed oil is almost as diverse as the cultures that use it. Still, the more common ailments include asthma, diabetes, cough, headache, diarrhea, and improving immunity.

Of the more interesting uses, it is supposed to reduce flatulence, increase milk supply in nursing mothers, and eradicate worms. At high doses, it has been used to induce abortion. When applied topically, it heals blisters, eczema, and swollen joints.

Recently, scientific studies have reinforced these claims. Nigella has been shown to stimulate the immune system, fight infection, act as an anti-inflammatory, lower blood sugar and blood pressure, alleviate asthma and allergies, and eradicate certain worm infections. It can even help in the battle with cancer.

In this realm of cancer, where I tread very lightly, thymoquinone, one of the components thought to have significant bioactivity, has been found to be effective against many types of cancer, including breast, colon, pancreatic, liver, lung, fibrosarcoma, prostate, and cervix cancer cell lines and in animal models of lung, kidney, skin, colon, and breast cancer.[30]

Like any and all plants, the list of chemical or molecular components is not only vast but also variable depending on specific conditions. There are probably hundreds if not thousands of compounds within the nigella plant.

Not unexpectedly, the percentage of ingredients varies with the geographic distribution, time of harvest, and cultivation methods; but overall, the seed composition is 20-85% protein, 40% fat, 7-94% fiber, and 30% total carbohydrates.

Of all of the components, the most bioactive are considered thymoquinone, dithymoquinone (nigellone), p-cymene, thymol, carvacrol, and alpha-hederin. Thymoquinone is the most abundant and is regarded as the most efficacious.

**THYMOQUINONE**

Other components include nigellicimine, nigellidine, nigellicine, thiamine, riboflavin, pyridoxine, niacin, folic acid, proteins, eight or nine essential amino acids, and significant levels of iron, copper, zinc, phosphorus, and calcium.

Over the counter, nigella is available as both a solid and an oil, which can be either fixed or volatile. *Fixed* mixtures contain mostly large molecules that do not vaporize, including omega-6 fatty acids and linoleic acid, which constitute roughly 50% of the oil. The *volatile* oil, meanwhile, contains many more aromatic compounds, including nigellone, thymoquinone, thymohydroquinone, dithymoquinone, thymol, carvacrol, α and β-pinene, d-limonene, dcitronellol, p-cymene, carvacrol, t-anethole, 4-terpineol and longifoline.

The thymoquinone concentration is around 3.5–8.7 mg/g.

The List...

## 1) DNA ALTERATIONS: 1

Studies looking at the epigenetic effects of black seed oil have focused mainly on treating various forms of cancer, and it is inherently difficult to extrapolate this to longevity issues. That being said, thymoquinone, in cell cultures, has demonstrated a decrease "in the level of HDAC1 (a Histone deacetylase) and an increase in the p21 expression that confirms its ability to induce epigenetic modulation."[22]

As well, in a leukemia model, thymoquinone was identified as a novel DNA hypomethylating agent.[25]

## 2) MITOCHONDRIA: 2

There is reasonable evidence that several of the components of nigella are antioxidants and free radical scavengers.[14] In fact, screening of multiple oil samples demonstrated the presence of four elements that all had respectable radical scavenging properties, including thymoquinone, carvacrol, tanethole and 4-terpineol.[31] And while several studies position thymoquinone as the most active agent in most medicinal categories, thymohydroquinone was shown to be a more potent antioxidant overall.[6]

Examining the effect on endogenous antioxidants, it is clear that nigella is beneficial. In a diabetic mice model, after one month of daily ingestion, the seed precipitated a substantial increase in superoxide dismutase, glutathione-s-transferase, and catalase.

Meanwhile, the gene expression of catalase and superoxide dismutase activity was also demonstrated in an obese rat model. Confirming the correlation, the increased catalase gene expression was concomitant with an increase in the hepatic catalase.

One of the routes to increase these endogenous antioxidants is through the nuclear factor-like 2 protein that binds with the antioxidant responsive element (ARE). This system was activated in the rats given thymoquinone.[14]

Lastly, the seeds also can increase cellular NAD+ levels. This was observed in a microglia cell culture, the degree of which seemed to be directly correlated to the concentration of the black seed oil.[33]

## 3) PATHWAYS: 2

Most, if not all of the studies in this category, involve rodents consuming thymoquinone and not the plant in its entirety. Regardless, in several separate studies, oral thymoquinone activated both SIRT1 and AMP Kinase in microglial cells, hepatic and liver cells.[5,33,34] It has also been reported that nigella stimulates AMP Kinase in muscles and the liver.[8]

## 5) IMMUNE SYSTEM: 3

There is significant evidence in cell cultures, mice, and people that nigella and its component, thymoquinone, have anti-inflammatory components.

In cell cultures, thymoquinone abolishes the expression of proinflammatory cytokines TNF-α and IL-1β, reduces the synthesis of monocyte chemotactic protein-1 and COX-2, and reduces the transport of NF-κβ from the cytosol to the nucleus.[17]

Thymoquinone in other cell cultures also suppresses the transcription of NF-κβ-controlled proinflammatory genes such as TNF-α, IL-6, IL-1β, as well as COX-2 and iNOS.[33]

Moving up the food chain, mice on nigella demonstrate a reduction in TNF-α, IL-1β[5], while diabetic rats on thymoquinone show a decrease in the COX-2 enzyme activity in pancreatic β- cells.[14]

In fact, rodent studies abound demonstrating the anti-inflammatory properties of nigella. Overweight rats taking thymoquinone had reduced proinflammatory cytokines (IL-1β and IL-18)[5] and arthritic mice on thymoquinone had reduced serum levels of proinflammatory cytokines IL-6, TNF-α, IL-4, IL-13, and IL-10.[32]

Moving to human trials, patients with liver failure were given two grams per day of nigella. Serum levels of high sensitive C-reactive protein, TNF-α, and NF-κβ were all substantially decreased.[9]

In a second example, women with rheumatoid arthritis who took 500 mg two times a day noted an improvement in the number of inflamed joints, the incidence of morning stiffness, and improved disease activity.[36]

Moving on to the immune component, in cell cultures, nigella was able to boost immunity by augmenting T cell- and natural killer cell-mediated immune responses.[23]

And in humans, it increased the ratio of T-lymphocytes helper cells to T-suppressor cells by 72% and enhanced T-killer cells function and number.[12]

## 7) WASTE MANAGEMENT: 3

Much like the inflammatory category, black seed oil excels here as well, with a plethora of evidence in both rodents and humans.

For example, in both diabetic hamsters [14] and rabbits [2], oral thymoquinone decreases blood glucose levels.

In Type II diabetic humans, oral intake for twenty days decreased the HgA1C and fasting blood glucose.[36]

To solidify this information, several studies have reviewed and clumped the results together such that "The result of the meta-analysis shows that TQ (thymoquinone) has a statistically significant ameliorative effect on blood glucose, serum insulin and bodyweight of the animals." [8]

Another meta-analysis indicated: "that N. sativa can modulate hyperglycemia and lipid profile dysfunction with various potential mechanisms including its antioxidant characteristics and effects on insulin secretion, glucose absorption, gluconeogenesis and gene expression."[15]

In an attempt to determine how this occurs, it's back to lab rodents.

In diabetic hamsters, thymoquinone decreased the blood glucose by increasing insulin levels and inhibiting the synthesis of gluconeogenic enzymes; specifically glucose-6-phosphatase and fructose-1, 6-bisphosphatase.[14]

Meanwhile, in diabetic rats, thymoquinone increased the plasma glucagon-like peptide-1 (GLP-1) levels. This protein is a gut hormone that is synthesized and released by intestinal cells and is known to control glucose and ameliorate obesity.[19]

Clearly, nigella reduces glucose levels. Because it is also an antioxidant, it will come as no surprise that it reduces AGEs as well.[24] This has been proven to be true in test tubes as well as in cell cultures.[4, 24, 37]

Lastly in this category, there is some evidence that black seed oil can decrease the accumulation of lipofuscin over time. In an albino female rat model, rat chicks given nigella from the age of 18 to 20 months had less lipofuscin accumulation in the olfactory nerve cells than did the controls.[13]

## SYSTEMS KNOWN TO BENEFIT

### ADIPOSE TISSUE
There is some evidence that nigella can precipitate weight loss[30], as was demonstrated in a study of obese men taking three grams per day.[30] This process is thought to occur via an increase in the UCP-1 (uncoupling protein) gene expression. This protein essentially leads to less efficient mitochondria and thus increased caloric utilization. Studies have shown this to be true for thymoquinone in isolation, but the results are improved when the whole herb is used.[21]

### BLOOD PRESSURE
The evidence here is robust. Black seed oil can lower blood pressure, but the degree that the blood pressure is actually reduced is not that impressive. In reality, however, it is hard to truly determine as many of the studies used doses smaller than one might guess would be effective.[29, 30]

### LIPID PANEL
There is significant evidence in innumerable human studies that black seed oil can improve the lipid panel. The questions, generally, remain in the details as to how much to take and what populations benefit the most.

Examples in healthy people abound. Giving five cc/day of black seed oil to healthy adults for eight weeks induced significant decreases in fasting blood cholesterol, low-density lipids, and triglycerides. In another study of healthy volunteers given two grams per day, cholesterol levels decreased as well after only two weeks.[30] Healthy females, given both the crushed seeds and the oil, decreased their triglyceride and cholesterol levels.[30]

Folks with poor lipid profiles benefitted as well. Patients in this category decreased their LDL-C and increased their HDL-C.[30] Thus the take-home message is clear; at least one to two grams of black seed oil daily can improve just about anyone's lipid profile.[28] Specifically, this includes serum triglycerides, total and LDL cholesterol, and HDL cholesterol.[27]

## MALE VIRILITY

In a somewhat unusual but actual randomized, double-blinded, placebo-controlled clinical trial, infertile Iranian men were given black seed oil. After two months of treatment, they were cured. Their sperm counts, sperm motility and morphology, and semen volume were all greatly improved.[18]

## SKIN

Whereas topical black seed oil does very little for longevity, it has positive effects on the skin. Studies have demonstrated that it helps wound healing by increasing collagen formation and increasing the rate of epithelialization. It also decreases the white blood cell count and limits tissue damage.[10] In addition, it has proven to have clinical utility for mild to moderate psoriasis.[1]

## HOMOCYSTEINE

While it isn't a common problem, elevated homocysteine can be a risk factor for cardiac disease. Thus, the reduction or control of the amino acid level may be useful. Treatment with nigella significantly reduced the levels of plasma homocysteine in cultured heart tissue by $43 \pm 8\%$.[11]

## BRAIN

Both in teenage boys (aged 14 to 17) and in the elderly, black seed oil was found to improve cognition, memory, and attention. For the teens alone, mood and anxiety were also miraculously improved.[30] This may be due to increased serotonin in the brain as was demonstrated in a rat study after treatment with the oil.[26] It may also be secondary to the inhibition of the enzyme acetylcholinesterase, allowing for higher brain levels of acetylcholine.[7]

Regardless of why this occurs, this property is well appreciated in Bangladesh, where residents imbibe the herb with either honey or boiled mint for memory improvement.[7]

## BIOAVAILABILITY

As with most natural agents, the bioavailability is not great. It is slowly absorbed from the GI tract and is quickly eliminated. In fact, the half-life is only 217 minutes following oral intake.[14]

## DOSE

Whereas there have been countless human studies, the actual dose is still a bit up in the air, most likely resulting from the different formulations. For example there is the straight nigella seed, the various oils, and then the isolated thymoquinone.

For the most part, the best outcomes are seen with one to two grams of seed per day. In Type II diabetes, over three months, one gram per day increased high-density lipoprotein cholesterol (HDL-c) levels, while the two or three grams per day significantly decreased serum levels of total cholesterol and triglyceride as well as low-density lipoprotein cholesterol and increased plasma HDL-c.[36]

And according to Akbar in 2018: "In most clinical trials, a daily dose of two grams of black seeds was found to optimally achieve health benefits, including effects on blood glucose, lipid profile, inflammation, gastrointestinal symptoms and the central nervous system (CNS)."[3]

Therefore, the present dose suggestions are one to two grams per day for either the black seed or black seed powder, or the fixed oil. The recommended tincture dose is three to five cc, three times a day.[35] At present, I could find no recommendations on the dose for straight thymoquinone.

## POTENTIAL SIDE EFFECTS

In terms of side effects, these were rare and usually tolerable in the vast majority of the studies.[30] Nevertheless, some folks report nausea, bloating, and a burning sensation, as well as a slight increase in liver and kidney enzymatic markers.[30]

Be careful with pregnancy; however, high doses can be dangerous and might precipitate abortion.

# CHAPTER 8
# CENTELLA ASIATICA
## (1.2.2.0.2.2.2)

BENEFITS:

BRAIN  SKIN  HAIR

*"C. asiatica (gotu kola) is a reputed plant species for its traditional use in Ayurvedic and Chinese medicines, and its positive effects on brain aging have been generally attributed to its two major triterpene saponosides; asiatic and madecassic acids as well as their heterosides; asiaticoside and madecassoside, respectively."*

~ E. Orhan 2012

Centella asiatica is yet another plant species that millions of folks know about around the world and, in fact, frequently serve with dinner as a side salad.

While we Americans generally dine on tasteless and nutritionally deficient Iceberg lettuce, other cultures appear to be doing much better in this category. Centella has been consumed and utilized by innumerable cultures for millennia, and with good reason. There is not only historical and cultural belief in its power, but objective scientific evidence that it not only can help with general health, but it can improve both your brainpower as well as your skin.

This perennial, herbaceous creeper belonging to the family *Umbelliferae* was already in use three thousand years ago in southeastern Asia, mainly in China, India, and Pakistan. It has more recently become popular in equatorial and southern Africa, the Philippines, Sri Lanka, Java, Indonesia, and Madagascar.

The plant comes to us through Indian, Ayurvedic medicine as well as traditional Chinese medicine, where it is considered one of the official *Miracle Elixirs of Life*.

And of course, as is true for most valuable plants that have been in use for thousands of years, different cultures have used it in different ways. In India, it is a paste applied to the skin, or a juice rubbed on the forehead to alleviate headaches. It's a cool drink on a warm afternoon in China, while in Sri Lanka, it's added to porridge and curry. But the most prevalent use is in a salad, as found on Indonesian menus. In Myanmar cuisine, centella is the main constituent in an onion salad mixed with crushed peanuts, bean powder and seasoned with lime juice and fish sauce.

As a medicinal substance, it treats cholera, jaundice, diarrhea, syphilis, measles, toothache, smallpox, lupus, rheumatism, varicose veins, neuralgia, and leprosy. It is known to reduce anxiety, depression and to make people smarter, although the appropriate phrase is improve cognition or enhance memory. It is also used to *purify the blood,* but I'm honestly not sure what that means.

Centella is also an aphrodisiac as well as a feminine birth control method. [which may or may not be considered convenient]

Following the trend of many cultures... many names, I have identified over forty different titles for the plant.

The prominent or more recognizable include the following:

| | |
|---|---|
| China and Sri Lanka: | *Gotu kola* |
| Bengali: | *Thankuni* |
| Hindi: | *Mandookaparni* |
| Malay: | *Pegaga* |
| Telugu: | *Bekaparanamu* |
| English: | *Pennywort* |

Like every other plant we have considered, centella is composed of numerous components. But again, just a few have standout, key qualities.

There are several elements that we have seen before. For example, the major flavonoid compounds are catechin, quercetin, naringin, apigenin, and chlorogenic acid - all of which are incredibly beneficial in their own right.

But the essential players seem to be the triterpenoid saponins known as centelloids, which account for between 1% to 8% of all C. asiatica constituents.

The most important of these compounds are asiaticoside and madecassoside.[18]

Other centellosides making contributions include brahmic acid, madasiatic acid, terminolic acid, centellic acid as well as their glycosides: brahminoside, madasiaticoside and centelloside.

The List...

## 1) DNA ALTERATIONS: 1

In a study seeking to identify telomerase activators in human cells, centella was compared to other substances, including astragalus extract, TA-65, oleanolic acid, and maslinic acid. The centella extract outperformed the competitors and increased telomerase activity 8.8 fold as compared to the controls. Of note, TA-65 demonstrated a two-fold increase, while oleanolic acid demonstrated a 5.9 fold increase.[54]

As of yet, there is no evidence in animal models, but I'm confident someone is looking into this!

## 2) MITOCHONDRIA: 2

Centella serves as both a primary antioxidant as well as a free radical scavenger.[34, 61] Compared to other extracts, it was roughly equivalent to grape seed extract, vitamin C[12], α-tocopherol[62], rosemary, and sage.[47]

Of note, the leaves have the highest phenolic content and the highest antioxidant activity.[47]

There is also a plethora of evidence in rodents that centella increases the production of endogenous antioxidants.

Rats and mice, from a wide range of ages, different genders, and with and without diabetes[15] were supplemented with the full extract, or the individual components including asiatic acid[15] or madecassic acid.[15] Most of these studies demonstrated an increase in glutathione[15, 23] or both glutathione and catalase.[8, 23, 34]

Most, however, demonstrated increases in several, including catalase, superoxide dismutase, and glutathione peroxidase.[8, 17, 63]

Lastly, in a relatively rare mechanism of action, centella precipitates an increase in the genes responsible for mitochondrial biogenesis. Specifically, within the brain cells of the hippocampus, cortex, and cerebellum, genes coding for the electron transport chain (cytochrome B, NADH dehydrogenase 1, cytochrome C oxidase 1, and ATP synthase 6) are activated.[31]

## 3) PATHWAYS: 2

Again, utilizing cultures as well as rodent models, asiatic acid was discovered to be an AMP Kinase activator and an inhibitor of the mTOR pathway.[27]

In addition, asiatic acid increased the expression of SIRT1 in mice.[55]

## 5) IMMUNE SYSTEM: 2

There is a plethora of evidence that centella possesses anti-inflammatory properties. While some studies utilize the intact extract, others use either asiatic acid or madecassoside. Regardless, the results are generally the same.

In mouse models, asiatic acid decreases TNF-α and IL-1β levels, COX-2, and NF-κB[17], while in human gingival fibroblast cultures, asiatic acid significantly inhibits LPS-induced IL-6 and IL-8.[11]

The other important saponin, madecassoside, was tested against a collagen-induced arthritis rat model. Results demonstrated a systemic down-regulation of inflammatory cytokines and an up-regulation of the anti-inflammatory cytokine IL-10.[59]

In diabetic mice, madecassic acid dose-dependently reduced interleukin IL-1β, IL-6, and tumor necrosis factor-α in the heart and kidneys.[15]

When using the intact extract, the pattern is the same; a reduction in cytokines, including NF-κB, TNF-α, IL-1β, IL-6, and COX-2.[28, 37]

## 6) INDIVIDUAL CELL REQUIREMENTS: 2

It turns out that centella is exceptional for the nervous system, both in cell cultures and live rodents. When the extract was introduced to budding nerve cells, there was a "robust induction in neurite outgrowth and neuro-filament expression."[21]

And in male rats that imbibed centella in their drinking water, injured nerves "demonstrated more rapid functional recovery and increased axonal regeneration (larger calibre axons and greater numbers of myelinated axons) compared with controls, indicating that the axons grew at a faster rate."[50] [There is far more information in this category in the systems section, keep reading.]

There is also a study suggesting a benefit in the bone category, with evidence that centella decreases the activity of osteoclasts.[14]

## 7) WASTE MANAGEMENT: 2

As we have seen in other categories, there are a wealth of rodent studies confirming a positive result, but a dearth of human studies.

In this instance, innumerable diabetic rat and mouse trials demonstrate a reduction in blood glucose.[6,15,25,35,36,41,43]

There is also good news in the glycation category. Several experiments have demonstrated that centella can effectively reduce the overall load of AGEs.[30]

*"The extract has shown a clear prevention of the phenomenon of glycation, targeting in a pleiotropic and complementary way the biochemical and cellular bases of skin aging."*[30]

## SYSTEMS KNOWN TO BENEFIT

Centella provides clinically relevant benefits all over the body, including aiding the immune system [47], alleviating fatigue [1], reducing anxiety, and even reducing fat accumulation.

Several human studies have confirmed this last finding. Participants took sixty mg of the dry extract once a day for ninety days in one particular study, and for "the patients taking the gotu kola extract, the diameter of fat cells (or adipocytes) decreased in both studied regions of the body with a predominance of positive changes in the gluteofemoral region."[4]

These findings were repeated in young, overweight women where the treatment had a "positive influence upon physical slimming, and body circumference reduction."[22]

Another interesting study, at least for women, was in postmenopausal lady rats. Oral centella improved the condition of their vaginas. [Yes, someone actually studied this.] Specifically, it increased the collagen content and precipitated a proliferation of vaginal wall tissue, thus making the organ more sturdy.[40]

The major highlights of centella, however, revolve around two very important, key systems; the brain and peripheral nervous system as alluded to earlier, and the skin.[8]

## BRAIN

As mentioned earlier, individual nerve cells in culture exposed to centella grow new nerve endings and sprout new connections.

*"At the in vitro level, C. asiatica promotes dendrite arborisation and elongation, and also protects the neurons from apoptosis."* [26]

When given oral centella, not only do the peripheral nerves improve for both juvenile and adult rats, but the brains do as well. Hippocampal CA3 cells, i.e., the cells involved with memory, have increased dendritic arborization with increased intersections and branch points.[44]

A second improvement within the brain involves neurotransmitters. Gohin et al. noted increases in serotonin, norepinephrine, and dopamine in the cortex, hippocampus, and thalamus after therapy with centella.[8]

The improvements in nerve structure and biochemical makeup seem to translate into many improved, testable behaviors in rodents, mainly concerning memory.[8,34]

This brainpower seems to extend to humans as well. There have been many studies, ranging from middle-aged adults to the elderly, that demonstrate improvements. Doses between 500 to 750 mg daily for two to six months consistently have shown cognitive-enhancing effects.[34]

## SKIN

In Asian medicine, C. asiatica has been used for hundreds of years to treat minor wounds, scratches, burns, hypertrophic wounds, and eczema.[4]

What exactly does centella do? To start with, all of the same properties discovered with other organ systems remain true for the skin.

For example, in the mitochondria category, rats studies have shown that the topical application of centella increases both enzymatic and non-enzymatic antioxidants.[4,5]

As well, inflammatory processes improve with topical applications. [28, 37, 48]

Centella also can limit MMP's; asiaticoside has shown "potent inhibitory activity against hyaluronidase, elastase and MMP-1."[32]

Turning to cellular effects, centella promotes fibroblast proliferation[5,45], and stimulates the production of collagen, hyaluronic acid, and fibronectin.[4,5,57]

In fact, there is a plethora of evidence demonstrating that both asiaticoside and madecassoside promote collagen synthesis.[4,5,57] However, madecassoside is the only compound that specifically promotes collagen Type III production, while asiaticoside improves the synthesis of collagen Type I. [30,60]

We even know the mechanism by which asiaticoside induces the synthesis of Type I collagen. In human dermal fibroblast cells, it phosphorylates Smad 2 in addition to the activation of tumor growth factor β receptor I kinase.[4] [A great piece of trivia for cocktail parties]

In animal models, a 10% gotu kola cream increased collagen levels in the skin of female rats by an average of 78%, while a 2.5% cream increased collagen levels, on average, 21%.[57]

In humans, topical C. asiatica extract (0.5%) was applied for six weeks on twenty volunteers. An increase in collagen density was reported in 70% of the patients and a significant decrease in wrinkle depth and volume was observed.[30]

It turns out that all of these cellular mechanisms also lead to improved wound healing.[29,60]

*"The asiaticoside-rich hydrogel formulation exhibited 40% fast wound healing*

*without any skin irritation as compared to untreated group. Thick epithelial layer and keratin formation can be found, while granulation tissue, fibroblasts and collagen were formed moderately."* [52]

Another thing that tends to bother people as they get older is the dreaded dry skin phenomenon. While people have tried to combat this issue by imbibing several gallons of water a day, the real answer lies in the production of hyaluronic acid and other molecules that bind and retain water molecules in the skin. As one might guess, centella can help solve this problem.

Aquaporin-3, filaggrin, loricrin, and involucrin are four of the most studied proteins expressed by epidermal keratinocytes that have a role in hydration for either their moisturizing or barrier capabilities. Madecassoside application precipitated an increase in these molecules in a skin model.[48] At the same time in the dermis, hyaluronic acid is also increased by treatment.

In a group of twenty-five volunteers using twice a day application for four weeks, a 5% cream improved skin moisture by increasing the skin surface hydration state and decreasing transepidermal water loss.[45] In Type 2 diabetics, whose skin is notoriously dry, a prospective, double-blinded, randomized, controlled study compared oral centella extract plus a 1% topical cream versus topical only for twenty-nine days and demonstrated that the combination therapy was the most effective.[24]

As if these skin benefits weren't enough, centella is also helpful in improving both stretch marks and cellulite. In a study examining oral centella (Centellicum® 225 mg 3x day) versus abdominal wall stretch marks, women at least six months postpartum were compared to women using a standard anti-stretch cream. After six weeks, the oral treatment proved to be more efficacious.[16]

Meanwhile, the topical application of Madecassol® applied four times a day for four months clearly showed a beneficial effect on inhibiting the progression of cellulite, with a significant improvement in the skin of 85% of the participants.[4, 5]

On a personal note, this is a key ingredient in the swamp juice that I mentioned in the aloe chapter. Now you know why!

## HAIR

Turning to hair, the benefits keep coming...
Centella improves the viability of dermal papilla cells and increases the expression of genes related to hair growth. More simply, centella can increase hair production.[7,46,52]

In a human study of hair lotion containing centella extract (Centerox®), volunteers with mild to moderate alopecia used 2 mL daily over a period of one to two months. The study demonstrated a 41% decrease in hair loss in the washing test and a slight increase in hair strength in the pull test.[53]

## POTENTIAL SIDE EFFECTS

The most common adverse events are gastrointestinal upset, flatulence, nausea, headache, dizziness, a decreased appetite, sedation, and rash if used topically. In addition, there have been three reported cases of hepatitis. Chronic treatment may block pregnancy or precipitate abortion and can also reduce sperm count.[31,58]

Long term supplementation can also lead to a buildup of metabolites, so most experts recommend cycling the herb. Specifically, although I have seen no precise science to back this up, recommendations are six weeks on, two weeks off.[8,34]

Interestingly, centella has poor oral absorption, slow delivery, and low bioavailability. On the other hand, studies, at least in rodents, have demonstrated that the active components do in fact, get into the body, and accumulate - mainly in the brain, stomach, and skin.[3]

## DOSE

The dose is yet another interesting question. This varies hugely from study to study and also depends on how the herb was processed.

Improvements in venous insufficiency, for example, seem to do best at 60 to 180 mg daily. Cognitive function was optimal at 750 mg daily in one study and at three grams in another.[39] Other studies suggest that a typical daily dose should be 600 mg.

Other preparations consist of the individual components; for example, you can get 10 mg madecassol tablets that are supposed to be taken three times daily.

In terms of creams, the choices are equally vast in range. You can get the complete herb as a topical, or you can get the components separately.

Therefore, it is challenging to recommend absolute doses. Thus, as usual, it's up to you, the reader, to determine the best elixir and dosing schedule.

# CHAPTER 9

# CHLOROGENIC ACID

(1.2.2.0.2.0.3)

BENEFITS:

FAT

HEART

*"Chlorogenic acid is an important and biologically active dietary polyphenol, playing several important and therapeutic roles such as antioxidant activity, antibacterial, hepatoprotective, cardioprotective, anti-inflammatory, antipyretic, neuroprotective, antiobesity, antiviral, anti-microbial, anti-hypertension, free radicals scavenger and a central nervous system (CNS) stimulator."*

~ M. Naveed 2018

Chlorogenic acid, unlike many of the exotic agents we have investigated, isn't very exotic. In fact, it's probably right under your nose or in a mug on your desk. Chlorogenic acid (CGA) is in potatoes, artichokes, eggplants, blueberries, and apples. The vast majority, however, is in coffee, especially green coffee bean extracts. And if you are seeking a more exotic source, there is always Yerba mate from South America.

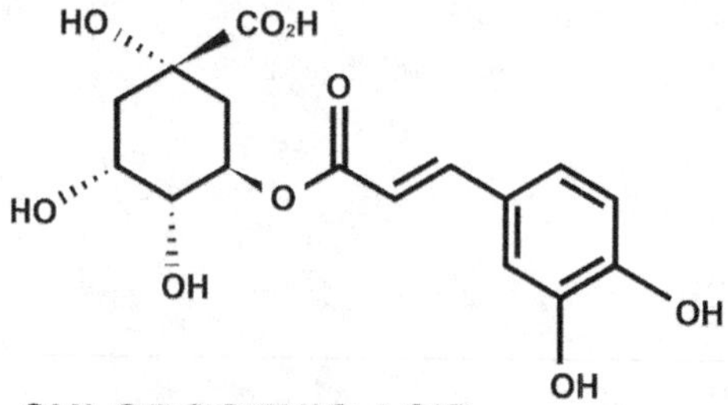

**CHLOROGENIC ACID**

**QUINIC ACID**  **CAFFEIC ACID**

A phenolic compound from the hydroxycinnamic acid family, chlorogenic acid's chemical structure consists of two pieces, a quinic acid and a caffeic acid.

**PHENOL**

It has several chemical names that chemists bicker over. Historically it was 3-CQA 3-caffeoylquinic acid (before 1976), but is currently known as 5-caffeoylquinic acid (5-CQA) as per guidelines of the International Union of Pure and Applied Chemistry (IUPAC).[25]

Structurally, CGA is related to a group of polyphenol compounds consisting of esters formed by hydroxycinnamates, i.e., caffeic acid, ferulic acid, and p-coumaric acid.

Of note, polyphenol compounds share a common phenol structural group, an aromatic ring linkage with at least one hydroxyl substituent.

I only point this out as there are innumerable polyphenols in the longevity arsenal. Knowing that a phenol molecule is pretty simplistic and easy to envision helps dispel the notion that the chemistry is too complex and difficult to comprehend. Once you can picture one molecule, you can easily picture a few of them stuck together. Alas… a polyphenol.

The List...

## 1) DNA ALTERATIONS: 1

Epigenetic modification is a relatively new science. We know that methylation changes over time can both be necessary for growth and development but can also precipitate premature aging and disease. Therefore, we hesitantly yet hopefully assume that agents that tend to be good for us display epigenetic effects that are positive as well, but remember - this is an assumption.

That being said, chlorogenic acid inhibits DNA methyltransferase.[27,45,49] This inhibition is primarily due to the increased formation of S-adenosyl-L-homocysteine, a known potent inhibitor of DNA methylation.[27]

There is also preliminary evidence, or at least a tenuous connection, between coffee consumption and increased telomere length.[19]

In a prospective study of 4,780 female nurses initiated in 1976, researchers discovered a significant, linear association between caffeine consumption from all dietary sources and telomere length after adjusting for potential confounders.[28]

Granted, there are no cellular studies confirming this, but it's a start.

## 2) MITOCHONDRIA: 2

Like many of the agents in our arsenal, chlorogenic acid is an antioxidant and scavenges free radicals.[8,26,33,40,42]

*"CGA has been shown to act as a scavenger of superoxide radicals, hydroxyl radicals, and peroxynitrite in a concentration-dependent manner in vitro."*[48]

Chlorogenic acid also can increase endogenous antioxidant levels, notably by increasing nuclear factor-erythroid factor 2-related factor 2 (Nrf2).[4,24]

More specific studies have confirmed increases in glutathione and superoxide dismutase.[13,14,37]

## 3) PATHWAYS: 2

At least in cell cultures and rodent models, there is ample evidence that chlorogenic acid activates AMP Kinase.[20,35,40]

*"The results of this study showed that CGA activated AMPK, leading to subsequent beneficial metabolic effects, such as suppression of hepatic glucose production and fatty acid synthesis."* [29]


There is also some preliminary evidence in cell cultures, i.e., human umbilical vein endothelial cells [43] and in a hepatocyte cell line [44], that chlorogenic acid activates SIRT1 as well.

## 5) IMMUNE SYSTEM: 2

In innumerable models of some irritating or inflammatory disease inflicted upon rodents, chlorogenic acid proved to be a potent mediator. [9,10,21,41,47] In a model of gout, a disorder where uric acid crystals accumulate and lead to severe joint pain, chlorogenic acid inhibited the production of proinflammatory cytokines including interleukin-1β, interleukin-6, and tumor necrosis factor-α.[29] This pattern was continued in models of liver and kidney injuries,[13] mouse mastitis,[38] and mouse colitis,[46] as chlorogenic acid reduced inflammatory cytokines.

## 7) WASTE MANAGEMENT: 3

As a reminder of why excess glucose is detrimental, hyperglycemia significantly enhances the generation of reactive oxygen species, it depletes superoxide dismutase activity, and it increases the production of AGEs.[36]

Luckily, chlorogenic acid reduces blood sugar, which is achieved through two mechanisms. First, glucose absorption from the GI tract is reduced through the inhibition of glucose-6- phosphate translocase (G-6-pase).[34] Once in the bloodstream, the same enzymatic effect controls the level of blood glucose being produced in the liver, by affecting both hepatic glycogenolysis (the breakdown of the molecule glycogen into glucose) and gluconeogenesis (the generation of glucose from non-carbohydrate carbon substrates). This works as G-6-Pase is the key enzyme that catalyzes the final steps in these pathways.[29, 34]

Some evidence suggests that G-6-Pase gets inhibited, while other studies demonstrate a decreased production of the enzyme.[29] The second mechanism of glucose reduction is through the inhibition of α-amylase and α- glucosidase. Whereas the relative magnitude of this reduction varies from study to study, it does contribute to the hypoglycemic effect.[29, 31]

This has been clearly demonstrated in diabetic mice, where after twelve weeks of intervention, the fasting plasma glucose and glycosylated hemoglobin levels were significantly decreased.[20]

Standard coffee is generally used as the testing agent in people. This muddies the results a bit, but we can extrapolate. We know, for example, that the regular consumption of coffee is associated with a lower risk of Type II diabetes mellitus in both sexes, multiple geographical locations, and at various levels of obesity.[29] In fact, the daily consumption of three to four cups of decaffeinated coffee containing high concentrations of chlorogenic acid reduced the risk for Type II diabetes by 30%.[29]

Using Coffee Slender®, forty-five subjects between 20 to 50 years of age ingested the high concentration CGA coffee and had significantly lower plasma glucose levels noted thirty minutes after ingestion.[40]

The other beneficial component of chlorogenic acid in this category is its ability to block the formation of AGEs.[2,3,5,7,16,18,22,23]

In several studies, chlorogenic acid was more efficacious than aminoguanine, the gold standard, in inhibiting both the formation of AGEs as well as the cross-linking of AGEs to collagen.[23,39]

## SYSTEMS KNOWN TO BENEFIT

The most notable contribution chlorogenic acid seems to make is improving metabolic syndrome. This disease is a composite of a few interrelated issues, including obesity, high blood pressure, diabetes, and dyslipidemia that all fall into the category of too much fat and too many calories. The good news is that chlorogenic acid precipitates positive changes in all of these categories.

## ADIPOSE TISSUE

First and foremost, chlorogenic acid decreases full body and visceral fat in both rodents and people. In fact, in an overweight human study, after twelve weeks of a diet with chlorogenic augmented coffee, 80% of weight reduction was due to the loss of body fat.[40]

## LIPID PANEL

There are several ways that chlorogenic acid can improve the lipid panel:

1) Chlorogenic acid has a strong inhibitory effect on the activity of hepatic and pancreatic lipase. Thus, less lipid gets absorbed from the gut in the first place.[29,33]

2) Chlorogenic acid can inhibit the enzymes necessary for the production of lipids. It blocks β- hydroxy-β-methyl glutaric acyl-coenzyme A reductase (HMG CoA reductase), the rate-controlling enzyme of the mevalonate pathway, responsible for cholesterol biosynthesis.[29,40]

Chlorogenic acid also inhibits fatty acid synthase and acyl-CoA:cholesterol acyltransferase, two enzymes necessary for adipose production.[11]

3) Chlorogenic acid increases the breakdown of lipids more efficiently by increasing fatty acid β- oxidation activity.

Thus, overall, one sees decreased triglyceride and cholesterol levels and reduced low-density lipoproteins.[11]

## BLOOD PRESSURE

Innumerable human studies exist trying to confirm any effect on systemic blood pressure. In a composite study examining a collection of randomized clinical trials, the evidence suggests that although moderate, chlorogenic acid does precipitate statistically significant reductions in systolic and diastolic blood pressures.[33]

## COAGULATION

The last effect of note is that chlorogenic acid moderately inhibits platelets. *"Chlorogenic acid in a dose-dependent manner (0.1 to 1 mmol/L) inhibited platelet secretion and aggregation induced by ADP, collagen, arachidonic acid and TRAP-6, and diminished platelet firm adhesion/aggregation and platelet-leukocyte interactions under flow conditions."*[17]

Do not get overly concerned; however, the inhibition is less than an equivalent dose of aspirin.

## METABOLISM

The metabolism of chlorogenic acid is not straightforward. About one-third of the ingested agent is absorbed in the small intestine and gets into the bloodstream. The remainder reaches the colon and gets metabolized by colonic bacteria. The resulting compounds, such as hippuric acid, caffeic acid, and ferulic acid have positive effects as well.[12,40]

## POTENTIAL SIDE EFFECTS

Life is generally not without a few issues, and chlorogenic acid is no different. As well, my evaluation would be incomplete if I didn't mention any drawbacks.

First, there is one 2012 report suggesting that chlorogenic acid induces high levels of topoisomerase I- and topoisomerase II-DNA complexes in cells, making it potentially cytotoxic. This effect was more prominent in lung cancer cells than normal fibroblast cells, supporting previous findings that it might selectively kill cancer cells. Regardless, there could be potential for harm in this regard.[6]

The other thing to note is that a high chlorogenic acid intake is associated with elevated serum homocysteine levels. One study found that a two-gram intake, comparable to 1.5 liters of strong coffee, raised homocysteine concentrations in postprandial plasma by 12% and in fasting plasma by 4%.[32] [That's a lot of coffee! Meanwhile, recall that black seed oil reduces homocysteine levels]

As chlorogenic acid also reduces the production of cholesterol, depending on how much coffee you drink, make sure you augment your CoQ10 intake.

So, how does one consume chlorogenic acid? You can get it through regular coffee, green coffee bean extract, or even *Yerba mate*.

Derived from the leaves of the *Ilex paraguariensis* plant, Yerba mate is consumed as tea in Argentina and the southern Americas. Chlorogenic acid is the primary polyphenol in Yerba Mate, and on average, each leaf contains roughly 90 mg. The leaves are chocked full of other agents as well, including other polyphenols, xanthines (caffeine and theobromine), purine alkaloids (caffeic acid, 3,4-dicaffeoylquinic acid, and 3,5-dicaffeoylquinic acid), flavonoids (quercetin, kaempferol, and rutin), amino acids, minerals (phosphorous, iron, and calcium), and vitamins (C, B1, and B2).

## DOSE

Examining a multitude of studies, it becomes immediately apparent that there is no optimal dose. If there was a known limit, Starbucks might go out of business.

If one had to make an educated guess, I would have to limit intake to somewhere between 500 mg to 1 gram daily to maximize benefits while avoiding any potential side effects.

If you are going to go the coffee route, pick the least roasted variety as roasting can break down the chlorogenic acid. Therefore, green coffee beans have the most chlorogenic acid, followed by a light roast. The dark roast clearly has the least.

# CHAPTER 10

# CISTANCHE DESERTICOLA

(2.2.0.0.2.2.2)

BENEFITS:

BONE　　BRAIN　　HAIR　　IMMUNITY　　VIRILITY

*"Athough Cistanches has been used as an elixir for thousands of years, scientific research on Cistanche plants started in the1980s, and since then, studies have demonstrated tremendous utility in hormone regulation, improving the immune system, being neuroprotective, antiinflammatory,having anti-fatigue properties and promoting bone formation."*

*~ T. Wang 2012*

*Cistanche deserticola* is an extraordinarily clever lifeform - it's a plant with no chlorophyll, and it gets all of its nutrients and water from other plants. Considered a holoparasitic plant, it is entirely parasitic on the roots of a desert shrub.

Growing only in Chinese deserts, it is found mainly in Inner Mongolia, Gansu, and Xinjiang, attached to the roots of sand-fixing plants, such as *Haloxylon ammodendron, H. persicum, Kalidium foliatum,* and *Tamarix.*

One cannot blame the plant for being parasitic, however, because life in this desert is challenging. [I actually applaud its cunning]

The environmental conditions are severe: an extremely arid climate, depauperate soils, significant temperature variations, intensive sunshine, and very little rainfall.

Other than the fact that I find this plant really cool, the actual reason for its inclusion is that daily consumption of cistanche is associated with longevity in China and Japan.[14] It is also believed to be a favorite of Genghis Khans.

Wild deserticola has been historically overexploited, primarily due to its host plant being utilized locally as firewood. Therefore, the plant we are interested in has suffered by association and is on the verge of extinction. Luckily, in order to meet increasing demand, cistanche is now cultivated on a more commercial basis.

The plant is called Rou Cong-Rong in Chinese, derived from the stem's meaty nature (rou), while the effects are calming (congrong). It is also referred to as suosuo dayun, or as *Ginseng of the desert.*

Coming to us by way of Chinese medicine, the plant has been used for about 2,000 years, first recorded during the Eastern Han Dynasty, and is used to strengthen Yin and boost Qi.

It is also a component of an ancient herbal formulation known as the *Youth Returning Pill* or Huan Shao Dan in Japan and Taiwan.

In addition to delaying age and strengthening Yin, cistanche historically has been used for geriatric constipation, to protect semen, treat kidney disease, treat tetanus and alleviate fatigue.[27] [basically pooping and procreation]

The composition of cistanche is exceptionally complex and varied, including both volatile compounds and more than a hundred isolated non-volatile compounds.[20]

Chemical groups from the stem include phenylethanoid glycosides (PhGs), iridoids and iridoid glycosides, lignans and lignan glycosides, polysaccharides, free amino acids, and minerals.

A total of twenty different phenylethanoid glycosides have been identified, including echinacoside, acteoside, and isoacteoside, which are also polyphenolic compounds.[6,18]

In addition, some of the phenylethanoid glycosides are phenylpropanoid glycosides, molecules known to expedite the repair of DNA.[27]

There are several other species in the genus *Cistanche* in China aside from deserticola: *Cistanche salsa, Cistanche tubulosa,* and *Cistanche sinensis.* Of note, the chemical contents and pharmacological effects of *Cistanche tubulosa* are similar to *C. deserticola,* and it is frequently used as a substitute.[14,27]

The List…

I believe, with time, these ratings will improve as information about cistanche is presently limited by a paucity of human studies.

## 1) DNA ALTERATIONS: 2

There appears to be good and very recent evidence that cistanche benefits telomeres. There is no human evidence as of yet, but the rodents seem to be doing well.

In a mouse model, isolated polysaccharides increased telomerase activity in the heart and brain, lymphocytes, and macrophages.[35]

Acteoside, one of the phenylethanoid glycosides, was also found to increase telomerase activity in the heart and brain tissue of mice when given at a dose of 40 mg/kg for two weeks.[16]

Lastly, in a mouse model using *C. tubulosa,* treatment for two weeks increased telomerase activity in the heart and brain.[1]

## 2) MITOCHONDRIA: 2

*"There are two main flows by which the cistanches scavenge the free radicals, namely directly involving in the removal of free radicals or blocking their production and regulating the antioxidant enzymes related to the free radical metabolism in vivo, such as SOD, CAT and GPX."* [21]

To date, researchers have identified nine compounds that possess significant free radical scavenging[26], tubuloside A, syringalide A, 3'-α-rhamno-pyranoside, cistanoside A, and cistanoside F, all of which have more robust free radical scavenging activity than α-tocopherol.[30]

In addition, cistanche increases endogenous antioxidants. In rat hearts, it increases mitochondrial glutathione levels [23], and in countless animal studies, the phenylethanoid glycosides at various doses increase superoxide dismutase.[3,15,17,25,26]

One of these, echinacoside, enhanced superoxide dismutase activity in the serum and brain in wild-type worms and also increased their lifespan.[26]

In an added mitochondrial bonus feature, treatment of both cultured cells and entire rats with cistanche increased mitochondrial ATP production. This was associated with an increase in mitochondrial complex I and III activity, with complex I benefiting to a greater degree. This is important as complex I is considered a rate-limiting step in regulating oxidative phosphorylation.[11,12]

## 3) PATHWAYS: 0
In neuronal cell culture, echinacoside increased the expression of SIRT1.[4]

Not yet good enough for points, however

## 4) QUALITY CONTROL: 0
There is limited evidence that Cisanoside B increases the rate of DNA repair.[22]
[same, still sketchy on the details]

## 5) IMMUNE SYSTEM: 2
As an anti-inflammatory, cistanche reduces levels of tumor necrosis factor-α, interleukin-1β [26], and IL-6.[36]

Cistanche is also a potent immune-stimulant, acting on several aspects of the immune system. At the cellular level, cistanche polysaccharides enhance the phagocytic activity of macrophages in a dose-dependent manner.[28]

In a mouse study, the extract significantly reduced the degree of mucosal hyperplasia and intestinal helicobacter infections thought to be due to an increased number of splenic macrophage and natural killer cells.[10]

In prematurely-aged mice, there was a significant increase in naive T and natural killer cells in blood and spleen cell populations following four weeks of treatment. This increase in the peripheral naive T cell pool was associated with a concomitant extension of lifespan and a reduction in the frequency of tumor formation.[36]

In addition to an increase in T and B cells, treated mice have also demonstrated improved titers of IgG, IgG1, and IgG2a [34,38], while a close relative, *Cistanche salsa* increases the production of IgM and IgG 5-fold.[19]

## 6) INDIVIDUAL CELL REQUIREMENTS: 2

Cistanche has definite benefits on bone quality, decreasing the activity of osteoclasts and increasing that of osteoblasts.[24]

*"C. deserticola extract has been proven to promote the bone formation of cultured osteoblasts through increasing alkaline phosphatase, bone morphogenic protein (BMP)-2, osteopontin(OPN) mRNA expression and bone mineralization in vitro."*[27]

First, the agent promotes osteoblastic production from mesenchymal stem cells.[27,33] There is then an increase in the activity of these osteoblasts, demonstrated by increases in alkaline phosphatase, bone morphogenetic protein (BMP)-2, osteopontin mRNA expression, and bone mineralization.[13,27]

As a rodent example, of which there are many, in postmenopausal, osteoporotic mice after twelve weeks of treatment with cistanoside A, the animals "exhibited significant anti-osteoporitic effects, evidenced by enhanced bone strength, bone mineral density, and improved trabecular bone microarchitecture."[31]

## 7) WASTE MANAGEMENT: 2

Many studies in this category have utilized *Cistanche tubulosa* as well as deserticola; luckily, both species precipitate a reduction in blood glucose levels.

For example, in diabetic mice, C. tubulosa significantly suppressed elevated fasting blood glucose and postprandial blood glucose levels, and improved insulin resistance and dyslipidemia.[1,26]

Acteoside, isolated from deserticola, was found to improve glucose tolerance in starch-loaded mice[1,26], as did isolated polysaccharides that were determined to be anti-hyperglycemic and hypolipidemic.[26]

## SYSTEMS SHOWN TO BENEFIT

### ANTI-FATIGUE [37, 39]
Given to mice for three weeks, swimmer-mouse athletes demonstrated an increase in the time to exhaustion. In addition, the serum creatine kinase, lactate dehydrogenase, and lactic acid levels were decreased compared with the control, while the hemoglobin increased significantly. Researchers concluded that the treatment enhanced swimming capacity by "decreasing muscle damage, delaying the accumulation of lactic acid and by improving the energy storage." [2]

[The key is finding googles that fit]

### LIFESPAN
Cistanches appears to extend lifespan in senescence-accelerated mouse models [36], flies (*Drosophila*) [21], and worms. [26]

### MALE VIRILITY
Men have been using cistanche for thousands of years, hoping to become more manly. Rumors suggest that this may be at the heart of Genghis Khan's prolific procreation. [In a 2003 genetic study, 16 million men were found to be descendants]

*"C. deserticola has been used to cure erectile dysfunction and other impotence conditions and to boost male sexual activity by enhancing the production of testosterone and protecting sperm."* [26]

In a eunuch rat model, animals were treated for four weeks after castration. Cistanche therapy successfully shortened erectile latency and prolonged erectile duration to minimize the adverse effects of castration.[9] Studies have also demonstrated that echinacoside can increase sperm count, sperm motility and attenuate poor sperm quality and testicular toxicity in rats through the up-regulation of steroidogenesis enzymes.[26]

### BRAIN
There is good evidence that cistanche can improve nerve function, leading to enhanced mental performance.

In cell cultures, cistanche increases nerve growth factor (NGF), preventing neuronal death, promoting neurite outgrowth, supporting synapse formation, and enhancing memory function.

In mice, this increased NGF localizes in the brain, increasing neuronal cell differentiation, neurite length, and synapse formation in the mouse hippocampus.[5] These same mice demonstrated enhanced learning and memory.[5]

In fact, several rodent models all demonstrate mental improvements. Acteoside improved memory in a senescent mouse model.[21]. At the same time, phenylethanoid glycosides precipitated gains in the learning and memorizing ability of rodents [15], and stressed-out rodents were less stressed out, with reduced serum concentrations of corticosterone after treatment.

## SKIN & HAIR

There is also some good news when it comes to hair and skin.

In skin cell cultures, *Cistanche deserticola* extract decreases the activity of tyrosinase, collagenase, and elastase, thus helping to preserve skin with advancing age.

In a double-blinded, placebo-controlled clinical trial, researchers compared hair density and diameter in patients with and without treatment. A statistically significant increase in both measurements was recorded after sixteen weeks.[1]

## DOSE & POTENTIAL SIDE EFFECTS

Cistanches has been considered a safe traditional medicine in China for 1000s of years (Shen Nong's Herbal Classic). Common adverse clinical reactions include nausea and vomiting, abdominal pain, and dizziness.[7, 8]

The recommended daily dosage for *Cistanche tubulosa* extract is 100-400 mg/day, but it is difficult to find any recommendations for *Cistanche deserticola*. Although the Chinese pharmacopeia recommends up to 9 gm daily, my best guess would be somewhere between 500 to 1,000 mg daily.[32]

# CHAPTER 11

# COLLAGEN

(0.2.0.0.2.3.1)

BENEFITS:

HEART     SKIN     JOINTS

*"Without collagen a human being would be reduced to clump of cells interconnected by a few neurons."*

*~ S.N. Deshmukh 2016*

Unlike many of the agents we have talked about, collagen isn't an exotic agent from a foreign country, nor does it taste good in food, even though people love to put it into smoothies. Collagen isn't fancy, and it isn't all that interesting. It just is, but it's pretty important.

Collagen is one of the most prevalent proteins in nature and especially in animals, where it is both abundant and ubiquitous. In fact, it comprises about 30% of all proteins, at least in humans. It's present in all fibrous tissues, such as tendons and ligaments, as well as in the cornea, cartilage, bones, skin, and blood vessels. It resides in most organs and tissues that have structure and even in the gelatinous substances of the body.

As people, we notice collagen, or the lack thereof, in the skin primarily. Within the dermis, collagen, elastic fibers, and hyaluronic acid are the major structural constituents of the extracellular matrix, and collagen constitutes greater than 70% of the dry weight of the human dermis.

So we are composed of a plethora of collagen, but why are we talking about it here? Because the addition of oral collagen or its components brings innumerable benefits, both to the visible skin and the invisible organs that lie beneath.

*"These beneficial effects include antioxidant, anti-aging, anti-osteoporotic and anti-osteoarthritis, anti-inflammatory, anti-tumor, wound healing, anti hypertensive and anti-atherosclerotic, anti obesity and hypoglycemic effects."* [32]

## WHAT IS COLLAGEN?

In a simplistic sense, collagen is essentially a rope. It consists of linear chains of amino acids, increasing in complexity as they become intertwined, creating thicker and thicker structures.

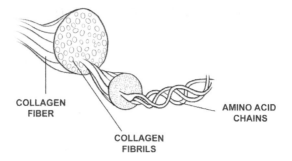

Three individual amino acid chains, called alpha chains, get twisted together at the most basic level. These, in turn, get combined with other, similar triad chains, forming first a collagen fibril and then joining together to become a collagen fiber.

These rope-like chains provide tensile strength, stability, and structural integrity to tissues and organs.

We know the most about collagen in the skin, where it resides in the dermis, assembled into a meshwork of fibers. Several dermal layers can be identified, and closest to the epidermis, the papillary dermal layers host thin collagenous fibers adjacent to the basement membrane. A bit deeper, the reticular dermis consists of thicker, well-organized, coarse collagen bundles that interface with the subcutaneous fat. Most dermal collagen forms a *'basketweave'* structure, with perpendicular collagen fibers intersecting at right angles.

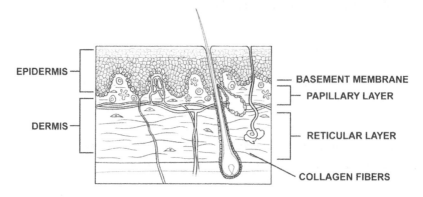

## SYNTHESIS

Collagen is synthesized mainly by mesenchymal cells and their derivatives, such as fibroblasts, chondrocytes, osteoblasts, and odontoblasts. In the skin, the fibroblasts are the stars in this arena.

The most basic component, or the alpha-amino acid chain, consists of three repeating amino acids, demarcated as Glycine - X-Y, where the first amino acid in the triplicate is always glycine, and the X and Y tend to be prolines and hydroxyprolines, although this is not always the case. Interestingly, hydroxyproline is unique only to collagen, as is hydroxylysine. Regardless, the triplet pattern is repeated over and over for thousands upon thousands of repetitions, creating a left-handed helix alpha chain.

As these complexes combine, three alpha chains then coil around each other in a right-handed helix. Interactions between the individual amino acids stabilize this trifecta. The ever-present glycine forms a hydrogen bond with amide groups in adjacent chains. As well, the hydroxylation of proline and lysine (when it is present) allows the formation of intramolecular hydrogen bonds that provide additional stability.

Several things determine the differences in collagen type, most importantly the specific amino acids in the repeating triad. But it is also determined by the type or types of alpha chains, which can be three identical alpha chains or two or three different alpha chains.

Regardless, all collagen fibers are synthesized in the cell's endoplasmic reticulum as precursors, or procollagens, consisting of the three intertwined amino acid chains. In addition, extra polypeptides are attached at both ends that will eventually be trimmed off.

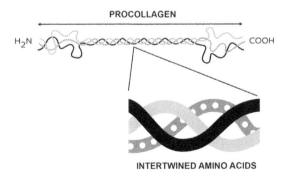

These pro-collagen strands are processed into tropocollagen strands and then get cross-linked to add stabilization. To get an idea of size, tropocollagen molecules are about 300 nm in length and about 1.5 nm in thickness. Several tropocollagen molecules are then clustered and cross-linked, finally creating the collagen fiber. The resulting molecular chain is now a hydrophilic or water-loving protein.

## TYPES

So far, twenty-eight different types of collagens have been identified in vertebrates, categorized into eight subfamilies. These collagens are grouped according to their structure, function, and tissue distribution and are designated by Roman numerals in the order of their discovery (I-XXVIII). [Clearly, they were first discovered by scientists during the late Roman empire]

## TYPE I[17]

Type I collagen is a fibrillar collagen and constitutes more than 90% of the organic mass of bone. It is also the essential collagen in tendons, skin, ligaments, cornea, and most interstitial connective tissues. In bones, it is responsible for biomechanical properties such as load-bearing and tensile strength.

To complicate matters, there are two subtypes of Type I collagen, one with identical alpha chains and one with two identical α1 and one α2 chain. These are located on genes Col1A1 and Col1A2.

I repeatedly say that we don't appreciate things until they fail, and this falls into that category. Mutations in Type I collagen are associated with a multitude of diseases, especially Osteogenesis Imperfecta and Ehlers–Danlos syndrome, bone and connective tissue diseases respectively.

## TYPE II

Type II collagen is composed of homotrimers, meaning it has three identical α1 chains and is the primary collagen in mammal cartilage, intervertebral discs, and in the vitreous humor of eyes.

Interestingly, Type II collagen has anti-inflammatory properties and reduces the destruction of other collagens within the body.

### TYPE III
Type III is also a homotrimer, widely found in conjunction with Type I collagen, and is prevalent in skin, blood vessels, and lymphoid tissue.

It is vital for elasticity as well as wound healing.

### TYPE IV [1]
Type IV is primarily in the skin and mainly within the basement membrane as a component of the dermal-epidermal junction.

The other essential collagen in skin is Type VII, a component of the anchoring fibrils which attach the epidermis to the dermis.

In fact, there are six types of collagen in the skin; Type I comprises 70%, with Type III at 10%, and trace amounts of Types IV, V, VI, and VII.

## COLLAGEN RECEPTORS
In addition to collagen providing a structural framework around the body, it can also serve as a messenger service. We know this as there are a variety of cellular receptors that collagen can bind to that precipitate various cell behaviors.

For example, several structurally diverse transmembrane receptor families recognize the collagen triple helix, including integrins, discoidin domain receptors, glycoprotein VI, and leukocyte-associated immunoglobulin-like receptor-1. These collagen receptors regulate a wide range of behaviors, including cell adhesion and migration, hemostasis, and even immune function.[22]

## COLLAGEN BREAKDOWN
Like any infrastructure system, including roads, bridges, and pipes, collagen fails over time and must be replaced. Thus the body has a built-in system for chronic remodeling, including the need to dissolve old collagen to make room for new collagen.

Collagen degradation is a multi-step process that relies on the activity of extracellular proteases to break down the ECM collagen and the subsequent movement of the pieces into cells for lysosomal degradation of the fragmented fibrils.

As seems reasonable, the enzymes that dissolve collagen are called collagenases, and of course, as there are many types of collagen, there are many types of collagenases. These enzymes are grouped with other deconstructing enzymes and are referred to as matrix metalloproteinases or MMPs. The roughly twenty-eight enzymes, or matrixins, are metalloproteinases that are calcium-dependent, zinc-containing endopeptidases. [I refer to MMP's throughout the book, so pay attention here!]

Thus collagenase I, a.k.a. MMP-1, hydrolyzes Type I, II, III, VII, VIII, and X collagens as well as gelatin. Of note, MMP-1 hydrolyzes Type III collagen molecules faster than Type I. The enzyme is locally produced, primarily by keratinocytes, fibroblasts, and macrophages.

Collagenase 2, a.k.a. MMP-8, also hydrolyzes I, II, III, VII, VIII, X, aggrecan, and gelatin, but it degrades Type I faster than Type III collagen. This enzyme is limited to specific granules within polymorphonuclear neutrophil cells.

Even though we do not actually contain jello in our bodies, we do have enzymes that dissolve it, i.e., gelatinases. [Stay tuned for more info on gelatin]

Gelatinase A, or MMP-2, degrades collagen Types I, II, III, IV, VII, X, fibronectin, elastin, and of course gelatin, thus the name.

Gelatinase B, or MMP-9, has a high affinity for gelatin but can also cleave collagen Types IV, V, and elastin. Gelatinase B is produced by eosinophils, macrophages, keratinocytes and is also stored in the PMN neutrophil granules.

Stromelysins are also in the family of MMPs and degrade proteoglycans, basement membranes, laminin, and fibronectin in addition to collagen.

Stromelysin-1, or MMP-3, degrades collagen Types II, IV, IX, X, XI, gelatin, proteoglycans, fibronectin, laminin, and elastin. The enzyme can also activate other MMPs such as MMP-1, MMP-7, and MMP-9.

The list of MMP's goes on and on, the gist of which is that they are helpful in the reconstruction and renovation of tissue.

Too much deconstruction is obviously not a great thing, so the body also has a control system for the MMPs, which are the TIMPs, or tissue inhibitors of metalloproteinases, of which there are four.

Not to get ahead of myself, but as you could expect, the MMPs get a bit out of hand with both age and inflammation, resulting in more damage than reconstruction.

## CHANGES OVER TIME

As one might expect, there are many avenues by which collagen fails with aging. "With the passage of time, aging elicits the reduction of 1% of collagen content yearly." [24]

1) Collagen production declines.

2) The ratio of collagen types shift.[3, 9] As an example, Type III collagen synthesis decreases with age resulting in a skewed Type I/III ratio and alters skin tension, elasticity, and healing.

3) Increased destruction of collagen.[23]

4) The collagen that is present declines in quality, becoming fragmented and disorganized.[7, 29]

Overall, the general outcome is that collagen fibers lose thickness and strength; tissue becomes thinner, weaker, and less supple, and dermal thickness is reduced.

As if the breakdown of collagen is not bad enough already, it is accelerated by sunlight, smoking, environmental pollution, alcohol abuse, menopause, and nutrient deficiency.[8]

As far as diseases, diabetes is especially detrimental to collagen as hyperglycemia reduces standard collagen production and it induces non-enzymatic glycosylation of collagen and keratin. This leads to abnormally rigid collagen and increased tissue breakdown.

Diabetic skin also has elevated levels of MMP-1 and MMP-2, a higher percentage of crosslinked collagens, and the collagen fibrils are fragmented and disorganized.[2]

The question then arises, after all of this collagen lecture, can we improve the status of our collagen? And the answer appears to be yes. In addition, exogenous supplementation seems to have other benefits as well.

... *"molecular results also indicate that the collagen drink could enhance the gene expression with respect to ECM protein synthesis, antioxidative enzymes, protein folding, DMR, and BER. In a word, the fish collagen drink exerts the synergetic effect on delaying skin aging by interfering with the aging parameters and compensating the oxidative damage and physiological loss in human fibroblasts."* [24]

The List...

## 2) MITOCHONDRIA: 2

There is evidence in mice [34,35] and obese rats [30,32] that collagen polypeptides have free radical scavenging capabilities, and collagen peptides with the highest total anti-radical activity contain around 14 to 15 amino acids.[37] The presence of hydrophobic amino acid residues within the peptide sequences is thought to explain these findings.[33]

In addition, there was a rise in the levels of superoxide dismutase and catalase in rats fed collagen fragments.[30, 32]

## 4) QUALITY CONTROL: 1

There is a bit, albeit not very much, evidence that collagen increases both DNA and protein repair mechanisms.

In cell cultures, collagen additives enhanced correct protein folding and mitophagy. [24]

*"In a word, DNA repair efficiency can be improved by the collagen drink."* [24]

## 5) IMMUNE SYSTEM: 2

In this category, we are going to go straight to the effect on MMP's.

The daily ingestion of collagen hydrolysate by four-week-old male rats altered the ratio of Type I and IV collagens in the skin as compared to controls, but more importantly, it decreased MMP-2.[39]

Likewise, fish-derived low-molecular-weight collagen peptides given to hairless mice promoted the recovery of collagen and elastic fibers in the skin after UVB radiation. They showed reductions in levels of MMP-3 and MMP-13 expression as well as the gelatinases, MMP-2 and MMP-9.[20]

A similar study in hairless mice exposed to UVB radiation demonstrated parallel reductions in MMP-2, -3, -9 and -13 after oral collagen peptides; in addition, there was a significant reduction in wrinkle formation, skin thickening, and trans-epidermal water loss.[28]

In terms of general inflammation, in obese rats, oral collagen fragments not only precipitated a decrease in body weight and a rise in superoxide dismutase and catalase, but it also lowered the serum concentrations of the proinflammatory cytokines IL-1β and IL-12 while increasing the concentration of the anti-inflammatory cytokine IL-10.[30]

## 6) INDIVIDUAL CELL REQUIREMENTS: 3

First off, we know that oral collagen is beneficial for fibroblasts. It promotes not only the proliferation of skin fibroblasts but also induces fibroblast migration. These fibroblasts then augment the production of collagen and other extracellular matrix components.

It is interesting that the oral collagen molecules do not embed themselves in tissues, but they actually induce the synthesis of brand new collagen.

In cell cultures, 0.125%, 0.25%, and 0.5% of collagen mixtures improved collagen production by fibroblasts by 11.1%, 17.9%, and 22.5% respectively.[24] In these same cells, the collagen mixture also enhanced the production of elastin; the same concentrations increased elastin production by 7.9%, 8.4%, and 10.9%.[24]

Meanwhile, in a double-blinded, human placebo-controlled study, a hundred and fourteen women aged 45 to 65 years were randomized to receive 2.5 g of the collagen peptide, VERISOL®, or placebo, once daily for two months. At its conclusion, researchers reported a statistically significant improvement in procollagen Type I (65%) and elastin (18%) compared to controls.[27]

## 7) WASTE MANAGEMENT: 1
Unexpectedly, it turns out that collagen peptides can even help control glucose.

In a mouse model, oral collagen improved glucose tolerance by inhibiting intestinal glucose uptake and enhancing insulin secretion.[18] Collagen supplementation in an obese rat model decreased fasting blood glucose levels and glycated hemoglobin.[30]

## SYSTEMS KNOWN TO BENEFIT

### SKIN
There have been innumerable studies looking to determine if oral collagen is beneficial, and in study after study, it has.[5, 10, 11, 16, 27, 31]

For example, oral supplementation with ten daily grams of Peptan®F (fish) or Peptan®P (porcine) in two different populations, i.e. one in Tokyo and one in Bordeaux, demonstrated that collagen peptides "very clearly and significantly increased the amount of glycosaminoglycans in the epidermis in a dose-dependent manner," "increase(d) collagen density in the human dermis" and "Peptan reduced collagen fragmentation in the human dermis."[4] In fact, oral supplementation with the fish collagen peptides increased the collagen density in the dermis by 9%.[4]

Reducing the intake to 2.5 and 5.0 grams of collagen peptides, women 35 to 55 years of age took part in a double-blinded, placebo-controlled trial for two months. The main finding was a statistically significant improvement in skin elasticity. This effect persisted even after the four weeks of follow-up. Also of note, no side effects were noted throughout the study.[26]

To take a birds-eye look at all of the available studies, researchers did a literature search to pool information. Eleven studies with a total of 805 patients were included in the review. Eight studies used collagen hydrolysate, 2.5 g/d to 10 g/d, for 8 to 24 weeks to treat pressure ulcers, xerosis, skin aging, and cellulite. The general conclusion was that "Oral collagen supplements... increase skin elasticity, hydration, and dermal collagen density."[10]

As a general consensus, the dose can vary between 2 to 10 grams daily, and the type of collagen can vary widely, but the results are pretty clear. Skin improves, and there are minimal to no side effects.

## EXERCISE

Collagen supplements can boost lean muscle gain, decrease recovery time after exercise, and improve damaged joints. But this is a book about longevity, not athletic performance, so we're not going to go into detail here.

## CARDIOVASCULAR SYSTEM

Collagen peptides are great for the cardiovascular system for several reasons. First, it lowers blood pressure. The antihypertensive effect of collagen has been reported in animal experiments as well as in clinical trials because it inhibits Angiotensin I Converting Enzyme (ACE).

In addition, collagen can block the development of atherosclerosis, at least in high cholesterol rabbits. It also inhibits platelet aggregation, which may help prevent excessive clotting.[32]

Lastly, in many rodent studies, collagen peptides reduce high fat-induced weight gain. It down-regulates serum levels of total cholesterol, triglycerides, and low-density lipoproteins and lowers blood glucose levels.[32]

## JOINTS

In several clinical studies, collagen has had positive effects on osteoarthritis, but less so for rheumatoid arthritis.[36]

*"The study demonstrated that collagen peptides are potential therapeutic agents as nutritional supplements for the management of osteoarthritis and maintenance of joint health."* [21]

## BONE

In ovariectomized mice on a diet of hydrolyzed collagen, the bone mineral density was significantly better than in the control group. There was also a significant and dose-dependent increase in alkaline phosphatase activity, a well-known marker of osteogenesis, and a decrease in osteoclast activity.[15]

The lack of human evidence in this category is stunning and shameful. There is no doubt that, at least in theory, collagen should benefit bone.[14]

## METABOLISM

Ingested collagen is broken down within the gastrointestinal system into di- and tripeptides, and then transported across the intestinal mucosa by the peptide transporter 1 (PepT1). In humans, studies have shown that di- and tripeptides containing hydroxyproline are detectable in the blood one to two hours after ingestion, peaking at two to four hours.

To date, more than thirty peptides (mainly dipeptides and tripeptides) have been identified in blood after gelatin and collagen intake, and Pro-Hyp is the most abundant collagen-derived peptide.[38] We also know that these peptides are transported throughout the body and can last in the skin as long as two weeks.[28]

Plus, unlike many of the agents we are familiar with, there is high bioavailability after oral ingestion. In rats, this hovers around 50%.[34]

## GELATIN

[This is going to seem like a side trip, but hang on and it will make sense] Gelatin is essentially jello, and who doesn't love the Jello jingle? "Watch it wiggle, see it jiggle,"...... But kudos to those folks that popularized jello, because it certainly wasn't easy.

Despite gelatin being around since the Egyptian pharos who used it as glue, it wasn't considered a food source until the French started consuming it in the 1680s. By the 1800s, the french utilized it as a significant source of protein, especially during the Napoleonic wars.

Charles Knox simplified the process of making gelatin in the 1890s, and his wife, Rose, published *Dainty Desserts,* a book of recipes using the Knox gelatin. The formula and the name Jell-O originated in 1895, but the product failed commercially secondary to the lack of funding. After changing hands a few times, Jell-O finally started to catch on in the early 1900s.

Jumping forward, the jello shot, made with vodka or rum, was invented in the 1950s.

And then oddly, in 2001, jello became the official Utah state snack food, apparently very popular with the Mormons. [We won't talk about Bill Cosby]

In reality, however, jello isn't that impressive or colorful.

Gelatin is simply a mixture of single or multi-stranded polypeptides, each with extended left-handed helix conformations containing 50 to 1000 amino acids, traditionally made from pigskin, bovine hide, bones, and other sources.

Because it comes from collagen that is partly broken down by hot water, gelatin contains mainly glycine, proline, and 4-hydroxyproline residues.

Thus collagen, gelatin, and hydrolyzed collagen are essentially the same things, just getting dismantled into smaller and small pieces.

Studies have in fact, shown that the bioavailability of hydrolyzed collagen is actually greater than collagen, sitting between 70 to 85%, which is not surprising as it is already partially digested.[38]

So, what type of collagen to take?

Collagen comes from numerous sources, including both land and sea creatures. Historically, it originated from cows and pigs, but marine sources such as fish and jellyfish have gained in popularity.

What's the difference?

Mammalian collagen contains significant amounts of hydroxyproline and hydroxylysine, and the total imino acid (proline and hydroxyproline) content is high. This is useful as the peptides more closely match what is in our own bodies.

On the other hand, collagen extracted from cows and pigs has had some challenges, with potential complications from outbreaks of bovine spongioform encephalopathy and foot and mouth disease, to autoimmune and allergic reactions. This is in addition to any personal choices stemming from religion or dietary restrictions.

In chemistry, an imino acid is any molecule that contains both imine and carboxyl functional groups. Imino acids are related to amino acids, which contain both amino and carboxyl functional groups, differing in the bonding to the nitrogen. *Of note, the Imino acid content of poultry is similar or slightly lower to that mammalian collagen.*

Today, many commercial collagens are extracted from marine life, mostly fish, sea urchins, sponges, and jellyfish. Fish, as a comparison, generally have less proline and hydroxyproline, but are higher in serine, threonine, and methionine. It turns out that the skin, bones, and scales of fish are quite rich in collagen, making this potential waste material both an eco-friendly and low-cost alternative.[12]

Thus, the choice of surf versus turf boils down to several things: desired amino acids and polypeptide composition and personal choice.

## DOSE

The choice of quantity is also a personal decision. Beneficial studies range from 2.5 grams per day up to 10. [On the other hand, Jell-O shots are certainly a fun alternative!]

# CHAPTER 12

# COENZYME Q10

(0.3.0.0.3.0.0)

BENEFITS:

HEART

> *"Coenzyme Q10 is one of the most significant lipid antioxidants, which prevents the generation of free radicals and modifications of proteins, lipids, and DNA."*
>
> ~ R. Saini 2011

I adore molecules that tell you exactly what they are without having to ask. CoQ10 is one such molecule.

It is a **Co**enzyme, a **Q**uinone, with **10** isoprenoid units. Simple, right?

Even better, because the quinone molecule is found in every plant and animal cell, it is ubiquitous and thus is called a ubiquinone. It is a *coenzyme* because of its unique ability to participate in chemical reactions while remaining in a relatively steady state within the cell.

More formally named 2,3 dimethoxy-5 methyl-6-decaprenyl benzoquinone, it is structurally similar to vitamin K, but it is not a vitamin as it is endogenously produced.

It has a molecular weight of 865 g/mol and exhibits limited solubility in fats and oils. Upon exposure to light and elevated temperatures, CoQ10 becomes progressively unstable, and its golden yellow color darkens to a deeper gold.

## HISTORY

First identified in 1940, CoQ10 was isolated in the mid-1950s by a research group at Liverpool University in the UK from the heart of a cow. A few years later, in 1958, Karl Folkers and Otto Isler in the US developed the first process for synthesizing CoQ10. However, it wasn't until 1975 that Peter Mitchell theorized that ATP synthesis was driven by an electrochemical gradient across the inner mitochondrial membrane. This characterization of biological energy transfer, in which CoQ10 was necessary, won him a Nobel Prize in 1978.[27]

Today we know that Coenzyme Q10 is a vital link in the electron transport chain, without which energy production is not possible. As we mentioned earlier, CoQ10 is ubiquitous in human tissues. Its level is variable, however, and the highest amounts can be found in organs requiring a lot of energy, such as the heart, kidney, and liver (114, 66.5, and 54.9 g/g tissue, respectively).[30] Therefore, it makes sense that a deficiency of CoQ10 would be associated with cardiovascular disease above all else.

Coenzyme Q10 is also known as Coenzyme Q, CoQ, CoQ10, Ubiquinone, Ubiquinone-Q10, Ubidecarenone, or Vitamin Q10; and it is composed of two pieces, a ring and a tail.

The ring is a benzoquinone ring, which can assume alternate redox states due to the existence of different possible levels of protonation, yielding three alternative CoQ forms:

*Ubiquinone: fully oxidized (CoQ)*
*Ubiquinol: fully reduced (CoQH$_2$)*
*Ubisemiquinone: partially reduced (CoQH) or ubisemiquinone*

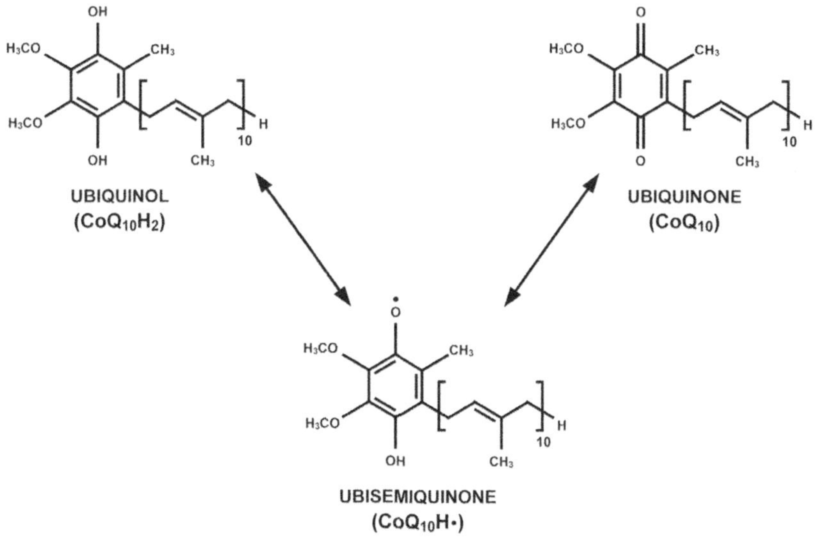

UBIQUINOL
(CoQ$_{10}$H$_2$)

UBIQUINONE
(CoQ$_{10}$)

UBISEMIQUINONE
(CoQ$_{10}$H•)

Tissues that are highly aerobic or have a high oxygen requirement contain more of the oxidized form (ubiquinone) than the reduced form; however, in the bloodstream, about 95% of CoQ10 is present in the ubiquinol or reduced form.

In fact, the ratio between ubiquinone and ubiquinol in the blood has shown to be a measure of oxidative stress.

The chain part of the molecule is a hydrophobic or water-fearing, isoprenoid chain of varying lengths, depending on the species.

It is the hydrophobic nature of this chain that allows it to get cozy within the lipid bilayer of the mitochondria, but it can also be found within the lipid bilayers of other cellular structures as well.

Of note, the number ten comes from the ten units or carbons in the isoprenyl tail, and interestingly, this number changes by species.

CoQ9 is the predominant form in rats and mice, while humans and other, long-lived mammals mostly have CoQ10 (we also have some 9).[34]

Meanwhile, *Escherichia coli* have eight, while *Saccharomyces cerevisiae* have six.

## SYNTHESIS

The creation of this large molecule is complex, requiring somewhere between eleven to thirteen genes for its biosynthesis.[11] The quinone structure is derived from tyrosine, the methyl groups are supplied by methionine via S-adenosyl methionine, and the isoprenoid side chain originates from the mevalonate pathway.[27]

The synthesis of endogenous CoQ10 also involves multiple compounds, including B-vitamins and vitamin C.

These nuclear genes send directions to the mitochondria, where the molecules are produced.

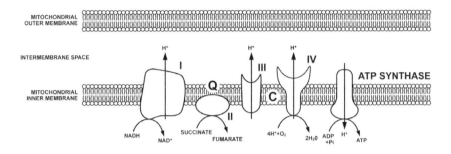

## WHAT DOES IT DO?

CoQ10 has two essential functions:
1.) It serves as an electron carrier in the ETC.

2.) It's a CoEnzyme for mitochondrial enzymes.

So, in the first category, CoQ10 is positioned between complexes I, II, and III, where it acts as a mobile electron carrier.

There are a ton of details that I am going to gloss over here, but in reality, unless you are a cell biologist, they aren't really that necessary to know.

*"Because human life cannot be sustained without this process, the need for ubiquinone is high."* [27] [I think that covers it]

In the second category, CoQ10 is important for:

1.) Sulfide detoxification

2.) Proline, arginine, and glycine metabolism

3.) Mitochondrial glycerol-3-phosphate dehydrogenase (G3PDH), which connects oxidative phosphorylation, fatty acid metabolism, and glycolysis

4.) Dihydroorotate dehydrogenase (DHOH), an enzyme involved in pyrimidine nucleotide biosynthesis, and thus key for DNA production

Numerous other functions of CoQ10 have been described as well, such as cell signaling, gene expression, and membrane stabilization. [It is also an antioxidant, but more about that a bit later]

## AGING

Why are we talking about CoQ10? Well, as per usual, things that we need and take for granted seem to decline with age.

During the aging process, and in some particular aging-related diseases, there is a significant reduction in the rate of CoQ10 biosynthesis.[26,27] In fact, the whole body content of CoQ10 is only about 500-1500 mg and this decreases with age.[30]

Whereas it is challenging to determine levels in the tissues of still-living humans, scientists have determined that the highest levels of CoQ10 occur roughly at the age of twenty and then decline. There are, however, discrepancies between studies about the levels of CoQ10 and aging, with some studies not identifying a consistent pattern.[7] The decline in CoQ10 is also not homogenous, at least in rats. In aged rodents, CoQ10 decreases in the heart, kidney, and specific muscles. In opposition to this, the concentration increases in the liver and is relatively constant in the brain, lung, and other muscles.[34]

The List...

## 2) MITOCHONDRIA: 3

CoQ10 has several roles in the mitochondria. As already mentioned, it is an essential element in the electron transport chain.

In addition, it's an antioxidant.

*"CoQ is one of the most powerful endogenously synthesized membrane antioxidants, being present in all membranes. Its antioxidant function efficiently protects lipids from harmful oxidative damage, but also DNA and proteins."*[34]

In such a role, it lends antioxidant protection to cell membranes and plasma lipoproteins, both in the mitochondria, but also in other lipid membranes. Embedded in low-density lipoprotein (LDL) particles, it can reduce lipid peroxidation and confers health benefits against cardiovascular diseases.[34]

Next, in its reduced form, ubiquinol, it can recycle and regenerate other antioxidants in the body[26], such as ascorbate and α-tocopherol.[34] Additionally, in this category, CoQ10 increases endogenous antioxidants.

Forty-three patients with at least 50% stenosis of one major coronary artery or that were undergoing percutaneous transluminal coronary angioplasty, were given either placebo, 60 mg, or 150 mg/ day for three months. In the high dose group, the plasma concentration increased significantly, and importantly, they had substantially higher catalase and superoxide dismutase activity than the placebo group.[18]

In another study of patients with coronary disease, supplementation with 100 mg of CoQ10 three times a day in a controlled, randomized study resulted in a significant increase in extracellular superoxide dismutase activity in those deficient patients.[27]

At this point, we are going to remain in the mitochondria category but take a brief field trip and examine something called the mitochondrial permeability transition pore (mPTP).

[This is a good time to put on your thinking caps]

The mitochondrial permeability transition pore is a non-specific channel located in the inner mitochondrial membrane. It is a multi-component mega-channel that allows for the nonselective diffusion of molecules greater than 1500 Dalton in size. It is also voltage-gated, activated by matrix calcium overload and reactive oxygen species.

The channel serves a bit like a pop-off valve. When a mitochondrion gets a bit stressed, this pore opens just a little bit to relieve that stress. These transient, very brief openings are sometimes referred to as flickering, lasting for milliseconds. The activity serves a physiological role by allowing for a quick exchange of solutes between the mitochondrial matrix and the cytosol. This process is thought to convey intracellular messages and help with calcium homeostasis.[35] These short, infrequent openings of the mPTP are also believed to trigger protective pathways. [28, 29]

On the other hand, sometimes this channel opens for more extended periods, up to seconds, akin to opening the flood gates. There is a rapid deluge of important matrix metabolites across the barrier, including superoxides, hydrogen peroxide, calcium, NAD, and glutathione flowing out while protons flow in.
As a result, the mitochondrial membrane potential collapses, and oxidative

phosphorylation and mitochondrial metabolism fails. This severe decline in membrane potential precipitates a drop in ATP production and actually utilizes ATP in repair attempts, leading to an overall energy deficit.

This event also causes the mitochondrial matrix to swell, and if the situation gets prolonged, the outer mitochondrial membrane eventually ruptures, releasing its components into the cell.

These agents, now in the cytosol, precipitate extensive damage to proteins, nuclear DNA, ion channels, transporters, and membrane phospholipids. Eventually, prolonged pore openings in enough mitochondria within the cell can lead to complete cell death.

The extended opening is also associated with bursts of reactive oxygen species that overwhelm the cell's antioxidant systems, resulting in extensive DNA damage. This damage precipitates an increase in PARP1 activity in an attempt to initiate repairs. This then leads to a decline in NAD+, a necessary substrate for the repair mechanisms.

Meanwhile, the NAD+ that has been released into the cytoplasm is then hydrolyzed by NADase, a.k.a. CD38, in the inter-membrane space, further reducing cellular stores of NAD.

This reduction in NAD becomes rate-limiting to many mechanisms, including the sirtuins and especially the mitochondrial-based SIRT3. This sirtuin generally controls the mitochondrial scavenging ability, which is already stressed beyond the capabilities of the cell. In an ironic twist, the expression of CD38 increases with age, which further enhances the destructive effects of mPTP opening during aging.

So what exactly stresses out mitochondria? It turns out that many things do… free radicals, oxidative stress, adenine nucleotide depletion, high phosphate concentrations, NAD deficiency, and excessive matrix calcium concentrations.[25] [Problems at work, issues with the spouse...]

The calcium homeostasis is thought to go haywire with aging as oxidative damage to calcium transporters and channels destroys the calcium balance.[28,29] The free radicals that activate the mPTP opening also create a negative spiral, further increasing ROS production. This is in addition to the free radicals that increase with age in the first place.

*"Thus, the increase in ROS production seen as a byproduct of aging initiates mPTP opening, but mPTP opening leads to further ROS production (H2O2) via proapoptotic signals."* [25]

This combination, especially the calcium overloading and the reactive oxygen species, contributes to the increasing frequency of pore opening.

In addition, the frequency of pore opening is age-dependent. The older you get, the more they open.

The pore complex itself is known to consist of many pieces and parts, but the only unequivocally identified regulatory component is cyclophilin D (CypD).[5]

This component is a peptidyl-prolyl cis-trans isomerase, translocated into the inner mitochondrial matrix during high matrix calcium conditions [22] and sensitizes the rest of the pore complex to free radicals and calcium. [Overwhelmed yet? It's almost over]

Functionally, CypD blocks an mPTP inhibitory phosphate-binding site, thus turning the channel on. Not unexpectedly, the concentration of CypD has been shown to increase with age, another reason explaining the enhanced mPTP opening over time.[28]

Luckily, SIRT3 confers protection in this department; it deacetylates CypD and can inhibit mPTP opening. Unluckily, with the levels of NAD declining, this ability is arrested.[17]

*"The various pathways that control the channel activity, directly or indirectly, can therefore either inhibit or accelerate aging or retard or enhance the progression of aging-driven degenerative diseases and determine lifespan and healthspan."* [29]

So how does one fight off this dreaded long-term, pore opening?
The most obvious thing to do first is to minimize the triggers by decreasing free

radicals or increasing available antioxidants.

*"Again, while the above therapies do not interact directly with the mPTP, they do reduce ROS levels or production within the mitochondria, which leads to inhibition of mPTP opening."* [17]

Treatment with Mito-TEMPO, a mitochondria-targeted ROS scavenger, blocks mitochondrial ROS generation and mPTP opening.[36] as does sulfuraphane,[14] astaxanthin,[3,30] and the chronic administration of melatonin.[3]

Nicotinamide adenine dinucleotide deficiency is another trigger; therefore, exogenous NAD can suppress the frequency and duration of mPTP opening.

Metformin also has been shown to inhibit the full activation of mPTP, but thankfully not the beneficial, transient opening of mPTP.[28] The exact mechanism of this is not clear, however. The same holds for spermidine and spermine. They inhibit pore opening but with no definitive etiology.[16,29]

It is also possible to directly inhibit the channel. Cyclosporine, for example, is a non-selective inhibitor of cyclophilins and inhibits pore opening. Because cyclosporin itself is a potent immune suppressant, this medication is not feasible for our purposes here, but molecular relatives might be a reasonable option at some point.

In terms of practical applications, control over these mitochondrial pores is considered vital to survival in ischemic cardiac disease.[3,10] In fact, the lengthy opening is thought to be the decisive factor in transitioning from a reversible to an irreversible myocardial injury.[20]

The pores also play a major role in diabetes and multiple neurodegenerative diseases.

Ok... back to CoQ10...

It turns out that the permeability transition pore hosts a ubiquinone binding site where different forms of CoQ10 are able to stabilize the pore in the closed conformation.[34]

## 4) QUALITY CONTROL: 0

There is a possibility of increased DNA repair in this category, but there is a paucity of detail. In a single study from 2001, lymphocytes with DNA damage but enriched with CoQ10 demonstrated speedier repair.[33]

The only other DNA connection is that CoQ10 is a coenzyme for Dihydroorotate dehydrogenase, an enzyme involved in pyrimidine nucleotide biosynthesis.

## 5) IMMUNE SYSTEM: 3

By contrast, in this category, there is significant evidence that CoQ10 reduces tumor necrosis factor in humans.[6,12,19,31,37]

Other studies have demonstrated a reduction in IL-6 and CRP as well. Not surprisingly, patients with higher baseline IL-6 levels see the greatest improvement.[12,31] Meanwhile, other studies have demonstrated no difference in IL-6 at all.[1,6]

These differences probably reflect variations between studies regarding dose ranges, patient types, and treatment times.

## SYSTEMS KNOWN TO BENEFIT:

### CARDIOVASCULAR DISEASE

Because the heart requires a lot of energy to continue contracting and pumping every minute of every day, it comes as no surprise that maintenance of the cardiac mitochondria is of the utmost importance. Therefore, the most obvious need for CoQ10 is clearly in patients with cardiac challenges. And in fact, "the severity of heart failure correlates with the severity of coenzyme Q10 deficiency." [2]

There is also emerging evidence that people with heart failure have elevated levels of reactive oxygen species, and coenzyme Q10 may help to reduce these toxic effects secondary to its antioxidant activity.[2]

In the Q-SYMBIO trial looking at moderate to severe heart failure, a two-year, prospective study using 300 mg/day in 420 patients demonstrated an improvement in symptoms and reduction in major cardiovascular events. Although short-term functional endpoints were not statistically different from placebo, CoQ10 significantly reduced the primary long-term endpoint—a major adverse cardiovascular event—which was observed in 15% of the treated participants compared to 26% of controls. Their conclusions: "Long-term CoQ10 treatment of patients with chronic heart failure is safe, improves symptoms, and reduces major adverse cardiovascular events."[23]

This study was repeated in Europe and reported in 2019 with similar results. "The therapeutic efficacy of CoQ10 demonstrated in the Q-SYMBIO study was confirmed in the European subpopulation in terms of safely reducing MACE (major adverse cardiovascular event), all-cause mortality, cardiovascular mortality, hospitalization and improvement of symptoms."[24]

The other improvement noted in several cardiac studies was that at doses above 150 mg/day, the activity of catalase and superoxide dismutase was elevated.[18,27]

As an additional benefit, a few studies have demonstrated a small but statistically significant decrease in both systolic and diastolic blood pressure in patients with Type II diabetes and dyslipidemia.[32]

## STATINS

Knowing that statins have tremendous value in the medical armamentarium, they do come with some drawbacks.

First, they are associated with a reduction in CoQ10 levels. Statins reduce cholesterol production by inhibiting the mevalonate pathway, which also produces ubiquinone. Thus, the more statin, the less CoQ10.

*"Statins inhibit 3-hydroxy-3-methylglutaryl–coenzyme A reductase (HMG-CoA), a rate-limiting step that converts HMG-CoA to mevalonate in the production of cholesterol."*[27]

Second, statins are associated with some unpleasant side effects, specifically called SAMS or statin-associated muscle symptoms. These can range from minor muscle aches to more severe muscle pains, severe cramps, muscle weakness, and, in rare instances, rhabdomyolysis. ( this is pretty rare but occurs most frequently in women and older adults). SAMS is hypothesized to be a result of mitochondrial failure precipitated by CoQ10 deficiency.

Therefore, it is advisable to take exogenous CoQ10 in order to ameliorate SAMS while taking statins.[26,27] Other medications also potentially lower

CoQ10 levels, including β-blockers[27], the anti-depressant amitriptyline, and some hypoglycemic drugs.[27]

## EXERCISE

The jury is still out as to whether or not CoQ10 improves exercise capacity.[8]

*"The previous studies investigating the effects of CoQ10 supplementation on physical performance in humans have found negative effects, no effect, positive effects, decreased exercise-induced muscular injury in athletes, and positive effects on aerobic and anaerobic threshold."*[4]

## MALE VIRILITY

One of the causes of idiopathic male infertility is thought to be associated with an increase in oxidative stress, and CoQ10 supplementation (200–300 mg/day) in men with infertility does, in fact, improve sperm concentration, density, motility, and morphology.[15]

## LIFESPAN

In *C. elegans*, exogenous CoQ10 prolongs life span, but there are mixed results in rodent studies.[34]

## METABOLISM

Unfortunately, the bioavailability of oral CoQ10 is terrible. Studies and estimates differ, but only 6% permeates the GI tract and gets into the bloodstream and organs in rats. Looking at it a different way, roughly 60% of the oral dosage is excreted in the feces.[34]

Luckily, there are ways to improve this a bit. For example, since CoQ10 is lipophilic, absorption increases if it is ingested with a fatty meal, as it gets aided by secretions from the pancreas and bile in the small intestine.

Regardless, following absorption, CoQ10 is reduced to ubiquinol and transported to the liver, where it is incorporated into very-low-density lipoprotein particles and released into the circulation. Only a minute amount gets incorporated into high-density lipoprotein particles.

The maximum plasma concentration after oral intake is reached after six to eight hours, and has an elimination half-life of over thirty hours. Once in the body, it accumulates in the liver and skeletal muscle the most, with medium levels in the heart and the least in kidneys and the brain.

It also takes two weeks after consistent intake to reach steady-state levels in the plasma, but it hangs around for several months once treatment has ceased.

The key here is that oral intake, despite its limited bioavailability, does get into the body as long as the dose is 100 mg/ day or greater.

## DOSE

Having cited 100 mg/ day in the last sentence, there is no actual dose; there are just educated guesses followed by clinical trials at different doses from which we make more guesses.

That being said, in neurological diseases such as Huntington's disease, Parkinson's disease, and amyotrophic lateral sclerosis, high doses of 600 to 3,000 mg have been utilized.

A more standard dose, however, falls within the range of 100 to 400 mg/ day.

For convenience, CoQ10 comes in 30, 60, 100, 200, 300, 400, or 600 mg aliquots and is available as soft gel capsules, an oral spray, hard shell capsules, and tablets.

Beware, however, the clinical effect is not immediate and may take up to eight weeks.

Another key here is that the endogenous biosynthesis of CoQ10 does not seem to be negatively influenced by the oral, exogenous supplementation.

One also needs to be cognizant of the exact formulation. For example, the reduced form (ubiquinol) has better bioavailability than its oxidized form (ubiquinone).[26]

In addition, the relative bioavailability depends on the type and amounts of oil in the formulations and its delivery system. The order of decreasing bioavailability is nano-particulated, solubilized, oil-emulsioned, and finally powder.[34]

## POTENTIAL SIDE EFFECTS

The potential side effects are generally infrequent and reported as mild, but include decreased appetite, diarrhea, dizziness, dyspepsia, rash, and nausea/vomiting.

## PRODUCTION

The most common method of producing CoQ10 is yeast or bacterial fermentation. This method tends to produce the all-trans isomer, which is structurally identical to normal, human CoQ10. Alternatively, other synthetic processes produce a mixture of trans- and cis-isomers.

If you prefer the even more natural way of intaking CoQ10, the best sources are oily fish such as salmon and tuna and livers from other animals. Non-animal products also contain fair amounts of CoQ10, the most derived from broccoli, spinach, soybean/canola/palm oils, nuts, and legumes. Levels obtained from the diet however, are pretty low, estimated to be only 3 to 5 mg per day.

# CHAPTER 13

# DELPHINIDIN

(2.3.2.1.2.1.2)

BENEFITS:

BONE

SKIN

*"It exhibits one of the highest polyphenol content and antioxidant activities of all fruits including the most consumed berries, being particularly rich in anthocyanin delphinium and cyanidin."*

*~ D. Vergara 2015*

The best thing about delphinidin? It's color. It's a marvelous, purplish-blue. It's the pigment in the grapes that make Cabernet Sauvignon. It's the beautiful blue in the flowers of the delphinium genus. And it's the color of eggplant.

What is it really? Delphinidin is an anthocyanidin, a water-soluble flavonoid mostly found in berries and grapes. Because the molecule absorbs light in the visible wavelength range, the darker and more beautiful the fruit, the higher the concentration of the molecule. Therefore, the most significant amounts are in blueberries, bilberries, and maqui berries.

The ripe maqui berry, *Aristotelia chilensis*, with a deep-black purple color, in fact, has the highest concentration of delphinidin.[30] Indigenous to Argentina and Chile, this plant is at least three times richer in total polyphenols than in all other berries tested, including red grapes, cranberries, strawberries, raspberries, blueberries, and blackberries. As a result of this, the radical absorbance capacity (TRAP) of maqui berry juice is about three times higher than the corresponding values for other berries.[1]

Like nearly all other anthocyanidins, delphinidin is pH-sensitive, i.e., a natural pH indicator, and changes from red in basic solution to blue in acidic solution.

Presently, maqui berries are commercially extracted to yield a polyphenolic extract, standardized to a minimum of 25% delphinidins, in addition to other anthocyanins, phenolic acids, and flavonols.

To clear up any confusion, anthocyanidins and anthocyanins are closely related but not identical. Both are red-blue plant flavinoids, but anthocyanin has an extra sugar molecule attached to it. Luckily, the maqui berry contains both molecules.

**DELPHINIDIN**

Delphinidin is the anthocyanidin, with no sugar group (above), and Delphinidin-3-glucoside is the anthocyanin with the added sugar.

**DELPHINIDIN-3-GLUCOSIDE**

In Chilean folk medicine, delphinidin is used for diarrhea, as an anti-inflammatory, and for treating fevers.

*"The currently available research on delphinidin points to significant photo-protective, antioxidant, anti-inflammatory and anti-aging virtues."* [30]

The List...

## 1) DNA ALTERATIONS: 2

Delphinidin inhibits the histone acetyltransferase activity of p300, which, like spermidine and fisetin, results in the up-regulation of autophagy.[24]

This inhibition of p300, in turn, inhibits the acetylation of p65, which is a subunit of NF-kβ, limiting the inflammatory response.[24]

Some studies also suggest that delphinidin can act as an "epigenetic demethylating agent of the Nrf2 promoter, (and) can activate the Nrf2-ARE pathway." [13]

Delphinidin also has a weak but measurable stabilizing effect on DNA and tRNA.[12, 17]

*"Delphinidin with a positive charge induces a more stabilizing effect on DNA duplex than quercetin and kaempferol."* [12]

Structural analysis has shown that delphinidin, as well as quercetin and kaempferol, all bind weakly to adenine and guanine in the major groove, and thymine in the minor groove as well as to the backbone phosphate group in DNA, improving structural stability.[12]

## 2) MITOCHONDRIA: 3

There is a plethora of evidence that delphinidin is a potent free radical scavenger.[12] It has the most activity of the flavinoids, effectively neutralizing peroxyl-, hydroxyl-, superoxide anion, and peroxynitrite-radicals, as well as singlet oxygen.[30]

*"At comparative concentrations, it is far more potent than ascorbic acid, catechin and quercetin."* [30]

In a head-to-head with astaxanthin (one of my all-time favorite molecules), officially Delphinol® vs. AstaReal®, there was "comparative antioxidant efficacy for the prevention of human plasma lipid peroxidation at same concentration level." Of note, the study identified a synergistic antioxidant effect of the two at a 1:1 ratio to prevent human plasma lipid peroxidation.[30] This may be because astaxanthin is fat-soluble, while delphinidin is water-soluble, so together, they may reach more cellular locations. [If you wanted to minimize your antioxidant supplement intake, this combination is probably the way to go]

Moving along, in cells, maqui berry extract suppresses light-induced photoreceptor cell death by inhibiting ROS production [27], and in obese mice, delphinidin supplementation successfully inhibits oxidative stress.[6] This agent also increases endogenous pathways by activating Nrf2.[15, 21]

Interestingly, the agent was cytotoxic at high concentrations in a cultured human umbilical vein endothelial cell model. However, low concentrations were protective and associated with an increase in intracellular glutathione.[7]

This molecule can also improve mitochondrial function. For example, delphinidin 3-O-glucoside can increase the activity of complex I within the electron transport chain.[25]

## 3) PATHWAYS: 2

Delphinidin is a known activator of sirtuins and AMP Kinase [10], as demonstrated in several mouse models.[3,5]

## 4) QUALITY CONTROL: 1

In multiple cell cultures, delphinidin has been shown to increase autophagy.[28]

*"Delphinidin activated cytoprotective autophagy to protect chondrocytes during oxidative stresses."* [15]

In fact, in human umbilical vein endothelial cells, delphinidin induced autolysosomal and autophagosome formation.[28, 31] This is no doubt related to the epigenetic effect on the p300 protein.

There, at present, are no human or even rodent studies examining this phenomenon. I have no doubt that over time, this rating score will be much higher.

## 5) IMMUNE SYSTEM: 2

Starting with human chondrocytes, delphinidin inhibits IL-1β-induced expression of COX-2 and the production of $PGE_2$.[8]

Moving to rodents, there are many mouse models where delphinidin inhibits the activity of NF-κβ.[6,18] This is hypothesized to occur through Fyn kinase, a member of the tyrosine kinase family involved in TNF-α-induced COX-2 expression. Delphinidin directly inhibits Fyn kinase activity.[9]

Additionally, in mouse epidermal cells, delphinidin suppressed UVB-induced COX-2 expression.[14] Lastly, in the flaky skin of mice, treatment resulted in a reduction in psoriasiform lesions, reduced infiltration of inflammatory cells, and decreased mRNA and protein expression of inflammatory cytokines.[22]

## 6) INDIVIDUAL CELL REQUIREMENTS: 1

In cell cultures, delphinidin seems to have beneficial effects on fat and bone. Delphinidin inhibits human mesenchymal stem cell adipogenesis, as well as improving osteogenesis and chondrogenesis.[23]

## 7) WASTE MANAGEMENT: 2

In a study of thirty-six prediabetic subjects aged nineteen to fifty, delphinidin significantly lowered postprandial blood glucose 60 and 90 minutes after rice intake, with a single 200 mg dose before food consumption.

In addition, the acute intake of delphinidin, in the absence of any carbohydrate exposure, lowered both postprandial fasting blood glucose and insulin one hour after intake in a dose-dependent and significant fashion.

*"A significant effect was obtained even for very low doses of Delphinol: only 60 mg of Delphinol, corresponding to 21 mg of maqui berry anthocyanins, was statistically significant for the basal glucose drop. A dose of 180 mg of Delphinol, corresponding to 63 mg of maqui berry anthocyanins, was effective in significantly decreasing fasting insulin levels."* [2]

## SYSTEMS KNOWN TO BENEFIT

### MUSCLE

In mouse muscles, delphinidin prevented muscle atrophy from muscle disuse (or just laziness) by preventing the degradation of skeletal muscle proteins.[19,20]

## BONE
Delphinidin suppresses the differentiation and function of osteoclasts.[19, 20, 30]

*"Delphinidin, one of the major anthocyanidins in berries, is a potent active ingredient in antiosteoporotic bone resorption through the suppression of osteoclast formation." [18]*

Although less significant than the effect on osteoclasts, in mesenchymal stem cells, delphinidin also exerts favorable effects on osteogenesis and chondrogenesis in addition to inhibiting adipogenesis.[23]

## SKIN
Various cell and murine models demonstrate that delphinidin protects keratinocytes by inhibiting matrix metalloproteinase activation, inhibiting the expression of inflammatory cytokines, and acting as an antioxidant.

These effects essentially contribute to the regenerative impact on the mechanical properties of cells exposed to UVB radiation [26] and may counteract the breakdown of dermal collagen.[30]

## PLATELET EFFECTS
Delphinidin-3-glucoside significantly inhibits human platelet aggregation in both platelet-rich plasma and purified platelets. This likely contributes to its protective role against thrombosis and cardiovascular events.[32]

## BRAIN
Whereas studies are lacking, delphinidin does cross the blood-brain barrier and probably has positive effects.[30]

## DOSE
The maqui berry, the richest known natural source of delphinidin, is available from many companies. Most studies, however, utilize Delphinol®, standardized to 25% delphinidin.

It has a rapid metabolism, but one of its degradation products is gallic acid, also a free radical scavenger. [7]

There are considerable variations in dose recommendations, from 60 mg to 1000 mg, with no reported side effects at any dose. The best bet is probably between 200 to 300 mg daily.

The one limitation at the moment with this agent is the lack of human studies. I have no doubt that with a tinge of time, researchers will discover more and more fantastic capabilities of this purple hero and we will all benefit.

# CHAPTER 14

# ECKLONIA CAVA

(0.2.1.0.2.2.3)

BENEFITS:

FAT HAIR SKIN

*"Ecklonia cava Kjellman is an edible seaweed, which has been recognized as a rich source of bioactive derivatives mainly, phlorotannins. These phlorotannins exhibit various beneficial biological activities such as antioxidant, anticancer, antidiabetic, anti-human immunodeficiency virus, antihypertensive, matrix metalloproteinase enzyme inhibition, hyaluronidase enzyme inhibition, radioprotective, and antiallergic activities."*

*~ I. Wijesekara 2010*


*Ecklonia cava*, also known as paddle weed, or kajime in Japan, comes to us from under the seas and halfway around the globe. [Depending of course, on where you are reading this]

Ecklonia inhabits warm, temperate sub-tidal waters off the coast of China, southern Japan, and Korea, but it also grows in the coastal waters of Australia, New Zealand, and South Africa. Regardless, the most famous Ecklonia is from Jeju Island in the Republic of Korea.

The plant itself is a large perennial brown alga of the family *Laminariaceae*, growing up to ten feet in length. Attached to a rocky substrate by multiple rootlike structures, each plant consists of one long stalk, usually three to six feet in length, with multiple blades growing from the stalk to form a single clump at the top, resembling a palm tree. These plants cluster together and are the major components of extensive, dense kelp forests.

This brown seaweed has historically been an essential part of the Japanese and Korean traditional diet. In addition, and the reason it appears here is its long-established place as a folk medicine and as a modern adjunct to longevity.

There are, of course, other marine brown algae, such as *Eisenia arborea*, *Ecklonia stolinifera* and *Eisenia bicyclis*, which have been studied for their potential biological activities. These seem to have many of the same properties, but *Ecklonia cava* is the star of the sea, probably because it has more phlorotannins than any other brown algae.

Unfortunately, the population of ecklonia has been shrinking, thought to be secondary to the rise in seawater temperature, along with increased grazing by sea urchins and herbivorous fish.

Like every other plant we have examined, ecklonia is composed of several molecules that are considered biologically important. These key components are the phlorotannins; unique marine, polyphenolic compounds which include eckol, dieckol, bieckol, [And of course, Heckle and Jeckle] phloroglucinol, phlorofucofuroeckol, and fucodiphlorethol. It also is rich in fucoidans, laminaran, fucoxanthin, fucosterol, vitamins, minerals, and proteins.

The List...

## 2) MITOCHONDRIA: 2

There is significant proof that the various components, as well as the extract in its entirety, have potent antioxidant and free radical scavenging activity.[2,4,16,25,26,55] Some studies make them comparable to vitamin C, while in others, ecklonia is superior.[16,25]

There is also limited evidence that eckol, one of the components, can increase the transcriptional activity of Nrf2, a critical regulator of binding to the antioxidant response element (ARE), a known pathway of increasing levels of endogenous antioxidants.[32]

*"Therefore, the effects of eckol on cell viability might involve dual actions: direct action on oxygen radical scavenging, as shown by DPPH radical, $H_2O_2$, OH radical scavenging, and indirect action through induction of anti-oxidative enzymes."* [23]

## 3) PATHWAYS: 1
While a few of the components contribute in this category, the lead player appears to be dieckol. Many studies have confirmed it's activation of AMP Kinase.[42, 68]

*"Our data demonstrate that dieckol inhibits lipid accumulation via activation of AMP signaling and cell-cycle arrest."* [9]

## 4) QUALITY CONTROL: 0
In one single study, there is preliminary evidence that triphlorethol-A, isolated from ecklonia, can activate the DNA repair system, at least in Chinese hamster lung fibroblast cells.[33]

## 5) IMMUNE SYSTEM: 2
In both cell cultures and mice, many of the components of *Ecklonia cava* have anti-inflammatory qualities; the standouts seem to be eckol, dieckol, bieckol, and fucoidan.

Eckol reduced TNF-α and NF-κβ in a human keratinocyte cell line[9], while dieckol suppressed NF-κβ activity.[11] Bieckol, meanwhile, down-regulated the production of TNF-α and IL-6.[66]

The fucoidans, in a macrophage cell line, reduced several pro-inflammatory cytokines, including TNF-α, IL-6, and IL-1β.[51]

Meanwhile, and not unexpectedly, the complete extract suppresses the production of tumor necrosis factor-α and NF-κβ in human endothelial cells and TNF-α, IL-1β, and IL-6 in human mast cells.[40]

The only study in mice demonstrated that eckol suppresses the production of interleukin-4, IL-5, IL-6, and IL-13 and NF-κβ.[14]

## 6) INDIVIDUAL CELL REQUIREMENTS: 2

*Ecklonia cava* can both decrease osteoclastic activity[39] and increase osteoblastic activity. If you recall, this means that less bone is dissolved while more is created.[1,2,57]

## 7 ) WASTE MANAGEMENT: 3

In both mice and rat models, *Ecklonia cava* and many of its components reduce blood glucose levels [30,48,52]

This occurs via several mechanisms.

First, absorption of carbohydrates from the gut is reduced by the inhibition of α-glucosidase and α-amylase, with activity noted to be more robust than the gold standard, acarbose.[35,48,49,58,69]

Secondly, the activity of several enzymes necessary for glucose processing is altered. In fat mice, phloroglucinol inhibits glucose production in liver cells by limiting phosphoenolpyruvate carboxykinase (PEPCK) and glucose-6-phosphatase, critical aspects of gluconeogenesis.[68]

In a second mouse study, glucokinase activity was enhanced, while glucose-6-phosphatase and phosphoenolpyruvate carboxykinase activities were significantly reduced.[50] In addition, the pancreas increased insulin production in both Type I and II diabetic, treated rodents.[21,30]

Lastly, many research studies attribute the glucose-lowering ability to its activation of AMP Kinase.[21,68] With all of these mechanisms in action, it is not surprising that human studies have demonstrated a decrease in serum glucose as well.[19]

For example, eighty pre-diabetic male and female adults enrolled in a randomized, double-blinded, placebo-controlled trial took 1,500 mg of dieckol daily. The study demonstrated a significant decrease in postprandial glucose levels after twelve weeks.[54]

In a second study, to drive the point home, ecklonia extract demonstrated the same reduction in postprandial glucose and insulin levels during the first thirty minutes compared with the placebo group.[41]

To round out the category, ecklonia extract and its components inhibit the formation of AGEs. Dieckol specifically had the highest inhibitory activity against fluorescent AGEs formation and showed about eighteen times inhibition when compared with aminoguanidine.[62]

I know what you are thinking...how does this compare to berberine and Metformin? It's a great question with no good answer as there have been no comparative studies. Feel free to trial them and report back!

## SYSTEMS KNOWN TO BENEFIT

*Ecklonia cava* holds many secrets, many of which are either unusual or just surprising. For example, it turns out that *Ecklonia cava* is a reasonably potent angiotensin I-converting enzyme or ACE inhibitor. ACE inhibitors are molecules that reduce blood pressure, and therefore E. Cava, at the proper doses, may reduce high blood pressure.[3]

Ecklonia can help people sleep,[7,8] protect hearing from loud noises,[6] decrease cholesterol levels,[10] and has anti-allergy qualities.[14,45,61,63] Ecklonia also has anticoagulant properties. It turns out that sulfated polysaccharide from E. cava strongly interferes with the coagulation cascade by inhibiting the biological activity of serine proteases II, X, and VII. In fact, at high doses, *ecklonia* "showed almost similar anticoagulant activity to that of heparin."[65]

I wouldn't be too concerned however, I've been on this for a while without any bleeding issues. That being said, I would shy away from high doses.

### ADIPOSE TISSUE

One of the most researched benefits of *Ecklonia cava* is fat reduction and consequent weight loss.

First, we'll start with the clinical trials. After twelve weeks of supplementation on a polyphenol extract from *Ecklonia cava*, ninety-seven overweight adults who were given either a high dose extract (144 mg), low dose (72 mg), or placebo were evaluated. Both the low and the high dose groups showed significant decreases in BMI, body fat ratio, waist circumference, and waist/hip ratio. This was in addition to improvements in total cholesterol, low-density lipoprotein cholesterol, total cholesterol/highdensity lipoprotein, cholesterol, and atherogenic index.[59]

These results were confirmed in another twelve-week, randomized, placebo-controlled, double-blinded study, where overweight subjects took ecklonia extract (Seapolynol, 360 mg/day). In the end, treatment significantly reduced body fat. In the abdominal obesity subgroup, the treated folks demonstrated significant decreases in total adipose tissue area, percent body fat and fat/ lean mass ratio.[47]

Tracing the actual mechanism has led to the discovery that ecklonia and its many components block fat production in a myriad of ways. But essentially, it has the power to control enzymes and pathways that build adipose tissue.

For example, Seapolynol, an extract of *Ecklonia cava*, down-regulates triglyceride synthetic enzymes such as diacylglycerol acyltransferase 1 (DGAT1) and GPAT3. In addition, it down-regulates the expression of Krueppel-like factor 4 (KLF4), KLF5, CCAAT-enhancer-binding protein β (C/EBPβ), C/EBPδ, and Protein C-ets-2 (ETS2), while KLF2, an anti-early adipogenic factor, was up-regulated.[18]

Dioxinodehydroeckol, another isolate from ecklonia, inhibits the differentiation of preadipocytes into adipocytes.[28]

As well, it "down-regulated the expression of peroxisome proliferator-activated receptor-γ, sterol regulatory element-binding protein 1 and CCAAT/enhancer-binding proteins in a dose-dependent manner."[37]

Realizing your brain just fogged over, the take-home message is clear; there are innumerable mechanisms by which fat production is blocked.[34,42,44,67] Of note, this is one of the areas that separates *Ecklonia cava* from the other brown seaweed; ecklonia activity is significantly more effective in inhibiting adipogenesis.[44]

## HAIR
Another cool thing about ecklonia is its effect on hair.[22]

Before getting into this discussion, there are just a few simple things that you need to know. Amazingly, hair, the hair shaft, and the cellular configurations are more complex than one might think. But here are the absolute basics.

First, hair grows in unique, repetitive cycles. Telogen is a resting phase, lasting up to three months. Anagen is the growth phase, lasting two to eight years, and catagen is the regression phase, lasting two to three weeks.

Second, hair growth is highly dependent on dermal papilla cells (DPCs), which serve as de facto stem cells at the base of the hair shaft. The real stem cells, the hair follicle and melanocyte stem cells reside in the bulge area of the hair follicle. Dermal papilla cells, however, are specialized fibroblasts and play a critical role in regulating hair follicle development through the secretion of diffusible proteins, such as insulin-like growth factor-1 (IGF-1), hepatocyte growth factor (HGF), vascular endothelial growth factor (VEGF), and transforming growth factor-β (TGF-β).

Third, known medications that increase hair growth, finasteride and minoxidil, are both known 5α-reductase inhibitors.

So how does *Ecklonia cava* affect hair growth?[5]

It turns out that ecklonia works through several processes. [This list is based on both complete extracts and specific components.]

1) Encourages the proliferation of dermal papilla cells

2) Elongates the hair shaft in cultured human hair follicles

3) Promotes the transition of the hair cycle from the telogen to the anagen phase [demonstrated in mice]

4) Increases the insulin-like growth factor-1 and vascular endothelial growth factor

5) Inhibits 5α-reductase activity [especially Dieckol]

6) Opens the K+ ATP channel

Thus, there are many indicators that ecklonia should help with hair growth. Stay tuned for any human studies for verification.

Interestingly, vitamin C also stimulates dermal papilla cells and promotes hair shaft elongation in vitro and in vivo animal experiments.[60]

## SKIN

Whereas there is minimal information about the effect of ecklonia on fibroblasts, there is some evidence that both fucoidan [56] and dieckol [64] can stimulate collagen production in human skin fibroblasts. [56]

The most significant evidence in this category, however, falls into the subcategory of skin preservation rather than construction.

For example, ecklonia as a whole, and its components, dieckol [56], and fucoidan [20,56] are all able to inhibit the expression of MMP-1. This is key as MMP-1 degrades Type I and Type III collagen. There is also evidence that phlorotannins from E. cava inhibit the expression of MMP-2 and -9. [20,56] [Remember the collagen chapter?]

Eckol and dieckol both strongly inhibit NF-κβ and AP-1, which is thought to be associated with their ability to inhibit MMP-1 expression. [71]

Ecklonia shampoo is also good for the scalp. In a human clinical trial, patient's scalps were observed for hydration and redness for one month. Treatment recovered the loss of cell viability and abnormal cell cycle distribution induced by urban pollutants and also attenuated pollutant-induced damage to the skin barrier by decreasing MMP-1. [29]

In another realm, dieckol is a tyrosinase inhibitor that is useful for the prevention of hyperpigmentation. [15,16]

The other skin benefit has to do with limiting inflammation. In a hairless mouse model, both topical and oral administration prevented skin carcinogenesis after prolonged UVB radiation exposure. This was accompanied by the down-regulation of inflammatory enzymes such as COX-2 and iNOS, which were amplified during the exposure. [9]

## DOSE

As per usual, the range of dose recommendations is broad. The lowest in the literature has been 72 mg daily, reaching up to 1500 mg per day.

The most effective dose was determined to be 360 mg per day. Remember, however, that every formulation is going to be a bit different. [41]

And in this case, there is real potential for side effects. For example, too much may lower the blood pressure beyond a desired level, and it may lead to bleeding issues as well.[70] Knowing this, I would stick to 300 to 360 mg per day.

# CHAPTER 15
# ELLAGIC ACID
## (1.2.2.2.2.1.2)

BENEFITS:

G.I. TRACT    HEART    SKIN

*"While ellagic acid's antioxidant properties are doubtless responsible for many of its pharmacological activities, other mechanisms have also been implicated in its various effects, including its ability to reduce the lipidemic profile and lipid metabolism, alter proinflammatory mediators (tumor necrosis factor-α, interleukin-1β, interleukin-6), and decrease the activity of nuclear factor-κβ while increasing nuclear factor erythroid 2- related factor 2 expression."*

~ J. Ríos, 2018

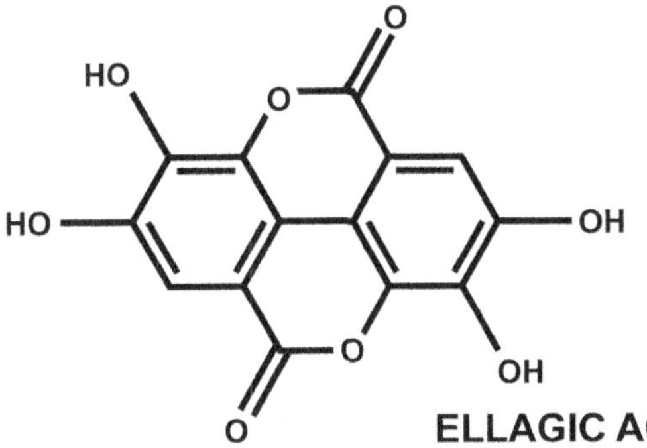

**ELLAGIC ACID**

Ellagic acid is simply a beautiful molecule. It's symmetric; it's elegant, sophisticated, intriguing. Like a beautiful woman, ellagic acid is not only attractive; it's also very complicated.

Discovered by Braconnot in 1831, it is a polyphenol present in many fruits and nuts but is exceptionally high in yellow raspberries, cloudberries, and pomegranates. It is also found in persimmons, regular raspberries, walnuts, and pecans. Of note, cognac contains a reasonable amount of ellagic acid as well. [see...sophisticated]

[Here's a bit of fun, useless information...]

*Cloudberries are soft, juicy, and golden yellow with a tart taste. They are a delicacy in Sweden, Norway, and Finland, as they can be cultivated in Arctic areas where few other crops are possible. In Sweden, cloudberries and cloudberry jam are used as a topping for ice cream, pancakes, and waffles, and in Norway, they are mixed with whipped cream and sugar into a dessert called multekrem or cloudberry cream. In the Alaskan Arctic, the Yup'ik mix the berries with seal oil, reindeer, or caribou fat and sugar to make "Eskimo ice cream" or akutaq.*

Back to business...

Ellagic acid is complex and confounding. It can go both ways in terms of solubility; it possesses a hydrophilic moiety with four hydroxyl groups and two lactones, along with a hydrophobic piece with two hydrocarbon rings.

Second, its metabolism can be a bit complex. Ellagic acid can be absorbed as it is or as an ellagitannin, which gets absorbed differently.

• Free ellagic acid can be absorbed in the stomach and proximal small intestine.

• Ellagitannins are hydrolyzed to release ellagic acid for absorption in the small intestine.

• The unabsorbed ellagic acid and ellagitannins continue on and are metabolized by gut microbes, generating urolithins in the colon, which are then absorbed.[15]

Thus, the bioavailability of ellagic acid is regulated by multiple factors, including the ellagic-to-ellagitannin ratio, the pH of the GI tract, and the gut microbiota composition.

ELLAGIC ACID → UROLITHIN D → UROLITHIN C → UROLITHIN A

As ellagic acid is known to have low bioavailability, comparable with that of resveratrol, the real functioning molecules may be the urolithins. These are created by the opening and decarboxylation of one of the lactone rings of ellagic acid, with the sequential removal of hydroxyls from different positions. [The urolithins are also relevant constituents of shilajit, which has its own chapter]

*"This poor bioavailability and the extensive gut catabolism suggest that urolithins rather than ellagitannins or ellagic acid may be the actual bioactive molecules."* [11]

The List...

## 1) DNA ALTERATIONS: 1

It appears that ellagic acid and its metabolites, the urolithins, have epigenetic effects.

First, ellagic acid increases histone deacetylase-9 expression.[15] It also decreases histone acetyltransferase activity in TNF-α-activated human monocytic cells, diminishing the inflammatory response and improving cell survival.[9] In addition, it inhibits histone arginine methyltransferase, an enzyme necessary for adipogenesis.[15]

Meanwhile, the metabolites urolithin A, B[11], and C also inhibit histone acetyltransferase activity.[15]

Of note, ellagic acid can bind directly to DNA, likely leading to the protection of binding sites from free radicals.[28]

## 2) MITOCHONDRIA: 2

Ellagic acid has high radical scavenging activity[14], and, in fact, is more potent than vitamin E.[28]

The presence of the four hydroxyls and two lactone functional groups enables ellagic acid to scavenge a wide variety of oxygen and reactive nitrogen species. Studies show that at physiological pH, ellagic acid in an aqueous solution can deactivate not only hydroxyl radicals but also peroxyl radicals, nitrogen dioxide, and peroxynitrite.[28]

Ellagic acid is also efficiently and continuously regenerated after scavenging two free radicals per cycle. This is advantageous, as well as unusual, in that it continues to have antioxidant activity even at low concentrations.[12]

Unfortunately, urolithins do not possess free-radical scavenging activity.[18]

In human dermal fibroblast cultures, ellagic acid up-regulates the Nrf2 signaling pathway [28], which explains the increased activity of superoxide dismutase, catalase, and glutathione peroxidase.[14]

In obese mice after supplementation for fourteen weeks, oxidative stress–induced endothelial dysfunction and atherosclerosis was attenuated through Nrf2 activation[15], and in lymphoma-prone mice, ellagic acid increased the expression and the activity of the antioxidant enzymes catalase, superoxide dismutase, glutathione peroxidase, and glutathione reductase.[28]

Turning to humans, supplementing the diet with pomegranate juice (1%) for 6–30 days in pregnant women significantly decreased oxidative stress and apoptotic cell death in the placenta.[15]

Lastly, in this category, ellagic acid appears to be an effective metal chelator, especially for copper[12], and iron. [28]

## 3) PATHWAYS: 2

In cell culture, a relatively low concentration of ellagic acid for thirty minutes induced AMP Kinase activation in fully differentiated 3T3-L1 cells.[15]

The urolithins A, C, and D, as well, induced AMP Kinase in cultured human adipocytes.[15]

*"Activation of AMPK is triggered by EA, as well as by urolithin A, C, and D in cultured human adipocytes."* [15]

In rats on a high cholesterol diet, ellagic acid induced AMP Kinase phosphorylation in the liver.[19] Additionally, urolithin A activated AMP Kinase and was shown to promote mitophagy via this mechanism.[32]

There is some limited evidence that ellagic can increase the activity of SIRT6,[27] as well as SIRT3.[17] Meanwhile, urolithin A increases SIRT1 in microglial cultures.[31]

## 4) QUALITY CONTROL: 2

In terms of DNA repair, female mice, on a diet supplemented with ellagic acid and dehydrated berries with varying ellagic acid contents, demonstrated a 3 to 8 fold increase in the expression of genes involved in DNA repair: xeroderma pigmentosum group A complementing protein (XPA), DNA excision repair protein (ERCC5) and DNA ligase III (DNL3).[1] Realizing there is no evidence for this in humans yet, I believe ultimately there will be and this may be crucial in the battle against DNA damage.

There is also evidence that urolithin A increases the activity of autophagy[31], as well as mitophagy.[32]

The only caveat here is that ellagic acid also may deter autophagy as it selectively blocks histone H3R17 methylation.[5]

## 5) IMMUNE SYSTEM: 2

Many rodent studies demonstrate that ellagic acid reduces gene expression of the proinflammatory cytokines IL-1β, IL-6, TNF-α, NF-κβ, while increasing the anti-inflammatory cytokines (IL-10). [20,28] In addition, it reduces histamine release from mast cells.[8]

To date, two studies have demonstrated ellagic acid's ability to decrease serum levels of IL-17 in experimental mouse models.[28]

Of note, urolithin A and B inhibit $PGE_2$ by reducing the expression of the two primary enzymes responsible for synthesizing prostaglandins under inflammatory conditions (mPGES-1 and COX-2). Ellagic acid, meanwhile, had no effect.[11]

## 6) INDIVIDUAL CELL REQUIREMENTS: 1
Although there is limited evidence, ellagic acid has been shown to precipitate a proliferation of neural stem cells through the Wnt/β-catenin signaling pathway.[21]

## 7) WASTE MANAGEMENT: 2
Ellagic has substantial activity in this category, with known ability to suppress the formation of AGEs[13], acting mainly on the inhibition of carboxyethyl lysine (CEL)[25] and carboxymethyl lysine (CML) formation.[26]

Because of this, ellagic acid helps prevent the loss of lens transparency in the eye through the inhibition of AGEs.[25]

In addition, ellagic acid can protect superoxide dismutase against glucose-induced glycation.[3] Interestingly, urolithin A, but not B, also showed significant anti-glycation activity with increasing concentrations.[11]

Ellagic acid is also valuable in the glucose category. In innumerable rodent models, treated diabetic animals have shown reduced blood glucose and HbA1c levels. [6,7,28]

This is thought to occur as "ellagic acid exerts anti-diabetic activity through its effects on pancreatic β- cells, increasing both their size and number, as well as on serum insulin and antioxidant status, all while decreasing blood glucose."[28]

Ellagic acid also has inhibitory effects on α-amylase, hexokinase, glucose-6-phosphatase, and fructose-1,6-bisphosphatase activities, thus reducing circulating sugar levels.[28]

Realizing that the rating on this agent isn't very impressive at the moment, I believe it suffers from the lack of looking. Perhaps with a little time and curiosity, researchers will be tempted into doing the studies necessary to prove its worth.

# SYSTEMS KNOWN TO BENEFIT

## METABOLIC SYNDROME
In many rodent studies, improvements have been documented in terms of improving the lipid panel [2,15,28,30], attenuating obesity[15], and improving metabolic syndrome overall.[15,28]

Middle-aged, overweight men between 45 to 55 years took 50 mg of ellagic acid for three months. Treatment "improved the levels of blood lipid metabolism with a 4.7% decline in total cholesterol, 7.3% decline in triglycerides, 26.5% increase in high-density lipoprotein, and 6.5% decline in low-density lipoprotein."[22]

## GASTROINTESTINAL TRACT
There is a reciprocal relationship between ellagic acid and the microbes in the gut. Ellagic acid alters the gut consistency, and these bacteria better metabolize the ellagic acid. For example, after four weeks of 1,000 mg of pomegranate extract, participants increased the diversity of their microbiomes and altered the ratio of *Firmicutes* to *Bacteroidetes*.

We know that healthy individuals generally possess microbiota that can generate more urolithins, except isourolithin A and urolithin B, two inactive metabolites that are more common in people with chronic illnesses. Of note, people with more urolithins tend to have fewer *Firmicutes*.

## SKIN
In skin models, topical ellagic acid reduces inflammatory cytokines[4,34] and edema.[23]

Topical ellagic acid also decreases matrix metalloproteinase production in human skin cells and hairless mice after UV radiation, helping to prevent collagen degradation.[4]

In addition, ellagic acid has skin whitening properties, inhibiting UV-induced skin pigmentation.[33]

> *"EA is thought to suppress melanogenesis by reacting with activated melanocytes and without injuring cells."*[29]

In a study of fifty-four multiethnic subjects, 0.5% topical ellagic acid and 0.1%

salicylic acid was applied twice daily for twelve weeks. In this double-blinded study, the test product demonstrated comparable tolerance and efficacy to that of 4% hydroquinone.[10]

The melanogenesis inhibition is thought to occur as ellagic acid binds copper, which is necessary for melanin-producing enzymes to act, and thus prevents the hyperpigmentation.[16]

Lastly, in a model of skin healing, 13% topical ellagic acid increased the tensile strength of incision wounds by 32%.[24]

## DOSE

Ellagic acid itself has poor oral absorption, and plasma levels peak between 30 minutes to an hour.

Urolithins, on the other hand, have a much higher absorption rate, presumably due to an increased lipophilicity, and are 25 to 80-fold more bioavailable. As a consequence, urolithins are present in the circulation in much higher concentrations. They take longer to reach peak concentration, i.e., up to twenty-four hours but persist in the circulation up to forty-eight hours.

There is a wide degree of variability between humans in the metabolism of ellagic acid as determined mostly by gut microbiota.[15] Therefore, the dose recommendations are very broad, somewhere between 30 to 850 mg of ellagic acid daily for an average adult.[15] Most studies use either 100 mg or 200 mg daily, which is probably the best recommendation.

The good news is that there seem to be minimal side effects, even at extremely high doses.[28]

# CHAPTER 16

# FISETIN

(3.2.2.2.3.3.2)

BENEFITS:

BONE    BRAIN    FAT    LUNGS

*"Fisetin, a newly characterized senolytic drug was recently demonstrated to significantly extend lifespan and healthspan, including musculoskeletal function, in progeroid mice."*

~ S. Hambright 2020

Fisetin is extraordinarily common in the natural, plant world, found in species as varied as the acacia tree, to the Venetian sumac, to everyday fruits. It was first isolated from the smoke tree in 1833, but today can be derived from apples, grapes, nuts, vegetables, herbal juices, wines, and teas. Strawberries have the highest content at 160 mg/gm; apples are next at 26.9 mg/gm. Unlike its chemical relative quercetin, there is very little fisetin in onions, coming in at a low 4.8 mg/gm.

Fisetin is also yellow. In fact, in traditional dying methods, it is the key component in a yellow dye called young fustic, created from the wood of the smoke tree. This is not to be confused with old fustic, which comes from a different tree, and is apparently more permanent. Regardless, this hydrophobic polyhydroxyflavone, a.k.a. 3,3′,4′,7-tetrahydroxyflavone, is a flavonoid and a polyphenol, and carries substantial health benefits.

Because it is anti-hyperlipidemic, an antioxidant, anti-inflammatory, anti-hyperglycemic, and has neurotrophic benefits, unlike many of the agents we have identified, there are many human studies either proposed or currently underway.

One might say that there is a fruitful body of scientific literature. At present (2022), the Mayo Clinic has recently begun a clinical trial aimed at the *Alleviation by Fisetin of Frailty, Inflammation, and Related Measures in Older*

*Adults* (AFFIRM-LITE) with fisetin administered orally in doses up to 20 mg per kilogram.[7]

Also, at the time of this writing, the same group under James L. Kirkland at the Mayo Clinic was looking into the value of fisetin in battling COVID-19 (SARS-CoV-2).

The List...

## 1) DNA ALTERATIONS: 3

In the epigenetics category, fisetin reduces histone acetyltransferase (HAT) expression at p300 (EP300) in human monocyte cultures. Therefore, like delphinidin and spermidine (which we will get to shortly), it can increase autophagy in addition to decreasing the inflammatory response.[16]

EP300 directly inhibits acetylation of several autophagy-essential and autophagy-related proteins, and indirectly stimulates deacetylation of tubulin by inhibiting α-tubulin acetyltransferase 1 (αTAT1). Therefore, the inhibition of EP300 precipitates the deacetylation of ATG proteins and increases acetylation of tubulin, thus stimulating autophagic flux.

Fisetin also improves DNA structural support.

In addition to the standard DNA spiral and the coiling around histones, another level of DNA complexity occurs in areas rich in guanine. Called a G-quadruplex, this configuration creates square-like structures that are stacked on top of each other. Biochemical studies have demonstrated that fisetin binds to this DNA structure, improving its structural integrity.[37] (Please refer to the Magnesium chapter for pics)

Stabilizing the G-quadruplex not only preserves DNA but is also thought to be protective against cancer. This is due to abundant G-quadruplex sequences, i.e., rich in guanine, that form in the promoter region of several oncogenes. The prevailing thought is that increasing the stability of these regions can decrease the incidence of malignancy. It is not clear exactly how this binding occurs, but the ligands are thought to stack externally along the G4 structure.[3]

## 2) MITOCHONDRIA: 2
Fisetin is a known free radical scavenger and is more robust than ascorbic acid.[38] In multiple studies, it has demonstrated activity against the superoxide anion and hydroxyl radicals. It has also been shown to protect against lipid peroxidation.[15]

In addition, fisetin induces the transcription factor Nrf2, thus up-regulating endogenous antioxidants, especially glutathione.[25] Lastly, it chelates metal ions.[28]

## 3) PATHWAYS: 2
In multiple cell culture models, fisetin increases the expression of SIRT1.[12, 17]

In a rat model of natural aging, six weeks of fisetin treatment up-regulated SIRT1 in the brain.[39] As well, in obese mice, treatment increased both AMP Kinase as well as SIRT1 in liver tissue.[21]

Our agent can also inhibit mTOR signaling.[1] This was demonstrated in cultured preadipocytes, where it "efficiently suppressed the phosphorylation of Akt, S6K1 and mTORC1 in adipose tissue."[14]

## 4) QUALITY CONTROL: 2
Due to its epigenetic manipulation of EP300 in aging rat models, fisetin up-regulates the expression of autophagy genes (Atg-3 and Beclin-1).[39] In a mouse model of Tylenol-induced liver injury, fisetin increased autophagy and inhibited the inflammatory cascade both in vivo and in vitro.[44]

## 5) IMMUNE SYSTEM: 3
There is significant evidence in both cultured cell lines and rodent models attesting to the suppression of the inflammatory response.

In a human basophilic cell line, fisetin suppressed the expression of TH2-type cytokines (IL-4, IL-13, and IL-5) by basophils. "Among the flavonoids examined, kaempferol and quercetin showed substantial inhibitory activities in cytokine expression but less so than those of fisetin."[11] In diabetic, neuropathic rats, fisetin treatment reduced interleukin-6 in sciatic nerves.[35]

Overall, fisetin decreases:

IL-1β, TNF-α, IL-6, COX-2, iNOS, MMP-3, and MMP-13 in bone models [46],

NF-κβ, IL-4, IL-5, and IL-13 in asthmatic mice [9],

TNF-α, IL-1β, and IL-6 in UVB-exposed skin of hairless mice [29],

and TNF-α, IL-1β, IL-6, and COX-2 in the colons of mice with colitis.[34]

Interestingly, fisetin is also a potent MMP inhibitor, inhibiting MMP-1, MMP-3, MMP-7, and MMP-9.[31]

In a human double-blinded, randomized placebo-controlled clinical trial, patients were given 100 mg of fisetin for seven consecutive weeks during chemotherapy for colorectal cancer. Fisetin significantly reduced the plasma levels of IL-8, hs-CRP, and MMP-7 levels as compared to placebo.[8]

## 6) INDIVIDUAL CELL REQUIREMENTS: 3

To start this conversation, it is essential to recall what senescent cells are. Some folks call them zombie cells; I like to think of them as grumpy old man cells. But the most important thing to understand is that these cells were once normally functioning then became abnormal and consequently toxic. This process is usually initiated by the accumulation of DNA damage, followed by the cells monitoring system arresting the cell in a quiescent or resting state. During this *time out*, the cell makes its best attempt at repairing the damage. If the repair process fails, the cell can either commit cell suicide, a.k.a. become apoptotic, or become senescent.

Senescent cells change morphology, become more plump and disorganized, the organelles become less functional, and the cell exudes the SASP, or Senescent-Associated Secretory Phenotype, which are essentially inflammatory cytokines. These cytokines are toxic both locally as well as systemically. Locally, neighboring cells become affected by these cytokines, becoming senescent themselves. [The one bad apple in the barrel analogy]

Senescent cells tend to accumulate at areas of tissue injury, areas of chronic inflammation, and especially after insults such as chemotherapy and radiation. Luckily, there is clear evidence that clearing out these cells is beneficial, and presently innumerable researchers are searching for molecules that can eradicate these cells. Unfortunately, at the moment, the only readily available options are quercetin, Danasatib, and fisetin.

Quercetin is easy to acquire; Danasatib is less so as it is actually a chemotherapy agent.

Thus, fisetin, a relative of quercetin, is only one of the two viable options available at this time. In a prematurely-aged mouse model, weekly doses of fisetin initiated at three months of age attenuated bone loss density, which was attributed to its senolytic capabilities.[10]

In a comparative study of various flavonoids, fisetin was the most potent senolytic in progeroid and old mice. Both acute and intermittent treatments with fisetin "reduced senescence markers in multiple tissues, consistent with a hit-and-run senolytic mechanism." As well, in normally-aged mice, fisetin "restored tissue homeostasis, reduced age-related pathology, and extended median and maximum lifespan."[43]

Fisetin has also been shown to reduce senescence in a subset of murine and human adipose cells, demonstrating cell-type specificity.[43]

Combined with Dasatinib, fisetin also is effective in monkeys.[5]

At present, human studies are pending.

## 7) WASTE MANAGEMENT: 2

In several rat studies, fisetin significantly decreased blood glucose levels and consequent glycosylated hemoglobin but elevated plasma insulin.[32,33]

In a new and exciting way of dealing with AGEs, fisetin also increases the level and activity of glyoxalase 1, the enzyme required for methylglyoxal removal, one of the major intermediate compounds implicated in AGE formation.[25]

## SYSTEMS KNOWN TO BENEFIT

### BRAIN

In cell culture, fisetin promotes the differentiation of nerve cells through the activation of extracellular signal-regulated kinases.[23,26]

But does it get into the brain? Yes, at least in mice. Following both intraperitoneal injection and oral administration, fisetin slowly disperses into the brain parenchyma.[18] Once in the brain, fisetin facilitates long-term potentiation in rat hippocampal slices and enhances object recognition.[23] It also has neurotrophic and neuroprotective properties in CNS neurons, enhancing memory and improving behavioral outcomes following ischemic stroke.[25]

*"We have recently identified and characterized the flavonoid fisetin as an orally active neuroprotective and cognition-enhancing molecule."*[25]

In mouse models of despair [that really is what they are called], fisetin produces an increase in serotonin and noradrenaline levels in the frontal cortex and hippocampus.[45]

## ASTHMA

In a plethora of rodent models, fisetin helps control asthma, reducing eosinophils and cytokines in the lungs.[9,13] It also attenuates lung inflammation, goblet cell hyperplasia, and airway hyperresponsiveness.[42]

*"These results suggest that fisetin may have potential as a therapeutic agent for the treatment of allergic diseases."*[13]

## BONE

Put in the simplest of terms, fisetin improves bone quality.

More specifically, it inhibits osteoclast differentiation and decreases the bone-resorbing activity of mature osteoclasts.[4,19,20,36]

These findings indicate that fisetin inhibits mature osteoclast formation by regulating NF-κβ activity via down-regulation of the p38-c-fos-NFATc1 signaling pathway.[2]

In addition to repressing osteoclasts, it also promotes osteoblast formation and activity.[19]

## ADIPOSE TISSUE

As a great beginning in this category, fisetin suppresses the early stages of preadipocyte differentiation.[14,17] In already formed adipocytes, or fat cells, treatment significantly inhibits lipid accumulation.[38] Finally, obese mice injected intraperitoneally with fisetin for ten weeks demonstrated decreased body weight and epididymal adipose tissue weight. The treatment also reduced liver lipid droplet and hepatocyte steatosis and significantly decreased fatty acid synthase.[21]

## DOSE

In the phase I/II randomized, Steadman Clinic and Steadman Philippon Research Institute (SPRI) trial, the proposed dose for eradicating senolytic cells will be 20 mg/kg for two consecutive days, followed by twenty-eight days off, then two more consecutive days. For an average 70 kg person, this works out to be about 1,400 mg.

Studies with other goals generally use 100 mg daily.[8, 41]

This wide range represents several things. Higher doses administered in a bolus fashion are believed to be more potent for treating senolytic cells, while other benefits can be derived from much lower, consistent doses.

Therefore, based on one's goals, there are several possible strategies. One could go with a smaller consistent dose or the bolus strategy. I believe that a combination of the two might be optimal, i.e., take a smaller dose on a regular basis, with monthly, elevated bolus doses. Following this strategy, use 1,200 to 1,400 mg on two consecutive days, once a month. On all of the other days, go with 100 mg.

Lastly, for those who want to follow a more natural route and eat fruit instead of swallowing capsules, a pound of freeze-dried strawberries yields roughly 72 mg of fisetin, thus at the minimum, one would need to eat around 1.5 pounds of freeze-dried strawberries daily for the smaller dosing strategy.

In comparison, a therapeutic dose of regular strawberries would be roughly fifty pounds daily.

# CHAPTER 17

# GANODERMA LUCIDUM

## (0.3.1.1.3.0.3)

BENEFITS:

ACCLIMATION  BRAIN  FAT  G.I. TRACT  IMMUNITY  LUNGS

*"A broad spectrum of Ganoderma lucidum's pharmacological actions have been established which include immunomodulation, anticancer, antidiabetic, antioxidant, antiatherosclerotic, antifibrotic, chemopreventive, antitumor, anticancer drug toxicity prevention, analgesic, anti inflammatory, antinociceptive, antimicrobial, hypolipidemic, hepatoprotective, antiandrogenic, antiangiogenic, antiherpetic, antiarthritic, antiosteoporotic, antiaging, antiulcer properties and estrogenic activity."*

*~ F. Ahmad 2018*

Considered an elixir of immortality, this spiritual mushroom has been popular in Asian countries for 2,000 years. The ancient remedy is esteemed for increasing energy and improving health and longevity throughout China, Korea, Japan, Malaysia, and Russia.

References to this mushroom date back to the early Han Dynasty, as described in the Shennong Bencao Jing. A thousand years later, during the Ming dynasty circa 1590 AD, the mushroom was credited with therapeutic properties such as enhancing vital energy, strengthening cardiac function, and increasing memory.

The fungal fame has continued into the 21st century, and in the State Pharmacopoeia of the People's Republic of China in 2000, it was noted to replenish qi, ease the mind, and relieve cough and asthma. It is also recommended for dizziness, insomnia, palpitations, and shortness of breath.

This fame has carried over to the United States, where September is national mushroom month. Oddly, national mushroom day is October 15th. [Think the guy that decided this was doing shrooms?]

Called lingzhi in China, the name represents a combination of spiritual potency and essence of immortality and is regarded as the *Herb of Spiritual Potency*[41], symbolizing success, well-being, divine power, and longevity. The common name, reishi, comes to us from the Japanese.

Before it was commercially cultivated, the plant was quite rare, and only the rich could afford the luxury. Today we are much more fortunate.

What is this precious mushroom? *Ganoderma lucidum* is a large, darkish-red shroom with a glossy exterior and a woody texture. The Latin word lucidus means *shiny* or *brilliant* and refers to the varnished appearance of its surface. It is an asymmetric fungus, with the stalk growing out of the side of the cap, and the top having a fan or kidney-bean shape.

Reishi comes in three colors, red, black and purple, has different characteristics, and is popular in different geographical regions. The black G. lucidum is popular in south China, whereas the Japanese prefer red.

Unlike our arctic, mountain-loving plants, G. lucidum thrives under hot and humid conditions, with many wild varieties found in subtropical regions.

Because this divine fungus is rare in nature, artificial growing techniques have blossomed. Since the 1970s, cultivation has been achieved by propagating it on wood logs, sawdust, and even cork residues.[41]

As the shroom has become more readily acquired, it has also become a popular food source, consumed globally, usually as a tea. [although it doesn't taste very good - triterpenoids tend to be rather bitter]

Another important thing to note is that our mushroom seems to be having a bit of an identity crisis. Historically, *Ganoderma lucidum* was considered the true name for the reichi shroom. However, it has come to light more recently that G. lucidum and G. lingzhi are actually two closely related, but different mushrooms and that G. lingzhi is the true mushroom that we need to be concerned with. They look very similar, at least to untrained fungal examiners, but there are a few morphological differences. The most important distinction however, is content: G. lingzhi has a higher triterpenic acid content and is therefore more bitter in taste. Unfortunately, most scientific literature is not that discriminating so, at least for this review, any study using either mushroom has been included.[14a]

Moving along...
By weight, the fungus is 90% water. The remaining 10% consists of protein, fat, carbohydrate, fiber, and vitamins and minerals, with potassium, calcium, phosphorus, magnesium, selenium, iron, zinc, and copper accounting for most of the mineral content.[6]

The most active biomolecules are most likely the terpenoids or triterpenoids and the polysaccharides, especially the beta-glucans.

Terpenoids are a large class of organic compounds, including terpenes, diterpenes, and in fact, many-unit terpenes. They are molecules composed of linked isoprene units, generally having the formula $(C_5H_8)n$. Triterpenes are a subclass consisting of three terpene units with the molecular formula $C_{30}H_{48}$; they may also be thought of six isoprene units. [Beta-carotene is a tetra-terpene]

In addition to mushrooms, both animals and plants produce triterpenes and there are significant triterpenes in ginkgo, ginseng, and rosemary. However, more than one hundred different triterpenes have been reported in G. lucidum, and among them, more than fifty are unique to this fungus. The vast majority are ganoderic and lucidenic acids, but other triterpenes such as ganoderals, ganoderiols, and ganodermic acids have also been identified.[4]

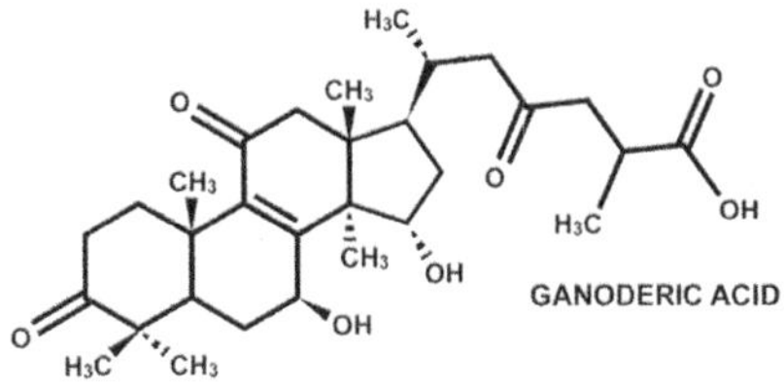

GANODERIC ACID

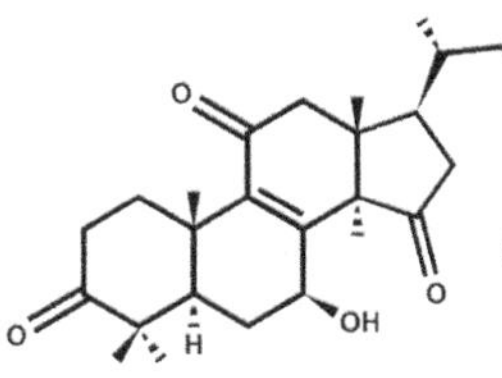

LUCIDENIC ACID

These terpenes all have four rings (only three of which are the typical six carbon variety) and are associated with anti-inflammatory, anti-tumorigenic, and hypolipidemic activities.[41]

The polysaccharides, most importantly the beta-glucans, are high molecular weight polysaccharides that are anti-inflammatory, hypoglycemic, anti-ulcer, anti-tumorigenic, and immuno-stimulating.[41]

Other bioactive compounds include peptidoglycans, ergosterols, phenols, proteins, and amino acids, with lysine and leucine being predominant.[41] [Leucine has its own chapter]

The List...

## 1) DNA ALTERATIONS: 0
Reishi has potential epigenetic effects, as seen in an Alzheimer's rat model, where alcohol extracts from G. lucidum altered some methylation regulators, including Histone H3, DNMT3A, and DNMT3B in brain tissue.[21]

Unfortunately, not enough evidence here to appear on the scoreboard. [sorry]

## 2) MITOCHONDRIA: 3
In this category, reishi can act as a free radical scavenger itself or up-regulate the production of endogenous antioxidants.

The primary free radical scavengers of this plant are the terpenes.[50] These molecules possess strong free radical scavenging capacity as well as significant ferric-reducing activity and are highly effective in reducing lipid peroxidation.[34]

Second in this category, the amino-polysaccharide fraction can "protect against oxidative damage induced by ROS." It "showed a dose-dependent inactivation of hydroxyl radicals and superoxide anions." [22]

In several rodent models, reishi increases endogenous antioxidant levels, especially superoxide dismutase.[16] In gerbil models, for example, it protects the brain against cerebral ischemic/ reperfusion damage by significantly improving the activity of superoxide dismutase [49]. In a mouse pancreatitis model, reishi increased not only superoxide dismutase but the total antioxidant activity. [23]

Turning to people, in a very small study of seven volunteers who took a single dose of 3.3 grams, "Results show that intake of G. lucidum caused an acute increase in plasma antioxidant power." [42]

Within the mitochondrial structure, reishi seems to have several beneficial effects as well. In the electron transport chain, aged male rats taking G. lucidum once daily for fifteen days had enhanced mitochondrial complex I and II activity in the brain.[2] In another aged rat study, it enhanced the Krebs cycle dehydrogenases, in addition to the mitochondrial electron transport chain complex IV activity.[35] Lastly, in a mouse model, treatment up-regulated the mitochondrial DNA copy number, suggesting some aspect of control over mitochondrial biogenesis.[47]

## 3) PATHWAYS: 1

There is evidence that reishi activates AMP Kinase in preadipoctye cell cultures [38], rat skeletal muscle cells [17], and mouse skeletal muscle.[47]

Unfortunately, this is all there is for now.

## 4) QUALITY CONTROL: 1

In the DNA repair category, beta-glucan, one of the polysaccharides, demonstrated positive effects in young Swiss albino mice exposed to whole-body radiation.[28]

## 5) IMMUNE SYSTEM: 3

In terms of inflammation, there are limited studies. One of the few was in a mouse cell line where triterpene extract suppressed TNF-α and IL-6.[11]

In the category of immunity, however, research is plentiful, and reishi boosts the immune system in a multitude of ways.

*"Ganoderma lucidum: a powerful medicinal mushroom has been found to possess immune-modulating and immune-potentiating capabilities and has been characterized as a wonder herb."* [5]

Reishi induces IgA secretion in rat small intestines,[20] IgA and IgG in chickens,[25] and it inhibits histamine release from rat mast cells (especially Ganoderic acids C and D).[19] It also enhances the proliferation and maturation of T and B lymphocytes, splenic mononuclear cells, natural killer cells, and dendritic cells.[41,44] Most of these effects are thought to be mediated by the polysaccharide components, and especially the beta-glucans.

In humans, the shroom is already being used clinically. In cancer patients, it is used to boost immune systems during chemotherapy.

Surprisingly, however, there are few studies in healthy people. There is however, one in Columbian children. In this randomized, double-blinded, placebo-controlled clinical study, researchers gave yogurt containing β-glucans to asymptomatic, three to five-year-olds for three months. They were found to have significantly higher absolute counts of peripheral blood total lymphocytes, CD3+, CD4+, and CD8+ T cells.[14]

In summation: The immuno-modulating effects of ganoderma are extensive, including promoting innate immune function, humoral immunity, and cellular immunity.[44]

## 7) WASTE MANAGEMENT: 3

This is a standout category for the shroom as well, decreasing blood glucose levels and AGE production. In diabetic rats, treatment decreases blood glucose levels, serum AGEs, and RAGE levels.[8] In fact, there are countless different rodent studies with innumerable variables, all attesting to the hypoglycemic effect of reishi.[45,46,48] This is thought to be due primarily to three polysaccharides, ganoderans A, B, and C.[41] Ganoderan B was also reported to increase plasma insulin, decrease hepatic glycogen content, and modulate the activity of glucose-metabolizing enzymes in the liver.[41]

Various proteoglycans, proteins, and triterpenoids have also been shown to have hypoglycemic effects. The triterpenoids, for example, have inhibitory activity on aldose reductase and α- glucosidase, suppressing intestinal absorption of glucose and the resulting postprandial hyperglycemia.[26]

In humans, by way of historical evidence, *Ganoderma lucidum* has been used to treat diabetes in China for centuries.[45]

This was confirmed in seventy-one Type II diabetic patients using ganopoly, a polysaccharide from G. lucidum. "The patients received either ganopoly or placebo orally at 1800 mg, three times daily for 12 weeks. The treatment of ganopoly significantly decreased the mean HbA1c, from 8.4 at baseline to 7.6% at 12 weeks." [12]

In addition to reducing blood glucose levels, G. lucidum polysaccharides decrease AGEs and attenuate the consequent myocardial collagen cross-linking.[27]

## SYSTEMS KNOWN TO BENEFIT

### LIFESPAN
In terms of lifespan, *Caenorhabditis elegans* is the only animal that has demonstrated positive results with the therapy.[33,43]

### ADIPOSE TISSUE
On a cellular level, the fat-blocking effects of reishi are really cool. First, it suppresses the expression of enzymes and proteins responsible for lipid synthesis, transport, and storage, i.e., fatty acid synthase, acyl-CoA synthetase-1, fatty acid binding protein-4, fatty acid transport protein-1, and perilipin.

Second, it inhibits adipocyte differentiation through the suppression of adipogenic transcription factors.[38] In an obese mouse model, developed by administering a high sugar and fat diet for sixty days, *Ganoderma lucidum* extract had significant, beneficial effects. Treatment reduced body weight, improved serum lipid levels, ameliorated damage to the gut microbiota, and activated the leptin regulatory pathways in the hypothalamus to improve metabolism.[10]

## BRAIN
In several rat studies, oral G. lucidum enhanced cognitive performance. Memory and learning-related protein markers substantiated this finding.[30] Other studies have confirmed improvements in neuronal apoptosis and brain atrophy.[21]

## ALTITUDE SICKNESS
"The fungus is attributed with numerous therapeutic properties and has contributed towards the amelioration of several pathophysiological conditions, including acute mountain sickness (AMS)."[4]

In cells exposed to low oxygen conditions, treatment with reishi extract reduced hypoxia-induced cell death and augmented the transcription factors HIF-1α and Nrf2, conferring tolerance to hypoxia.[18] In rodent studies of high altitude-induced stress, treatment had protective effects against mitochondrial impairment by maintaining mitochondrial number and morphology.[3]

## ALLERGIES
As if reishi didn't already have enough beneficial attributes, it also works against allergies by inhibiting histamine release from mast cells. Cyclooctasulfur,[37] and Ganoderic acids C and D are thought to be responsible.[19]

## GASTROINTESTINAL TRACT
Reishi also can alter intestinal microbiota. It increases the relative abundance of beneficial bacteria such as *Lactobacillales, Roseburia,* and *Lachnospiraceae*.[23]

In addition, it helps maintain intestinal barrier integrity and reduces metabolic endotoxemia.[7]

## WOUND HEALING
For one final trick, G. lucidum significantly enhances healing, evidenced by increased wound contraction, collagen accumulation, and total protein contents of wounds.[9,13]

## DOSE
Despite the mushroom being an intricate part of Chinese medicine for thousands of years, recommended doses have been challenging to track down. This is probably because there are innumerable variables; for example, reishi is available as a complete mushroom, as a powder, in capsules, and as a liquid

extract and is processed differently all over the world.

In general, however, 25 to 100 grams of the fresh mushroom can be consumed daily. Traditional practitioners recommend 0.5 to 1 gram daily for general health, 2 to 5 grams daily for chronic illness, and up to 15 grams of extract daily for serious illness. Officially, the Pharmacopoeia of the People's Republic of China recommends 6 to 12 grams daily of the extract.

Thus, as a general consensus, somewhere between two and five grams a day should be sufficient.

## POTENTIAL SIDE EFFECTS

The side effect profile is pretty limited, although some folks ironically report an allergic reaction to the mushroom. Other potential issues include dizziness, itchiness, rash, headache, and an upset stomach. Of note, one should avoid combining reishi with blood thinners, chemotherapy, or immunosuppressants unless specifically directed to do so by a physician.

And lastly, I wasn't really a huge fan of mushrooms, but this one grew on me.

# CHAPTER 18
# HYALURONIC ACID
## (0.2.0.0.2.0.0)

BENEFITS:

JOINTS

SKIN

*"HA is involved in many key processes, including cell signaling, wound reparation, tissue regeneration, morphogenesis, matrix organization and pathobiology, and has unique physico-chemical properties, such as biocompatibility, biodegradability, mucoadhesivity, hygroscopicity and viscoelasticity."*

*~ A. Fallacara 2018*

Hyaluronic acid is the new and latest cool thing in almost all skincare concoctions, in addition to appearing in a zillion other medically-oriented products. In fact, the global market of hyaluronic acid was 7.2 billion dollars in 2016 and is expected to reach 15.4 billion by 2025.

The name hyaluronic acid (HA) derives from the Greek word hyalos, meaning *glass*, because hyaluronic acid is transparent. As well, depending on the concentration and molecular weight, it has the rough consistency of superglue. [before it dries and your fingers stick together]

Widely distributed in almost all eukaryotic cells, hyaluronic acid is absent in fungi, plants, and insects. HA is essentially a molecular sponge; it allows tissues to be *mushy*, and, in general, bugs are more crunchy than mushy.

Humans contain about fifteen grams of the stuff, with roughly five grams present in the skin. The rest appears where there is thickened fluid; the vitreous of the eye, the umbilical cord, and synovial fluid. But it is also found in skeletal tissues, heart valves, the lungs, the aorta, and the prostate.[1] The substance was, in fact, first spotted in the eye of a cow by Karl Meyer and John Palmer at Columbia University in 1934.

In terms of its molecular composition, it's really just a long, unbranched polysaccharide, essentially a coiled, linear chain. It consists of repeating disaccharides of D-glucuronic and N-acetyl- D-glucosamine linked by β-(1–4) and β-(1–3) glycosides.

**HYALURONIC ACID**

HA belongs to a group of substances called mucopolysaccharides, in the glycosaminoglycan (GAG) family, along with chondroitin sulfate, keratan sulfate, heparin, and dermatan sulfate. All GAGs are characterized by the same basic structure, repeating disaccharide units of amino and uronic sugars. Hyaluronic acid is a bit different as it is not sulfated and is not synthesized by Golgi enzymes.[2]

The repeating unit can range in size from reasonably short to unbelievably long, assuming a variety of three-dimensional configurations depending on the local environment. In terms of size, the molecular weight ranges from 20 to 4,000 kDa.

This negatively charged molecule is extremely hydrophilic. This means that it loves water. It really loves water. In fact, hyaluronic acid is one of the most hygroscopic molecules known in nature; one gram of HA can bind up to six liters of water. Put another way, one molecule of HA can attract and bind six thousand water molecules. Thus, one can easily imagine HA as a molecular sponge.

In this regard, hyaluronic acid modulates tissue hydration and osmotic balance; it provides tissue lubrication and resistance to compressive forces and helps regulate tissue homeostasis.[1]

The second role hyaluronic acid plays is controlling signaling molecules, as well as being one itself. While in the extracellular matrix, HA forms a protective coating around most cells, where it interacts with the proteins that approach the cells, thus regulating cell adhesion, migration, and proliferation. The HA molecules also control the diffusion of nutrients and growth factors and intercede in response to tissue injury and inflammation.

In fact, depending on the molecular weight or size of the hyaluronic acid, the cell's location, and cell-specific factors, the binding between HA and its proteins can precipitate different outcomes. For example, HA can have both pro- and anti-inflammatory activities, it can sometimes promote and alternatively inhibit cell migration, and it can both activate and block cell division and differentiation.[5] [Highly Ironic!]

As an example, high molecular weight HA, usually above 1,000 kDa, tends to be anti-angiogenic and immunosuppressive. Alternatively, smaller polymers are considered distress signals and potent inducers of inflammation and angiogenesis. HA also acts by its interaction with receptors, called Hyaladherins, because they adhere or stick to hyaluronic acid. These HA binding proteins are widely distributed in the extracellular matrix, on the cell surface, in the cytoplasm, and in the nucleus.[1]

The most common of such is CD44 (Cluster of Differentiation 44), a transmembrane glycoprotein found on virtually all cells, except red blood cells, and regulates cell adhesion, migration, lymphocyte activation, and cancer metastasis.

CD44 interacts with a variety of ligands, including osteopontin, collagens, and matrix metalloproteinases (MMPs), in addition to hyaluronic acid. Interaction of HA with CD44 depends on many factors, including its concentration and molecular size.[1]

Meanwhile, RHAMM (Receptor for Hyaluronan Mediated Motility) or CD168 was the first isolated cellular hyaladherin. Located both on the outside and inside of cells, it is present in several cell types, including smooth muscle endothelial cells, and mediates cell migration via interactions with skeletal proteins.

The interaction of HA with RHAMM controls cell growth and migration by a complex network of signal transduction events and interactions with the cytoskeleton.

Notably, transforming growth factor-β1, a potent stimulator of cell motility, works by eliciting the synthesis and expression of both RHAMM and HA, which then initiates locomotion.

Other such receptors include the Intercellular Adhesion Molecule 1 (ICAM-1), the Lymphatic- Vessel Endothelial hyaluronan receptor 1 (LYVE1), and Hyaluronic acid receptor for endocytosis (HARE).

LYVE1 is a HA-binding protein expressed in lymph vascular endothelium and macrophages and controls hyaluronic acid turnover by mediating its adsorption from tissues to the lymph system. The Hyaluronic acid receptor for endocytosis (HARE), also known as Stabilin-2, binds not only HA but also other GAGs and is involved in the clearance of GAGs from the circulation.

## SYNTHESIS

Unlike the other glycosaminoglycans, hyaluronic acid is synthesized on the cell's outer membrane, by three transmembrane enzymes. These enzymes are all located on the inner side of the plasma membrane, with the active end facing into the cell. The newly constructed HA chain is exported out of the cell and into the extracellular space through a pore in the Hyaluronic acid synthase (HAS) structure.

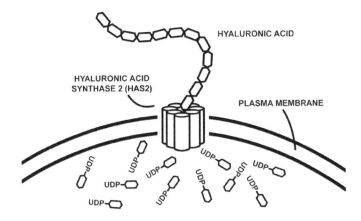

The difference between the several synthetases is the chain length they produce, creating either small, medium, or large polymers.

**HA Synthetase 1 (HAS1)**   **Located on chromosome: 19q13.4**
HAS1 exhibits the slowest activity and is the least active isoenzyme. But as with HAS2, it generates large-sized HA (from $2 \times 10^5$ to $2 \times 10^6$ Da).

**HA Synthetase 2 (HAS2)**   **Located on chromosome: 8q24.12**
HAS 2 represents the main hyaluronan synthetic enzyme in normal adult cells; it is more active than HAS1 and synthesizes HA chains greater than $2 \times 10^6$ Da.

HAS2 regulates tissue development and repair and is involved in inflammation, cancer, pulmonary fibrosis, and keloid scarring.

**HA Synthetase 3 (HAS3)**   **Located on chromosome: 16q22.1**
HAS3 is the most active isoenzyme and produces HA molecules with a molecular weight lower than $3 \times 10^5$ Da.

Despite being coded for on three different chromosomes, the three genes share 50 to 70% of their amino acid sequences, and they all have seven membrane-spanning regions and a central cytoplasmic domain.

These enzymes do, however, respond differently to different stimuli. For example, in human synoviocytes, transforming growth factor ß up-regulates HAS1 expression but down-regulates HAS3 expression.

Thus, the cells and the body as a whole can control the molecular weight and concentration of hyaluronic acid, creating microenvironments with the optimal degree of viscosity and viscoelasticity.

## HIGH MOLECULAR WEIGHT HYALURONIC ACID

High molecular weight HA is simply long chains of the same components, creating a less viscous but still fluid environment. The easiest place to imagine them hanging out is in joint spaces, acting as a lubricating agent and protecting articular cartilage.

But HMW HA also has beneficial roles in inflammation and immunosuppression, tissue injury, and repair. For example, it binds fibrinogen and controls the recruitment of inflammatory cells. It also limits the levels of inflammatory cytokines as well as the migration of stem cells.

Unfortunately, under some less than ideal conditions, such as asthma, pulmonary fibrosis, hypertension, chronic obstructive pulmonary disease, and rheumatoid arthritis, long-chain hyaluronic acid is cleaved into smaller-chain hyaluronic acid, which has proinflammatory and pro-angiogenic activities.

## LOW MOLECULAR WEIGHT HYALURONIC ACID

The low molecular weight molecules, or the shorter chains, sometimes do the opposite. They stimulate proinflammatory cytokines and chemokines, promote angiogenesis, and regulate growth factors to promote ECM remodeling.

The presence of the shorter chains also makes the medium they inhabit less viscous, thus enhancing cell motility. This is juxtaposed to the high molecular weight molecules that inhibit movement. [much like trying to swim in either a pool of water versus a pool of mud]

## PRODUCTION

When we compare cells to factories, hyaluronic acid is one of our best examples. It's a high turnover molecule, lasting only days once it has been created; thus, the manufacturing process is on high gear and never-ending.

One could also say that hyaluronic acid is the butterfly of the molecular world. HA in the skin lasts between twelve to twenty-four hours. In other internal organs, it lasts up to three days. Only in cartilage does it last up to three weeks. And if you take additional HA orally, it only lasts three to five minutes in the bloodstream. Therefore, out of the fifteen total grams of HA, it has been calculated that about five grams turn over daily.

The breakdown or degradation of HA is accomplished by two different mechanisms: one is specific, mediated by specific enzymes or hyaluronidases, while the other is nonspecific, destroyed by free radicals.

In skin and joints, 20–30% of the HA turnover occurs by local metabolism via these mechanisms. The lymphatic pathways remove the rest and transport it to lymph nodes, where it is internalized and catabolized by the endothelial cells of the lymphatic vessels.

## BREAKDOWN

In humans, six different hyaluronidases, enzymes that dissolve hyaluronic acid, have been identified.

**HYAL 1** was the first human hyaluronidase (HYAL) to be isolated and is the major HYAL in serum. The enzyme resides within lysosomes which degrade the HA into tetra-saccharides (800 Da).

Mutations in the HYAL1 gene are associated with a disease called Mucopolysaccharidosis Type IX.

**HYAL 2** is anchored on the external side of the cell surface, specializing in the demise of high molecular weight HA into products roughly 20 kDa in size, or twenty-five disaccharide units. This enzyme has very low activity in comparison to plasma HYAL1.

HYAL1 and HYAL2 are widely expressed in somatic tissues and act in concert to degrade HA chains.

Thus, on the cell surface, HYAL2 binds and internalizes HA in vesicles where the molecules are cleaved into 20 kDa fragments. Lysosomes further degrade these fragments by HYAL1 into tetra-saccharides.[5]

**HYAL 3** is expressed in bone marrow, testis, and lungs.

There are several other HYALs, but you get the picture.

In addition to enzymatic breakdown, HA is susceptible to damage from free radicals, produced in massive quantities during the inflammatory process. Therefore, an increased inflammatory state is detrimental to the status of hyaluronic acid.

*Agents that delay the free-radical-catalyzed degradation of HA are useful in maintaining the integrity of dermal HA and its moisturizing properties.*

This is thought to be one of the reasons that increased inflammation is bad for joints; there is a loss of viscosity in synovial fluid, consequent cartilage degeneration leading to increased joint stiffness and pain.

Time for the List...

## 2) MITOCHONDRIA: 2

Low molecular weight hyaluronic acid shows potent inhibition of lipid peroxidation, strong activity against hydroxyl radicals, and moderate scavenging of superoxides.[8]

In an immunosuppressed mice model, researchers gave LMW HA orally for seven days. Treatment reversed the CY-induced immunosuppression and significantly raised the activity of superoxide dismutase, catalase, glutathione peroxidase, and total antioxidant capacity.[8]

So, in essence, LMW HA possesses pronounced free radical scavenging and antioxidant properties.

## 5) IMMUNE SYSTEM: 2

This category is tricky, as low molecular weight molecules perform differently from high molecular weight ones.

For example, high molecular weight chains are anti-inflammatory, while low molecular weight molecules are proinflammatory. [3,7,17]

Enough said.

## SYSTEMS KNOWN TO BENEFIT

### SKIN

One of the most critical roles for hyaluronic acid is in skin, so we will focus there first.[10]

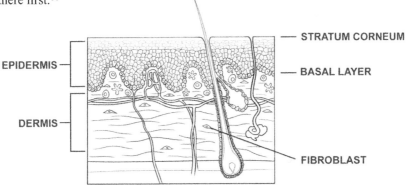

Fibroblasts produce hyaluronic acid in the dermis, which attracts fluid from the lymphatic and vascular systems. This water depot regulates osmotic pressure and ion flow and functions as a sieve, excluding specific molecules, enhancing the extracellular domain of cell surfaces, and stabilizing skin structures by electrostatic interactions.

Above, in the epidermis, hyaluronic acid is mainly in the extracellular matrix of the upper spinous and granular layers, whereas in the basal layer, HA is predominantly intracellular. The hydration is sealed into the epidermis by the more exterior layer called the stratum granulosum or corneum, a waterproof seal when intact.

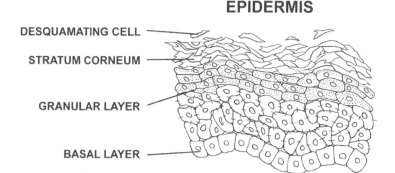

HA does more than just provide hydration for the skin; however, and as an indication of this, the various isoforms of HA are uniquely regulated by stimulating factors. TGF-β1, for example, differently up-regulates HAS1 and HAS2 in the dermis and epidermis.

As well, keratinocyte growth factor stimulates the expression of HAS2 and HAS3, which activates keratinocyte migration and promotes wound healing.

In fact, in multiple studies of wound healing, hyaluronic acid has proven to be invaluable. Of note, fetuses in utero do not scar secondary to elevated levels of HA. In adult mice models, HAS2 and HAS3 mRNA are significantly increased after a dermal injury, leading to increased epidermal HA. Thus, higher levels of HA are crucial to wound healing with scar reduction.

Unfortunately, glucocorticoids, commonly prescribed steroids, inhibit hyaluronic acid production, which contributes to the skin atrophy seen with steroid therapy.

As with everything else, good things decline with age, and so does hyaluronic acid. Epidermal deposits fall more dramatically than in the dermis, probably because the fibroblasts are located in the dermis, and HA must migrate into the epidermis.

The epidermal content of hyaluronic acid, in fact, decreases from 0.03% in 19-year-olds to 0.015% in 47 to 50-year-old women. This is further decreased to 0.007% in women by the age of 70.[6] Put more simply, the hyaluronic acid content in the skin of a 75-year-old person is lower than one-quarter of that in a 19-year-old.[14]

The other change is the reduction in the size of HA with aging, leading to a loss of hydration both in the dermis and epidermis, as well as atrophy and a loss of elasticity.

Part of this decline in HA is due to collagen fragmentation in the dermis. Precipitated by UV radiation and inflammation, collagen fragments contribute to the decrease in HAS expression.

In fact, in photo-exposed skin, nothing goes well. There is an increase in degraded hyaluronic acid, a decline in the production of new HA, and increases in HYAL 1, 2, and 3, which further degrade the chains. In addition, the hyaluronic receptors CD44 and RHAMM are significantly down-regulated.[15]

We know from radio-tagging that ingested hyaluronic acid does reach the skin. It appears about four hours after ingestion and lasts up to seventy-two hours. After oral intake, the amount in the skin is higher than in the blood, even after ninety-six hours. This extraneous HA doesn't stay long, but it does promote the growth of fibroblasts and increases endogenous HA production.[14]

We also know that the high molecular weight molecules get degraded into small polymers or LMW HA before being absorbed in the large intestine. Therefore, in terms of getting HA into the skin, the length of ingested chain does not really matter.

Human studies have confirmed all of this; oral HA does, in fact, help the skin. In several different randomized, double-blind, placebo-controlled trials, the amounts of HA ranging from 37 to 240 mg per day, ingested in a period of between four and six weeks, significantly improved cutaneous moistness.[5]

Topical application is yet another story as any hyaluronic acid must penetrate the protective epidermal barrier. In this instance, the smaller molecules are more successful, but the even newer and even smaller nano-molecules are the best.

## JOINTS

Put most simply, joints are a combination of cartilage at the end of bones and synovial fluid for lubrication. The primary cell in the cartilage is the chondrocyte, constituting 5% of the total mass. The rest is extracellular matrix, composed of complex networks of proteins and glycoproteins, all expressed by the chondrocytes.

Within the joint cavity, synoviocytes synthesize hyaluronic acid, which is essential for the biomechanics of normal synovial fluid. It is responsible for lubrication and viscoelasticity and can decrease pain by limiting prostaglandin and bradykinin synthesis, as well as substance P.

Specifically, hyaluronic acid molecules greater than 40 kDa produce an analgesic effect, and molecules between 860 and 2,300 kDa provide even better, longer-lasting analgesia by interacting with HA receptors.

Hyaluronic acid can also decrease the expression of MMP-1 and MMP-3 mRNAs, which are induced by IL-1β and other forms of inflammation in human synovial cells.[7]

Over time and with age, the concentration of HA and its molecular weight declines. Replacing this by direct injection into joint spaces has proven to be quite valuable in this regard, and histological evidence suggests that HA prevents the degradation of cartilage and may promote its regeneration. In fact, the HA appears to provide cartilage protection by the down-regulation of cytokines, free radicals, and proteolytic enzymes in synovial fluid.

But do you need to inject it? The efficacy of oral, 200 milligrams daily for one year, was evaluated in people with osteoarthritis. In addition to the hyaluronic acid, subjects also underwent daily quadriceps strengthening exercises. Improvements were more evident with the oral treatment, and this trend was more significant in subjects younger than seventy.[19]

In another three-month, double-blinded, randomized, and placebo-controlled study, forty overweight subjects with knee osteoarthritis took oral HA. There were statistically significant improvements in both pain and function. Analysis of the serum and synovial fluid samples demonstrated substantial reductions in the inflammatory markers. As well, there was a marked decrease in hyaluronic acid turnover within the joint.[13]

So, in reality, oral hyaluronic acid will get to the joint, but perhaps not as quickly or in the same magnitude as direct injection.

## EYES

Oral HA also gets transported to the cornea and can assist in corneal healing. Two hundred and forty mg of HA (390 kDa) were given to subjects for three months in a prospective, randomized controlled trial. They found that a combined oral and topical program more efficiently improved corneal epithelial wound healing and related symptoms than topical HA alone.[9]

## RANDOM

As a completely useless bit of trivia, if you are interested in preserving boar sperm, the addition of HA to the preservation concoction significantly improves sperm motility, sperm membrane integrity, mitochondrial activity, acrosomal integrity, superoxide dismutase, and catalase activity.[16] This also works for rooster sperm.[12]

## DOSE & METABOLISM

Studies with rats and dogs have demonstrated that oral supplementation accumulates in the skin, joints, vertebrae, and salivary glands four hours after administration and persists up to seventy-two hours.[7]

The main excretion route is via expired carbon dioxide, resulting in 76.5% of the dose being excreted by 168 hours after administration. In comparison, urinary and fecal excretion rates are 12% and 3.0% of the initial dose.[14]

Unfortunately, the bioavailability is low, as ingested hyaluronic acid gets quickly metabolized and excreted by the lungs.[14]

Once upon a time, commercially available HA originated from cow eyes or rooster combs. Today hyaluronic acid or hyaluronan is principally produced within a lab with microbial fermentation.

In terms of dose, the most widely accepted recommendation is 200 mg per day with no reports of significant side effects.[14]

# CHAPTER 19

# KAEMPFERIA PARVIFLORA

(0..3.2.0.2.1.2)

BENEFITS:

FAT  MUSCLE  SKIN  VIRILITY

*"Kaempferia parviflora rhizomes are used in folk medicine to improve blood flow and treat inflammatory, allergic, and gastrointestinal disorders. KP also reportedly exhibits beneficial activities, such as anticholinesterase, antiinflammatory, spasmolytic, gastric ulcer amelioration, antioxidative, and vasodilatory effects in addition to regulation of adipocyte differentiation."*

*~ S. Yoshino 2018*


*Kaempferia parviflora* (KP) hails from the tropical, exotic lands of the far east, originally Thailand, Malaysia, Sumatra, and Borneo where it is referred to as kra-chai-dam or krachaidum. In Japan, it is black galangal, black ginger, or black turmeric.

The purplish-black root or rhizome has been used for centuries as a tribal remedy, especially by the Hmong hill tribe, who believe it reduces perceived effort and improves physical work capacity.[34] It's popularity has grown beyond the tribes, however, and has become one of the top five herbal products produced in Thailand.[26]

The rhizome has been a traditional remedy for allergies, asthma, impotence, gout, diarrhea, and diabetes. This is in addition to the idea that it's just good for vitality and health in general. More recently, modern science has reinforced these original uses, demonstrating real utility as a treatment for allergies, inflammatory conditions, depression, infections, peptic ulcers, and as an aphrodisiac.[26] In addition, it's been proven to build muscle and diminish fat stores. There is even evidence that the rhizome has anti-cancer properties, but that remains beyond the scope of this book.

As is typical of any plant, it is composed of numerous chemical or molecular elements. Polymethoxyflavonoids constitute the bulk of these components, with contributions as well from polyphenols. In fact, more than twenty chemically identifiable constituents have been reported to have potent pharmacological effects.[26]

The polymethoxyflavonoids are thought to be responsible for most of the plant's therapeutic properties. Of note, these compounds are not found in other plants of the same *Zingiberaceae* species, such as *Curcuma longa*, the source of curcumin, that has outstanding therapeutic properties as well.[25]

Within the polymethoxyflavonoid category, 5,7-dimethoxyflavone, (5,7-DMF) and 5,7,4'- trimethoxyflavone constitute the two most essential components. 5,7-DMF is a known activator of AMP Kinase, while 5,7,4'-trimethoxyflavone is a potent AGE inhibitor. Other methoxyflavone substances are phosphodiesterase inhibitors, which enhance sexual performance.[26]

Of the polyphenols identified, gallic acid, apigenin, and tangeretin are the most prevalent.[12]

The List...

## 2) MITOCHONDRIA: 3

First off, *Kaempferia parviflora* (KP) has potent free radical scavenging activity. Of sixteen Thai plants with known activity, the highest antioxidant activity was detected in kaempferia.[1]

As well, in several cell cultures, the rhizome extract increased endogenous antioxidants, with an emphasis on catalase.[22] Additionally, in cell cultures, KP was able to induce mitochondrial biogenesis.[9]

At the human level, elderly volunteers who took either 25 or 90 mg of KP once daily for eight weeks enjoyed improved physical fitness and increased endogenous antioxidants.[36]

## 3) PATHWAYS: 2

Many studies have now validated KP as a SIRT1 activator,[9,24,25] and in test tubes, it was more powerful than even resveratrol.

*"K. parviflora extract elevated SIRT1 catalytic activity by eight - and 17-fold at 20 µg/mL and 100 µg/mL, respectively, compared with vehicle only. Two major polymethoxyflavonoids, 3,5,7,3',4'- pentamethoxyflavone and 5,7,4'-trimethoxyflavone, were isolated from this extract and are four and five fold more potent than resveratrol, hitherto the strongest known natural SIRT1 activator."*[25]

In addition, there is building evidence that KP, particularly the component 5,7-dimethoxyflavone, activates AMP Kinase in culture[33], as well as in mice.[3] In these rodents, this is thought to have contributed to increased skeletal muscle and running endurance.

Unfortunately, we are lacking any human studies in this category. This agent probably deserves a 3, but as of yet, we don't have enough evidence.

## 5) IMMUNE SYSTEM: 2

The anti-inflammatory properties of KP have been well documented in multiple cell cultures. Specifically, the expression of nuclear factor kappa β, interleukin-1β, interleukin-6, IL-8, and cyclooxygenase-2 are all reduced. [6,7,21]

A study by Tewtrakul, et al. indicated that 5-hydroxy-3,7,3',4' tetramethoxyflavone is primarily responsible for the anti-inflammatory activity.[29]

## 6) INDIVIDUAL CELL REQUIREMENTS: 1

*"In our screening study of medicinal plants on antioxidant and antiosteoporotis activities, the methanolic extract of K. parviflora exhibited significant antioxidant and antiosteoporosis properties."*[31]

There is preliminary evidence that KP can prevent cells from becoming senescent, at least in an $H_2O_2$-treated fibroblast model and in a mouse model.

*"KPE significantly increased cell growth and suppressed senescence-associated β- galactosidase activation. KPE inhibited the expression of cell-cycle inhibitors (p53, p21, p16, and pRb) and stimulated the expression of cell-cycle activators (E2F1 and E2F2)."*[22]

In addition, KP effectively attenuated the senescence-associated inflammatory responses or the SASP.[22] Generally, senescent cells stimulate neighboring cells by secreting inflammatory cytokines, including IL-6 and IL-8. Under these conditions, normal adjacent cells become dysfunctional and senescent themselves, contributing further to aging and pathology. Thus, the reduction of the SASP is beneficial.

Again, we are lacking good human studies here.

## 7) WASTE MANAGEMENT: 2

There is significant evidence that KP has anti-glycation activity, at least in test tubes.[4,25,39]

*"The anti-glycation activity of K. parviflora extract was observed to be seven times more effective than aminoguanidine. 3,5,7,3',4'-Pentamethoxyflavone and 5,7,4'-trimethoxyflavone showed the strongest anti-glycation activity among the tested polymethoxyflavonoids."* [25]

In a novel mechanism to decrease blood glucose levels, KP interferes with glucose pumps in the kidney. These cellular pumps, specifically the sodium-glucose co-transporter 2 (SGLT2) and the facilitated glucose transport 2 (GLUT2), reabsorb glucose from urine, ensuring that more glucose remains in the circulating bloodstream available to the body. As KP inhibits these pumps, it essentially increases the amount of glucose lost in the urine and decreases the glucose remaining in the circulation.[32]

This reduction was demonstrated in non-obese, diabetic mice where after eight weeks, KP successfully lowered blood glucose levels.[17]

## SYSTEMS SHOWN TO BENEFIT

### SKIN

Let's start with the anti-inflammatory nature of KP. Stress, i.e., sunlight or UV light or such, activates the inflammatory response in the skin. This precipitates the production of matrix metalloproteinases or MMP's, which dissolve collagen and elastic fibers and prevent the production of new collagen. The resultant structural failure leads to skin droopage and wrinkles. [3,12,38,21]

Oral KP, at least in mouse models exposed to UV radiation, suppresses these MMP's and prevents collagen loss. At the same time, it activates the production of new collagen fibers.[38]

Studies have specifically demonstrated a decrease in the expression of MMP-2, MMP-3, MMP- 9, and MMP-13.[21] KP also promotes an increase in the genes for Type I, III, and VII collagen.[21]

Continuing with the skin conversation, there is some evidence that KP contributes to the improvement of acne. This is hypothesized to occur by modulating the growth of facial bacteria, affecting the oil processing in skin cells, and/or suppressing the inflammatory response.[7]

## ADIPOSE TISSUE
Many exciting things happen to fat under the influence of KP.

In cell cultures, KP reduces intracellular triglycerides. It shrinks lipid droplets in mature fat cells (adipocytes). Lipases, or enzymes to break down fat, i.e., triglyceride lipase and hormonesensitive lipase, are up-regulated[19], while enzymes that create fat are decreased.[3]

KP also promotes the population of brown adipocytes over white fat cells. Brown fat allows for thermogenesis, the ability to create heat through shivering, increasing caloric expenditure, and promoting fat loss.[10,15,24] "The KP extracts... have been demonstrated to increase lipolysis, suppress lipid accumulation, and decrease hypertrophy in mature adipocytes."[3]

So does this fat reduction extend to real beings? In innumerable mouse and rat studies, it does.

In normal mice,[13] obese mice,[13] diabetic mice,[17] and middle-aged rats,[42] KP decreased fat accumulation in a variety of tissues, including the liver, subcutaneous fat, and visceral fat.

In one particular study, mice were given 200 mg/ kg/day, the high dose in this study. Fat stores declined, on average, 40%.[13] Specifically, epididymal, subcutaneous, and perirenal adipose depots were reduced 31.6%, 46.4%, and 41.4%, respectively.[13]

There is, at present, only one human trial addressing this question. In this three-month randomized, double-blinded, placebo-controlled clinical trial, seventy-six males and females aged 20 to 65 were given 150 mg once daily.

"Compared with the placebo group, the active KPE group exhibited significant reduction in abdominal fat area (visceral, subcutaneous, and total fat) and triglyceride levels after twelve weeks."[41]

## MUSCLE

Evidence for an increase in muscle mass is primarily in cell cultures and mice. At the cellular level, KP precipitates a significant up-regulation of mitochondrial biogenesis regulatory genes in skeletal muscle tissue.[9, 33] This leads to an increase in protein synthesis and promotes myotubular hypertrophy with consequent increases in myofiber size in middle-aged (SAMP1) mice.[20]

In the same mouse study that demonstrated the fat reduction after 200 mg/kg/day, the researchers also noted that it "markedly increased the muscle fiber size, muscle volume, and muscle mass, resulting in the enhancement of muscle function, such as exercise endurance and grip strength."[13] [I'm not sure how you test the grip of a mouse]

## EXERCISE

Utilizing their new muscles, male mice administered KP for four weeks significantly increased their swimming time until exhaustion, their motility after swimming, and their grip strength.[33]

Twenty-four healthy humans given 30 mg/day for a month were also found to have enhanced physical fitness. This was measured by improvements in grip strength, leg strength, balance, endurance, and locomotor activity. KP intake also slightly improved fatigue after exercise in folks unaccustomed to physical exertion.[33]

Meanwhile, healthy, elderly volunteers, given either 25 or 90 mg/day, improved in the 30-second chair stand test as well as the six-minute walk test.[36]

## MALE VIRILITY

KP historically has been known as an aphrodisiac and as a cure for erectile dysfunction.[35] In boy rats given a single dose of KP and then exposed to sexy lady rats, the frequency of mounting, intromission, and ejaculation increased while the latencies of all their sexual behaviors decreased.[35] Additionally, male diabetic rats had an increase in sperm density and recovered their once abandoned sexual behavior.[14]

Moving to men, the erectile function of healthy 50 to 70-year-olds was examined after thirty days of treatment. Only thirteen men completed the trial (no clear reasons as to why), but there was a statistical improvement in survey questions referring to erectile function, intercourse satisfaction, and orgasmic function. In general, 61.5% of subjects noted the product improved their erections.[5]

In a more clinical, measured study, men received either 25 or 90 mg a day for two months.

"They found that Krachaidum (KP) at a dose of 90 mg/day significantly decreased the response latency time to sexual erotic stimulation and still showed significant changes during the delay period. In addition, after 1 month and 2 months, the Krachaidum group at a dose of 90 mg/day experienced a statistically significant increase in length and width of penis both in resting state and erection state compared with the placebo group."[26]

Because of this, there is now a product called KaempMax™, an ethanol extract of the K. parviflora rhizome. In a study of middle-aged men with mild erectile dysfunction, "the effects were not as pronounced as what might be seen with prescription medication, (however) most participants found them satisfactory."[27]

Lastly in the *be nice to men* category, there is some suggestion that KP might be useful for benign prostatic hypertrophy. Like most prescription medications aimed at this issue, KP exhibits some inhibitory activity on 5-alpha reductase.[16]

## DOSE

Unfortunately, KP suffers from poor bioavailability.[3,23] In mice, oral methoxyflavones reached a maximum blood level concentration in one to two hours, with a half-life of three to six hours.[23]

As with most longevity supplements, there is no recommended dose. Human studies have ranged from 25 mg up to 1,500 mg per day. Some folks recommend starting at around 200 mg per day and slowly increasing to 1000 to 1200 mg/day. The best bet is 200 mg/ day.

## POTENTIAL SIDE EFFECTS

The good news is that few severe side effects have been reported, even at high doses. [26,41] Of these, sweaty armpits and an elevated body temperature are most frequent. The most serious issue, however, is the possibility of tachycardia, a fast heartbeat, and even cardiac arrhythmias that can be harmful, especially to the elderly.

# CHAPTER 20

# LACTOFERRIN

(1.3.1.2.3.2.2)

BENEFITS:

BONE   BRAIN   IMMUNITY

*"Lactoferrin (Lf), a multifunctional iron-binding glycoprotein, plays a significant role in anti-inflammatory, antibacterial, antiviral, reactive oxygen species (ROS) modulator, antitumor immunity, and anti-apoptotic processes."*

*~ Y. Li 2012*

Lactoferrin is a protein common to all mammals as it comes from milk, or at least it was originally identified in breast milk in 1939. Since then, it has also been identified in most mammalian secretions, including saliva, tears, vaginal fluids, bile, and pancreatic fluids. In addition to the exocrine secretions, it lives within the lysosomes of neutrophils and macrophages, white blood cells that fend off disease. The locations here hint at what the protein does, or at least one of the things it does, i.e., it protects the body from pathogens. [Another concern with being a vegetarian...almond milk does not count]

The name also tells us that the molecule carries iron. Again, this hints at another aspect of its activity: it provides the iron necessary throughout our body to make red cells, fend off free radicals, and such.

The protein itself is a single, really long chain of amino acids, ranging from 679 to 692 individual amino acids, depending on what article you read. This chain clusters into two separate regions, connected by a small linking chain, resembling either a dumbbell or a bowtie.

The two lobes, named N and C, share between 33 to 41% homology, and the linking piece is a ten to fifteen amino acid chain taking the form of a three-turn, α-helix. The conformation grants several properties to the molecule as a whole. First, the link allows the molecule to be a bit bendy. Second, there are two subpieces within each lobe that create a gap or pocket for carrying iron. Because there are two lobes, every lactoferrin molecule can either carry no iron, one iron molecule, or two irons, referred to as apo, monoferric, or homo or diferric, respectively. Under normal physiologic conditions, lactoferrin is generally 10 to 25% iron-saturated. This carrying capacity allows for the safe circulation of metal in a non-toxic form.

Systemic lactoferrin generally originates from neutrophils as they degranulate during infection or inflammatory states. For this reason, measuring serum or colonic lactoferrin has traditionally been a measure of inflammation. Thus blood levels of lactoferrin are relatively low under normal conditions, varying from 0.02 to 1.52 µg/mL. However, once under 0.1 g/L, infection and anemia become problematic as the body cannot mount appropriate responses.

Back to the molecule… Lactoferrin is a positively charged or cationic glycoprotein. In addition to carrying iron, it is also decorated with sugars (the glycol part). This glycosylation is both species-specific and tissue-specific, with many of its activities dependent on these unique glycosylation patterns.

In humans, there are three primary sites (Asn 138, Asn 479, and Asn 624) where a sugar complexes onto the molecule, with several different sugar varieties or patterns possible. These different variations affect the protein's folding, immunogenicity, protein solubility, and resistance to proteolysis, which are likely responsible for the heterogeneity of their biological properties.

In comparison, bovine lactoferrin has four to five potential glycosylation sites, while mice are limited to one.

Lactoferrin plays many roles around the body, some of which require interacting with cellular receptors.

One of these receptors is intelectin-1 (ITLN-1), which has a high affinity to lactoferrin and controls several cellular functions, ranging from facilitating intestinal iron absorption to strengthening the immune system.

The second is the low-density lipoprotein (LDL) receptor-related protein (LRP), which is involved in the uptake of lipoproteins through an endocytotic-mediated pathway. It is also involved in cell migration, survival, motility, and differentiation. Of note, lactoferrin translocation across the blood-brain barrier is LRP-mediated.

In infants, the intended recipients of human breast milk and thus maternal lactoferrin, breast milk represents the primary source of the molecule. In the first days of life, this assists with the initiation, development, and eventual composition of the neonatal gut microbiota. It also allows for the acquisition and development of protective immunity against infectious-related pathologies during the early stages of life.

In adults, lactoferrin continues to provide beneficial effects:

*"As a bioactive protein in foods, especially in dairy products, lactoferrin exerts multiple bioactive functions including iron metabolism, immunomodulation, anti-inflammation, antioxidation, antibacterial, antivirus, and anticancer."*[5]

*"Although lactoferrin is known to be involved with immunoprotection, its functions are not limited to the regulation of innate immunity, but extend to iron transfer to cells, control of the level of free iron in blood and external secretions, interaction with DNA, RNA, heparin, and polysaccharides, and pronounced antimicrobial and antiviral activities."*[7]

The List...

## 1) DNA ALTERATIONS: 1

In this category, we are on the verge of knowing things but don't really have a full picture as of yet.

We know, for example, that lactoferrin gets into the cell's nucleus, binds to DNA, and acts as a transcriptional activator. What it actually activates is a little hazy.[6,8]

In the most recent paper: "This protein can act as a transcription factor regulating the expression of some genes. This review demonstrates that lactoferrin can directly and/or indirectly influence epigenetic mechanisms (DNA methylation, histone modification, chromatin compaction, and microRNA pathways) in different types of cells, in particular cancer cells."[30]

There is evidence, at least in some leukemia cancer cell lines, that lactoferrin is a DNA methytransferase inhibitor.[39]

Of note, there is also evidence that lactoferrin production is significantly reduced in many cancers. In prostate cancer, for example, the CpG island spanning the transcriptional start site of lactoferrin is hypermethylated, leading to a significant reduction in lactoferrin production.[27]

The lactoferrin gene, on the short arm of chromosome 3 (3p21.3), is also compromised in many types of lung cancer. "The expression of the LTF gene was absent in 16 (59%) of 27 small-cell lung cancer cell lines, 33 (77%) of 43 nonsmall-cell lung cancer (NSCLC) cell lines, and 7 (54%) of 13 primary NSCLC cell lines."[12]

## 2) MITOCHONDRIA: 3

By virtue of sequestrating iron, lactoferrin controls the physiological balance of reactive oxygen species production and elimination. Specifically: "LTF inhibits free ferric ($Fe^{3+}$) ions reactivity with superoxide molecules, thus limiting formation of ferrous ($Fe^{2+}$) salt and ground state oxygen. In turn, there is reduced reactivity of ferrous ($Fe^{2+}$) ion with hydrogen peroxide to form ferric ($Fe^{3+}$) salt, a hydroxyl radical, and an alcohol. The end result is that LTF protects against oxidative stress, in particular by limiting the production of hydroxyl radical and lipid peroxidation."[28]

Lactoferrin also contributes to the up-regulation of antioxidant enzymes. In an obese pediatric diabetic population, the addition of lactoferrin stimulated superoxide dismutase levels and Nrf2 expression.[22]

## 3) PATHWAYS: 1

There is limited evidence that lactoferrin activates SIRT1.

In the same obese, Type II diabetic pediatric population, three months of oral lactoferrin increased SIRT1 expression.[22]

There is also one study demonstrating the activation of AMP Kinase in human kidney proximal tubular cells.[9]

## 4) QUALITY CONTROL: 2

In these same human kidney proximal tubular cells, the increase in AMP Kinase induced autophagy, a finding confirmed in lactoferrin-treated mice.[9]

# 5) IMMUNE SYSTEM: 3

## IMMUNOMODULATION

*"Lactoferrin is a modulator of both the innate and acquired immune systems. Following the penetration of a microbe into a tissue, cells of the innate immune system release pro-inflammatory cytokines, including interleukins 1 and 6 (IL-1, IL-6) and tumor necrosis factor-alpha (TNF-a), which increase the permeability of blood vessels enabling the recruitment of circulating neutrophils to the site of infection. The release of neutrophil granule content creates very high local concentrations of lactoferrin. Apart from direct antimicrobial activity, lactoferrin interacts with cells of the innate immune systems as well as with cells of the adaptive immunity; regulating their recruitment, proliferation, differentiation and activation."* [25]

The immunomodulatory nature of this protein derives from its unique ability to *sense* the immune activation status of an organism and act accordingly.[17]

As its primary role, lactoferrin promotes both innate and adaptive immunity.

Prevalent in the secretory fluids protecting us from the outside world, lactoferrin is one of the key components of the innate immune system as a first line of immune defense.[28] At the site of any injury or attack, lactoferrin activates phagocytosis by polymorphonuclear leukocytes and macrophages and activates natural killer cells.

In the second category, it improves adaptive immunity in two fundamental ways.

1) It matures T-cell precursors into competent helper cells.[17]

2) It accelerates the differentiation of immature B cells into efficient antigen-presenting cells.[17]

In a human study, eight healthy males aged 30 to 55 took escalating oral daily doses for twenty-one days. There was a placebo capsule for seven days, followed by 100 mg of lactoferrin for seven days and 200 mg of lactoferrin for an additional seven days. Statistically significant increases were identified after 200 mg of supplementation in total T-cell activation, helper T-cell activation, and cytotoxic T-cell activation (as measured by CD8+).[23]

In an older population, sixty-two elderly participants took 300 mg of bovine lactoferrin for three months. This randomized, double-blind, placebo-controlled trial demonstrated "significant intergroup differences in peripheral blood lymphocyte subset ratios, neutrophil phagocytic function, and NK cell cytotoxicity."[14]

They concluded that: "immune competent cells of the innate immune system in elderly individuals are specifically activated by LF intake."[14]

Lactoferrin is also useful as a vaccine adjuvant. In a mouse study, both bovine and human lactoferrin, when added to the BCG vaccine (an attenuated strain of Mycobacterium bovis Bacillus Calmette Guerin vaccine), led to increased protection against subsequent challenges with virulent mycobacterial organisms.[10,17]

In addition to assisting the immune response to pathogens, lactoferrin is antibacterial and can negatively affect microbes via two mechanisms. The first is iron-dependent, leading to a bacteriostatic effect.[26] The second is due to the high affinity of lactoferrin to lipopolysaccharides which can destroy gram-negative organisms.[13] Of note, only the N-lobe is believed to increase antimicrobial activities.

The C-lobe, meanwhile, modulates inflammatory and immune responses through the down-regulation of numerous cytokines, interleukin-6, granulocyte-macrophage colony-stimulating factor (GM-CSF), and TNF-α.

Once this inflammatory cascade is initiated, the next trick is to extinguish it when it is no longer needed. Lactoferrin becomes crucial at this juncture.

To make this understandable, follow this progression for a second. An injury or insult leads to activation of the NF-κB signal transduction pathway within monocytes and macrophages. This, in turn, stimulates the production of inflammatory mediators, which then trigger the production of fresh neutrophils and monocytes from bone marrow. This activates circulating neutrophils which degranulate, releasing secondary mediators, including lactoferrin. Here is the key.....the released lactoferrin is the signal for the aggressive cascade to cease. By binding to receptors on the monocytes and macrophages, the system gets turned off.

Thus, lactoferrin serves as a marker of inflammation, but it's also useful as an anti-inflammatory.

This anti-inflammatory activity has been ascribed to several things.

First, its positive charge allows it to interact with negatively charged groups present on the surface of the immune cells, signaling pathways that induce a physiological anti-inflammatory reaction.

Second, lactoferrin enters cells, translocates into the nucleus, and helps to regulate proinflammatory gene expression.[34]

For example, lactoferrin down-regulates the expression of intercellular adhesion molecule-1 (ICAM-1) in a DNA-binding dependent manner at the transcriptional level in endothelial cells. Lactoferrin binds to the ICAM promoter region, acting as a physical roadblock so that NF-κβ cannot bind to the same site.[16]

Through this and other mechanisms, lactoferrin regulates inflammatory cytokine production, suppressing the production of TNF-α, IL-1β, IL-6, and IL-8 while enhancing levels of IL-10 and IL-4.[34]

In human articular chondrocytes derived from osteoarthritic articular cartilage, camel lactoferrin inhibited IL-1β-induced activation and nuclear translocation of NF-κβ. Lactoferrin also inhibited the mRNA/protein expression of COX-2 and production of $PGE_2$.[29]

In a very telling human study, 40 mg/day of lactoferrin showed differential effects on immune responsiveness. In participants that tended to secrete high amounts of IL-6, i.e., had too much cytokinin release, lactoferrin down-regulated this response. In under responders, lactoferrin up-regulated the immune response.[34]

## 6) INDIVIDUAL CELL REQUIREMENTS: 2
Apparently, milk really is good for your bones... [Who knew?]

In several recent studies in humans and experimental animals, dietary lactoferrin improved bone mineral density, bone markers, and bone strength.[25]

Both human and bovine lactoferrin promotes the differentiation and proliferation of osteoblasts and inhibits the formation of osteoclasts.[1,3,4,25] Of note, the degree of glycosylation and level of iron-binding do not seem to affect osteogenic potential.[4]

These osteogenic effects are considered profound, at least in cultures, where it proved more potent than epidermal growth factor, transforming growth factor, parathyroid hormone, and insulin.[25] [Lactoferrin and naringenin might be the winning pro-bono combo]

Of note, the intact or entire molecule is key for optimal osteoblastic activity. "In contrast, the equivalent osteoclastogenic activities of the recombinant human lactoferrin and its N-lobe suggest that this activity might be largely located in the N-lobe."[25]

In white rabbits undergoing unilateral tibial distraction osteogenesis, meaning researchers pulled on the bone to make it grow faster, oral treatment with bovine lactoferrin "not only accelerated bone formation at early stages of distraction osteogenesis but also promoted bone consolidation at late stages."[18]

Of note, lactoferrin also stimulates the production of primary chondrocytes or cartilage cells.[3]

## 7) WASTE MANAGEMENT: 2

As mentioned at the beginning of the chapter, lactoferrin is a glycoprotein. It gets glycated under normal conditions at normal locations, which helps define the molecule's role. At the same time, this molecule also gets abnormally glycated to create AGEs or Advanced Glycation Endproducts.

There are two ways to look at this phenomenon, one not very useful and one potentially quite useful.

In the not so useful category, abnormally glycated lactoferrin means that the molecule loses its standard three-dimensional structure and the ability to carry iron.[33]

It also loses its ability to interact with its receptors, and it may not gain entrance into cells. This non-enzymatic glycation occurs at specific sites on the molecule, for example, Lysine 206 and Lysine 534 in apo-transferrin or Lys103, Lys312, and Lys382 in holo-transferrin. In nondiabetics, non-enzymatic glycation of serum lactoferrin is roughly 1% to 2%. In adult diabetics, this increases to 5%. In kids, it rises from 4% in healthy children to 11% in those with Type I diabetes. AGE-laden lactoferrin can even be found in the tears of diabetics. As well, glycated lactoferrin in the breast milk of diabetic mothers is increased.[19]

On the other hand, as lactoferrin is a known binder of AGEs, extra, exogenous quantities of lactoferrin may be beneficial. As early as 1995, a 17 to 18 amino acid domain was identified within the lactoferrin molecule that had a high affinity to AGEs.[2]

This AGE–binding cysteine–bounded domain or 'ABCD' motif occurs on both sides of the molecule. Thus every lactoferrin molecule can potentially bind two AGEs. At the time, it was thought that this would be detrimental as it would prevent lactoferrin from effectively preventing infection, especially in diabetics.[19, 20] There was even a patent registered using this discovery to extract AGEs from patient's blood during dialysis.[2]

Whereas, I confess that this is my own hypotheses, and I have no actual evidence, I believe that supplemental lactoferrin might be advantageous here. In binding to AGEs and consequently being excreted, circulating lactoferrin may prevent AGEs from adhering to and destroying our permanent tissues.

Turning to other glucose issues, lactoferrin has hypoglycemic effects as well, primarily by increasing insulin levels.[28]

In an obese Type II diabetic pediatric population, sixty patients with the mean age of 14 ± 3 years received lactoferrin (250 mg/day, p.o) for three months in addition to their standard treatment. Researchers reported improvements in the HbA1c, BMI, and lipid profiles.[22]

## SYSTEMS KNOWN TO BENEFIT

### BRAIN
Systemic lactoferrin crosses the blood-brain barrier via receptor-mediated transcytosis and accumulates in the brain capillary endothelial cells. At the same time, some lactoferrin is produced in the brain, synthesized in the substantia nigra, particularly in dopaminergic neurons and microglia. Once in the brain, lactoferrin does several things.

It reduces the toxicity of iron. This may prove useful in many scenarios; for example, in the event of a brain bleed or hemorrhagic stroke, lactoferrin binds the iron released from red blood cells before it causes damage. Lactoferrin can also reduce lipid peroxidation and neurological inflammation.[40] In fact, lactoferrin has proven to be beneficial in models of several neurologic diseases, including Parkinson's and Alzheimer's disease.[24]

In an aged mouse model, spatial cognition was improved after lactoferrin administration, with more pyramidal cells detected in the hippocampus. LF reduced iron deposition, malondialdehyde and reactive oxygen species in the hippocampus and serum levels of IL-1β, IL-6, and TNF-α.[41]

## ADIPOSE TISSUE

In several types of cell cultures, lactoferrin increases the production of bone cells while decreasing the production of fat or adipose cells.[25]

In several mouse models, adding lactoferrin to the diet enhances weight loss, decreases overall adipose tissue, and ameliorates fatty liver tissue.[31,34]

## SKIN & HAIR

In an odd finding, iron-free lactoferrin acts as a hypoxia mimetic, stabilizing the HIF-a factor and improving outcomes in the face of low oxygen. Interestingly, increased levels of HIF-a are associated with improved skin repair and regeneration.[37]

Continuing on this path, lactoferrin is beneficial for the skin for many reasons. Not only does it limit the inflammatory response, but it also improves re-epithelialization. In fact, "Lactoferrin stimulates the proliferation and migration of fibroblasts and keratinocytes and enhances the synthesis of extracellular matrix components, such as collagen and hyaluronan."[35]

In hair, lactoferrin regulates the growth and cycling of hair follicles by promoting the proliferation of dermal papilla cells, specialized mesenchymal cells which are located at the base of the follicle.[10]

## GASTROINTESTINAL TRACT

Lactoferrin has an interesting and evolving relationship with the GI tract as a person ages. In infancy, lactoferrin is crucial. It helps determine the microbiome populations and affects the growth and proliferation of the intestinal wall cells or enterocytes. This controls the mucosal surface and consequently the absorption of a variety of nutrients.[36]

In adulthood, oral lactoferrin is mostly, although not completely, digested into smaller molecules. Luckily, some of these smaller products retain some activity. Lactoferricin, a 25-residue peptide, and lactoferrampin, a 20-residue peptide, display potent antimicrobial activity. There are lactoferrin receptors in the intestine however, so intact molecules that do successfully make it through the stomach are effectively absorbed.[34] In fact, ten minutes after oral intake in adult mice, the intact molecule can be found throughout the body. [25]

Unfortunately, lactoferrin disappears very rapidly, with a half-life of 15 minutes. [32,34]

## DOSE

Most exogenous lactoferrin is consumed as bovine lactoferrin, which is roughly 69% identical to the amino acid sequence of human lactoferrin. It has been commercially available for many years and is Generally Recognized As Safe (GRAS) from the United States Food and Drug Administration.

As per usual, dose ranges are wide, ranging from 40 mg/day to 300 mg/day. Some studies go as high as 3 gm/day. By far, however, the most common dose is 250 mg per day.[14]

Because lactoferrin gets released by white blood cells during times of inflammation, lactoferrin levels are frequently used as a marker for trouble. High levels in the GI tract are associated with inflammatory bowel disease or colon cancer.[13]

One might extrapolate from this that elevated levels are bad- and in general, high, naturally occurring levels certainly are. On the flip side, artificially elevating our levels is not bad, and in fact may be quite beneficial. The caveat, of course, is that you had better inform your primary care doctor if you choose to take exogenous lactoferrin.

Research into the lactoferrin molecule has, of course, been chiefly carried out on human and bovine milk. In fact, the idea of oral administration of bovine lactoferrin was first introduced in 1978 by the Morinaga Milk industry in Japan.

But even more attention-grabbing was the finding that camel milk contains the highest amount of lactoferrin, coming in at 230 times higher in camels than cows. Thus, in the middle east, camel milk is quite popular, and in Egypt, the milk is a treatment for hepatitis B and C.

Intrigued by this idea, I tried powered camel milk. It was moderately tolerable. I even attempted to make brownies out of powdered camel milk chocolate. Per my daughters, these were not tolerable.

Meanwhile, folks are so interested in lactoferrin and its therapeutic use that it has been made into an ice cream called "Recharge," designed to enhance the immune system in patients undergoing chemotherapy.

Therefore, regardless of how you take your lactoferrin, take your lactoferrin!

# CHAPTER 21

# LEUCINE

(0.2.3.0.2.0.1)

BENEFITS:

BRAIN  HEART  MUSCLE

*"Chronically increasing leucine intake via the consumption of an overall increase in dietary protein appears to be the most effective dietary intervention toward increasing or attenuating lean mass during aging."*

~ M. Devries 2018

Granted, the diet must provide twenty essential amino acids, but the most important for longevity, without a doubt, is leucine. [This statement is probably a bit controversial]

Why? Because amino acids are not simply precursors for protein synthesis, but they also function as signals regulating various metabolic processes.[23]

Because humans cannot synthesize leucine, by definition, it's an essential amino acid. It is readily available in many animal protein and dairy products such as milk, eggs, pork, beef, chicken, whole grains, vegetables, oats, wheat germ, garlic, and black fungi. But it was first isolated from cheese in 1819.

Leucine is an α-amino acid, meaning it contains an α-amino group (which is in the protonated −NH3+ form), an α-carboxylic acid group (which is in the deprotonated −COO− form), and a side chain isobutyl group, making it a non-polar, aliphatic amino acid. Leucine is also a branched-chain amino acid (BCAA), along with isoleucine and valine.

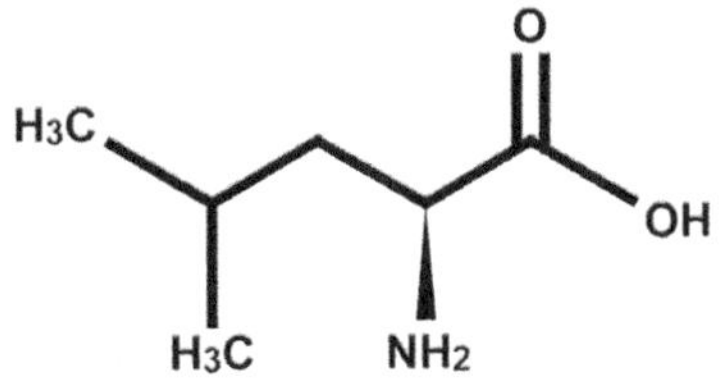

**LEUCINE**

*Amino acids are organic compounds that contain amino (–NH2) and carboxyl (–COOH) functional groups, along with a side chain (R group) specific to each amino acid.*

In addition to participating in protein synthesis, leucine is involved in energy production by way of its metabolites.

Catalyzed by BCAA transferase (BCAT), leucine generates α-ketoisocaproate (KIC) in the skeletal muscle and liver. This can then go one of two ways...

1) 90–95% becomes Isovaleryl -CoA, which further down the pathway becomes cholesterol, B-hydroxybutyrate, acetoacetyl-CoA, and acetyl-CoA. This is useful as the last two are intermediates of the tricarboxylic acid cycle, contributing to energy production.

2) 5–10% becomes β-hydroxy-β-methylbutyrate (HMB), ending up in the same place through a different intermediate.

In addition to this, the conversion of leucine to α-ketoisocaproate (α-KIC) creates glutamate from alpha-ketoglutarate. [Remember that AKG has its own chapter]

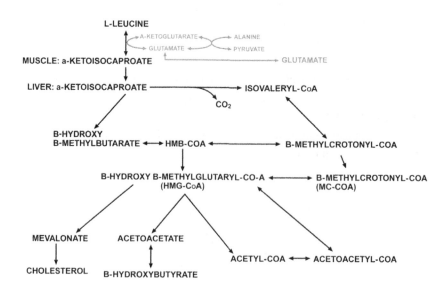

The List...

## 2) MITOCHONDRIA: 2

There is consistent evidence that leucine is beneficial for mitochondria, especially for mitochondrial biogenesis.[11,20] In cell cultures, i.e., cultured myotubes, "Leucine significantly increased mitochondrial content, mitochondrial biogenesis-related genes expression, fatty acid oxidation, SIRT1 activity and gene expression."[11,18]

Interestingly, both leucine and its metabolite β-hydroxy-β-methylbutyrate increase myotubular mitochondrial biogenesis. Consistent with this, HMB and leucine both stimulate the expression of mitochondrial regulatory genes, which suggests that HMB mediates these effects for leucine.[19]

This is valid in mice models as well as cultures, where BCAA supplementation increases mitochondrial biogenesis and sirtuin 1 expression in cardiac and skeletal muscle, but not in adipose tissue or the livers of middle-aged mice.[22]

## 3) PATHWAYS: 3

Leucine has a novel effect on sirtuins, activating SIRT1 in an allosteric fashion.

*allosteric: undergoing, or being a change in the shape and activity of a protein (such as an enzyme) that results from combination with another substance at a point other than the chemically active site*

This translates into requiring a much smaller dose of the amino acid for sirtuin activation and requiring far less NAD+. There is also evidence that this smaller dose of leucine can successfully activate sirtuins, before activating the mTOR pathway, which requires a higher dose.

More scientifically, "Leucine allosterically activates Sirt1, thereby reducing its Km for NAD+ by >50% and enabling Sirt1 activation by nonfasting levels of NAD+."[29]

In addition, "the optimal activation of Sirt1 occurs when leucine is increased from fasting levels of ~0.10 to 0.15 mM to a concentration of ~0.50 mM, and clinical dose-ranging studies with oral leucine administration demonstrate that this level is achieved with doses of 1–1.5 g leucine."[29]

So, in mice, BCAA's increase SIRT1 in cardiac and skeletal muscle.[22] In addition, in obese mice, the addition of leucine to the diet correlated with increased expression of SIRT1 and NAMPT (nicotinamide phosphoribosyltransferase) as well as higher intracellular NAD+ levels.[10]

In an innovative study, low-dose NAD+ precursors synergized with leucine to increase SIRT1 activity in adipocytes, hepatocytes, and muscle cells in mice. This combination also increased lifespan in *Caenorhabditis elegans* (worms).[29]

In another combination study with resveratrol and leucine, lower doses of both together synergistically activated SIRT1. "Although neither leucine nor this concentration of resveratrol exerted effects on SIRT1 activity, AMPK activation, or mitochondrial metabolism, the combination resulted in marked increases in SIRT1 and AMPK activity and mitochondrial metabolism in both skeletal muscle cells and adipocytes."[29]

Adding to this, it turns out that the metabolites HMB and KIC, as mentioned above, are also direct activators of SIRT1, and in fact, KIC is more efficacious than leucine at activating both the mTOR signaling and SIRT1.[30]

β-hydroxy-β-methyl butyrate (HMB), another metabolite of leucine, has been reported to activate AMPK synergistically with resveratrol in myotubes. This occurs, most likely, through SIRT activation as the primary event.[11]

Moving to the mTOR pathway, conversations are a bit challenging, as leucine is a known mTOR activator, and most longevity studies support the inhibition of mTOR.[12]

There is no doubt that many amino acids up-regulate protein synthesis via the mammalian target of rapamycin complex 1 (mTORC1). Leucine, however, is the most potent.[3,12]

In fact, in innumerable animal models, leucine and the BCAA's increased protein and muscle production, all acting through the mTOR pathway.[2,17]

In people, specifically, twenty-year-old male athletes, the addition of 5.6 g of BCAA after resistance training increased plasma concentrations of leucine (300 ± 96%), isoleucine (300 ± 88%), and valine (144 ± 59%), peaking thirty minutes post-ingestion. These BCAAs increased the phosphorylation status of mTORC1 signaling.[7]

Of note here, KIC is more potent than leucine at activating both the mTOR signaling and SIRT1.[30]

Therefore, the balancing game begins. Leucine stimulates mTOR and builds tissue, while mTOR inhibition helps with longevity.

## 5) IMMUNE SYSTEM: 2
There is good evidence that leucine is an anti-inflammatory agent.

Eccentric muscle contractions were induced by electrical stimulation in the anterior tibialis muscle of male rats, i.e., they underwent electrocution-induced exercise. Results demonstrated that oral branched-chain amino acids alleviated the expression of interleukin-6 and impeded the infiltration of inflammatory cells into the muscle. The take-home here was that "leucine-enriched amino acids accelerate recovery from muscle damage by preventing excessive inflammation."[8]

Jumping to human studies, adolescents and adults with Cerebral palsy were given leucine supplementation. Following ten weeks of oral leucine (192 mg/kg body mass), there was a 25.4% increase in strength and a 3.6% increase in muscle volume. Crucial for this category was the 59.1% reduction in CRP within the leucine group.[21]

## 7) WASTE MANAGEMENT: 1
There is mixed information in regards to leucine and blood glucose levels.

For example, long-term leucine supplementation, 2.5 grams three times a day in sixty elderly Type II diabetics during a 6-month intervention period, demonstrated no improvements in various indices of whole-body insulin sensitivity and glycemic control.[23]

On the other hand, when combined for twenty-eight days with low dose resveratrol (added to 2.2 grams of leucine), prediabetics had "significantly improved glucose dynamics, as demonstrated by improvements in glucose and insulin areas under the curves following oral glucose tolerance test in addition to 60-min glucose and insulin concentrations."[29]

# SYSTEMS KNOWN TO BENEFIT

## LIFESPAN
Branched-chain amino acids extend lifespan in yeast, worms, and mice.[22]

## MUSCLE AND CONNECTIVE TISSUE
There is no doubt that leucine is good for connective tissue and muscle.[15]

Young male rats with experimentally damaged connective tissue healed much faster following supplementation with leucine.[13] Additional BCAA's also improve regenerative muscle capacity in old rats[10], attributed to a decrease in inflammation and an increase in the number of proliferating satellite cells.[14] It also appears that leucine inhibits muscle atrophy, in part, through the down-regulation of proteolysis.[3]

Turning to people, in a meta-analysis examining sixteen studies and 999 subjects, [like they couldn't find just one more subject] leucine supplementation significantly "increased gain in body weight, lean body mass, and body mass index." The leucine supplementation was more effective in the subgroup with sarcopenia.[9]

In elderly men, meanwhile, there are mixed results. In a study of thirty healthy men on leucine supplementation for three months at 7.5 grams per day, there were no changes in skeletal muscle mass or strength.[23]

On the other hand, in a second meta-analysis of nine randomized controlled trials, researchers reported that "leucine supplementation is useful to address the age-related decline in muscle mass in elderly individuals, as it increases the muscle protein fractional synthetic rate."[25]

Therefore, it is not unreasonable to combat sarcopenia and frailty with leucine as age progresses.[5]

## METABOLIC SYNDROME
Dietary leucine supplementation is also appropriate for combating metabolic syndrome. *"The beneficial effects include increased loss of body weight, reduced white adipose tissue inflammation, improved lipid and glucose metabolism, enhanced mitochondrial function, and preserved lean body mass."*[26]

Low dose HMB, a metabolite of leucine, when given to diet-induced obese mice, demonstrated an increase in adipose SIRT1 activity, an increase in muscle glucose uptake, improved insulin sensitivity, a reduction in inflammatory stress biomarkers, and reduced adiposity.[30]

Some of this activity results from activating the mTOR system, such as inhibiting lipogenesis, promoting lipolysis and fatty acid oxidation, and increasing leptin secretion in adipocytes.[30]

*"Long-term low-dose supplementation with Leu reduces the body fat and the rate of fat production, increasing insulin sensitivity."*[30]

Leucine also reduces adiposity by increasing beige fat, white adipose tissue that functions as brown fat. Supplementation induces a nearly fourfold increase in the mRNA expression of uncoupling protein 1 (UCP-1), a brown fat-specific gene, in white adipose tissue capable of altering the tissue metabolism.[30]

In the trend of combo therapies, leucine was found to have synergistic effects with subtherapeutic doses of both metformin and sildenafil to amplify downstream effects of Sirtuin1 and AMP Kinase on energy metabolism and improve nonalcoholic fatty liver disease and obesity.[29]

## BRAIN

Leucine enters the brain from capillaries more rapidly than any other amino acid and then enters astrocytes. Within these cells, a mitochondrial branched-chain aminotransferase metabolizes the leucine, which provides 30 to 50% of all α-amino groups for brain glutamate and glutamine.[27]

## DOSE

The dose of leucine is tough to estimate, with a plethora of variables to consider. For example, bodybuilders or people craving muscle require more than others. Lower doses tend to augment sirtuin activation, while higher doses are more likely to activate mTOR.

That being said, recommendations for adults range from 42 mg/kg to 55 mg/kg daily, working out to be about two grams. In terms of optimal sirtuin activation, 1 to 1.5 grams seems to be the way to go.[1]

There are also recommendations on the upper limits, about 500 mg/kg, or roughly 35 g/day for an individual weighing 70 kg.[4]

## POTENTIAL SIDE EFFECTS

For those intense bodybuilders, too much can be harmful and can precipitate side effects such as delirium and neurological compromise which can be life-threatening

Of note, a high leucine intake may also cause or exacerbate symptoms of pellagra in people with low niacin status because it interferes with the conversion of L-tryptophan to niacin.

In terms of getting enough leucine through foods alone, whey protein has the highest concentration at about 10-12 g/ 100 gm. Soy protein is next at 8 gm/100 mg, followed by pea protein.

# CHAPTER 22

# MAGNESIUM THREONATE

(2.2.0.0.3.3.2)

BENEFITS:

BONE  BRAIN  HEART

> *"Because of the essential role of Magnesium in the stabilization of DNA, in defending the cell to the damage of ROS and in stimulating DNA replication and transcription, a Magnesium deficit may facilitate genomic instability, alter DNA repair, and reduce mitochondria functionality, thus facilitating an accelerated cellular senescence and aging."*
>
> ~ M. Barbagallo 2021

This particular agent is a bit incongruent with the others, but I think it has significant value regardless. First and foremost, most folks are magnesium deficient for one reason or another, and people spend a lot of cash supplementing their levels. Perhaps more to our concerns, however, magnesium threonate, in particular, crosses the blood-brain barrier and can improve neurologic function. Thus, this is an excellent tool to have available for people concerned about potential age-related cognitive decline.

We can thank Dr. Guosong Liu and other scientists at MIT for magnesium threonate, who worked very hard to develop something to maximize magnesium loading in the brain.

Why? Because it was well known at that time that increased brain levels could promote synaptic density and plasticity in the hippocampus. Getting magnesium into the central nervous system, however, was a problem.

Regular magnesium formulations have been in use for many years, treating migraine headaches, metabolic syndrome, diabetes, hyperlipidemia, asthma, premenstrual syndrome, various cardiac arrhythmias, and depression. It can also aid in the prevention of renal calculi and cataract formation.[16] Magnesium supplementation is helpful in these disease states, as the deficiency contributes to the illnesses. Suboptimal levels are also linked to osteoporosis, muscle spasms, respiratory diseases, and neurologic anomalies, including stress and anxiety.

*"Oral magnesium supplementation reduces the frequency, duration, and intensity of migraines by 41% compared to placebo at 15.8%."* [16]

In older adults, magnesium deficiency can present as hyper-emotionality, tremor, sleep disorders, and amnesic and cognitive disturbances that may be often overlooked or confused with standard age-related decline.

In addition, and for our purposes, long-term magnesium deficiency results in oxidative stress and chronic, low-grade inflammation, which, as we are aware, are linked to age-related disease and to the aging process itself.[4]

What exactly is magnesium?

It is the fourth most common mineral in the human body after calcium, sodium, and potassium and the second most common intracellular cation (positively charged) after potassium.

Magnesium is quite busy all over the body, both in and out of cells. For example, it's a cofactor in almost 500 enzyme systems and binds more than 3,500 human proteins. It is required for all energy-generating reactions and for both RNA and DNA synthesis. In the mitochondria, high concentrations are critical for the synthesis of ATP from ADP. Lastly, magnesium is an essential modulator of the N-methyl-D-aspartate (NMDA) receptor in the brain, which is involved in memory function and depression.

On average, a person contains 25 grams of magnesium with 55 to 65% in the bone, 27% in muscle, 19% in soft tissues, and less than 1% in the plasma.

Within the plasma, concentrations between 75–95 mmol/L are considered normal. Some researchers, however, believe that serum levels less than 85 mmol/l are actually deficient.[16]

As of 2017, nearly two-thirds of the western world's intake was considered inadequate.[16] Possible etiologies include diminished levels in processed foods, loss of nutrient value with various cooking methods, and reduced gastrointestinal absorption with concurrent vitamin D deficiency. Common medications also diminish the absorption of magnesium, including some antibiotics, antacids, and hypertensive drugs. Aging can also decrease absorption by 30%. As well, excess excretion can occur with diabetes and alcohol use.[16]

The List...

## 1) DNA ALTERATIONS: 2
Magnesium is essential for the maintenance of DNA structure.[11]

In addition to the standard DNA spiral and histone configurations, another level of DNA complexity occurs at areas replete with guanine. (Remember fisetin?) Called a G-quadruplex, this arrangement creates square-like structures that are stacked on top of each other. Magnesium enhances the stability of these structures, preventing damage to key DNA strands, including the telomeric regions.[11] In fact, telomere attrition is faster when cells are cultured in Mg2+- deficient conditions.[11]

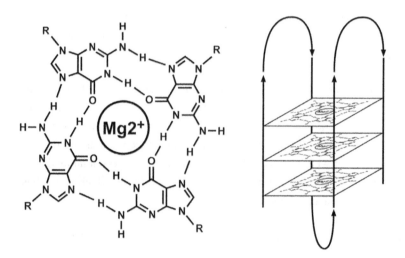

*"Telomeric chromatin structure and integrity is impacted upon by Mg2+ biochemistry."* [11]

There is also evidence that magnesium alters telomerase activity.[11] In animals subjected to twenty-one days of magnesium deficiency, telomerase levels become down-regulated.[11]

*Telomere Attrition: The main driver of telomere attrition is DNA replication. However, the amount of telomeric sequence loss can vary dramatically, from the theoretical minimum of less than ten base pairs per replication to an average of 50–200 bp, up to about one thousand bp.*

## 2) MITOCHONDRIA: 2

Magnesium is vital for the electron transport chain.[11] In a mouse model of premature aging, treatment with dietary magnesium improved the electron transport chain complexes I, III, and IV. It enhanced the coupled mitochondrial membrane potential and thereby increased ATP synthesis.[21]

## 3) PATHWAYS: 0

"mTOR demonstrates high sensitivity to Mg2+ oscillation, and it is therefore possible that Mg2+ acts as a regulator of the balance between anabolism and catabolism."[11]

That is truly all we know.

## 5) IMMUNE SYSTEM: 3

Magnesium deficiency is inflammatory, resulting in leukocyte and macrophage activation, the release of inflammatory cytokines and acute-phase proteins, and excessive production of free radicals.[12, 13]

In rats, the dietary reduction of magnesium to 25% and 50% of the requirement for three to six months increased substance P and tumor necrosis factor-α in tibial bone. Over more extended periods, magnesium deficiency in rats during aging was found to increase not only the inflammatory state but also oxidative stress leading to increased cardiovascular risks.[13]

In humans, a meta-analysis of seven cross-sectional studies with a total of 32,198 individuals found that dietary magnesium was significantly and inversely associated with serum CRP concentrations.[13]

## 6) INDIVIDUAL CELL REQUIREMENTS: 3

More information about magnesium and the brain will be discussed below, but I thought I would mention a few cool tidbits as a preview.

In the adult brain, neural stem cells (NSCs), located in the sub-granular zone of the hippocampus and the sub-ventricular area of the cortex, can self-renew as well as generate all neural lineage types. Many studies have shown that supplemental magnesium threonate can regulate this neural stem cell pool in the adult hippocampus. Both short and long-term treatments increase the number of hippocampal NSCs. In young mice especially, supplementation significantly enhanced hippocampal proliferation.

More importantly, chronic treatment in aged mice did not deplete the hippocampal neural stem cell reservoir but instead curtailed the age-associated decline in NSC proliferation.[7]

## 7) WASTE MANAGEMENT: 2

In this category, we have very significant correlations but no actual cause and effect data.

In a meta-analysis of fifteen studies in the *Cohorts for Heart and Aging Research in Genomic Epidemiology* (CHARGE), magnesium intake was inversely associated with fasting glucose and insulin.[16]

In addition, a cohort of 286,668 healthy individuals and 10,192 Type II diabetic patients revealed a significant, inverse association between magnesium intake and Type II diabetes. The authors recommended that increased consumption of magnesium-rich foods could reduce the risk of the disease.[14]

Similar observations were noted in the 2012-2013 Canadian Health Measures Survey. Both Type I and Type II diabetes were associated with lower serum magnesium levels when compared to individuals without the disease. In addition, body mass index, glycated hemoglobin, serum glucose, and insulin concentrations were negatively correlated with serum magnesium levels.[14]

## SYSTEMS KNOWN TO BENEFIT

### BONE

Magnesium is essential for bone as it interacts with calcium, vitamin D, and phosphate to control mineralization. This interaction determines absorption rates for calcium and phosphate and influences the synthesis of vitamin D. It is actually possible, at least in part, to treat vitamin D deficiency with magnesium.[14]

*"Magnesium is required for conversion of vitamin D into its active form which, in turn, supports calcium absorption and metabolism, as well as normal parathyroid hormone function."* [16]

Higher magnesium intake is associated with higher bone mineral density in elderly white men and women. Oral supplementation can suppress bone turnover in postmenopausal women, as well as decrease the risk of fractures.

Too much can also be detrimental, however, as excessively high magnesium levels appear to negatively affect bone health.[16]

## CARDIOVASCULAR DISEASE

In areas with low magnesium in the drinking water, there are higher incidences of atherosclerosis, ischemic heart disease, coronary vasospasm, hypertension, and sudden cardiac death.

More directly, both animal and human studies have shown an inverse relationship between dietary magnesium and atherosclerosis.[3]

Acting as a natural statin, magnesium can lower cholesterol and triglyceride levels. It also functions as a calcium channel blocker as well as a vasodilator in the microcirculation.[3,16] Thus, diets high in magnesium lower high blood pressure, improve atherosclerosis,[3,14] and reduce the incidence of sudden death.[16] Additionally, optimal levels of magnesium are essential for stabilizing heart rhythms.[14]

*"The consequences of Mg deficiency lead to oxidation and fragmentation of DNA and inflammation in cells of the cardiovascular system, phenomena characteristic of atherosclerosis, aging, and CVD."* [3]

## BRAIN

As was mentioned above, magnesium threonate is a fantastic adjunct to improving memory.[5]

In humans, standard magnesium formulations poorly infiltrate the brain. Elevating plasma magnesium levels by 300% only increases brain levels by 19%.

Magnesium threonate to the rescue; it permeates the blood-brain barrier and once in the brain, it is quite remarkable.

It is a NMDA receptor antagonist involved in memory function and depression.[22, 25]

It is a necessary coenzyme for the conversion of tryptophan to serotonin, a neurotransmitter involved in mood and mental health.[16]

It reduces neural inflammation,[6, 20, 22, 23, 27, 28, 24] and it improves CA3-CA1 synapses in the hippocampus.[1, 23]

Therefore, it is not surprising that magnesium threonate significantly improves learning and memory and reduces depression-like symptoms in innumerable rodent studies.[10, 15, 17, 19, 24]

Findings suggest that increased brain magnesium enhances both short-term synaptic facilitation and long-term potentiation, and improves learning and memory functions.[18]

In a human association study, the Qatar Biobank Study found that the incidence of suboptimal magnesium levels was 57.1%. They identified a direct association between serum magnesium and cognitive function. As well, low magnesium concentrations were associated with longer mean reaction times.[2]

Lastly, in a rare, randomized, double-blinded, placebo-controlled, parallel-designed trial in subjects aged 50 to 70 with cognitive impairment, magnesium threonate in the form of a brand called ClariMem® was given for twelve weeks.

The results speak for themselves:
*"After 6 weeks of treatment with MMFS-01, we observed improvements in executive function (TMT-B), and working memory (DigitSpan), both associated with the prefrontal cortex, and after 12 weeks, we observed improvement in episodic memory, associated with the hippocampus. These observations suggest that the mechanisms of action of MMFS-01 might work at different time scales in different brain areas."*[8]

# DOSE

The Institute of Medicine has set the upper limit of magnesium supplementation with no side effects at 350 mg/day,[16] and most recommendations range between 300 to 400 mg/ day.[14] Higher levels can be used when appropriate, but excess magnesium will usually result in bowel intolerance. Meanwhile, the recommended dose specifically for Magnesium Threonate ranges from 500 to 1,000 mg. Your best bet? 500 mg once a day.

The half-life of magnesium is forty-two days, so once a deficiency is corrected, the daily intake can be reduced.

There are various magnesium supplements available, including magnesium citrate, magnesium glycinate, and magnesium malate in addition to magnesium threonate. The choice should depend on the end goal, but magnesium threonate is the only way to go for brain effects.

## POTENTIAL SIDE EFFECTS

A common unpleasant side effect of oral magnesium supplementation is diarrhea.

Have caution; however, too much is not a good thing. Hypermagnesemia can lead to hypotension, bradycardia, and in extreme situations, coma.

**CHAPTER 23**

# NARINGENIN

(0.3.2.1.2.3.2)

BENEFITS:

BONE  BRAIN  FAT

*"Naringenin possesses various biological activities such as antidiabetic, antiatherogenic, antidepressant, immunomodulatory, antitumor, anti-inflammatory, DNA protective, hypolipidaemic, antioxidant, peroxisome proliferator-activated receptors (PPARs) activator, and memory improving."*

~ P. Venkateswara 2017

Naringenin comes to us from traditional Chinese medicine, historically from a fern called drynaria. In China, Taiwan, Vietnam, Thailand, and Laos, the rhizomes of this fern, also called gu-sui-bu, are used to treat bony injuries. The name, in fact, translates into *mender of shattered bones*. Today, gu-sui-bu is available as Qianggu capsules, marketed in Asia to help with osteoporosis. [Realizing this doesn't sound very exciting, anyone at risk for osteoporosis should take note.]

The most active ingredient, naringenin, is actually more prevalent in citrus fruits, especially grapefruit, bergamots, oranges, and tomatoes, substantially increasing its availability. [Naringenin is the reason grapefruit are so bitter]

Naringenin comes in many molecular forms, the most important being naringin, a glycated version. This is key as it is frequently commercially available in this form.

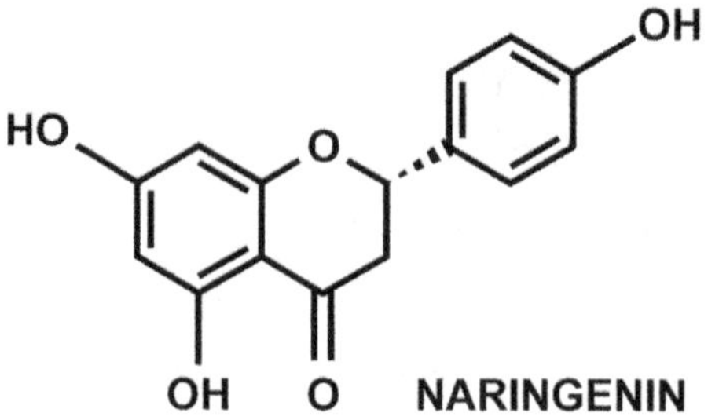

**NARINGENIN**

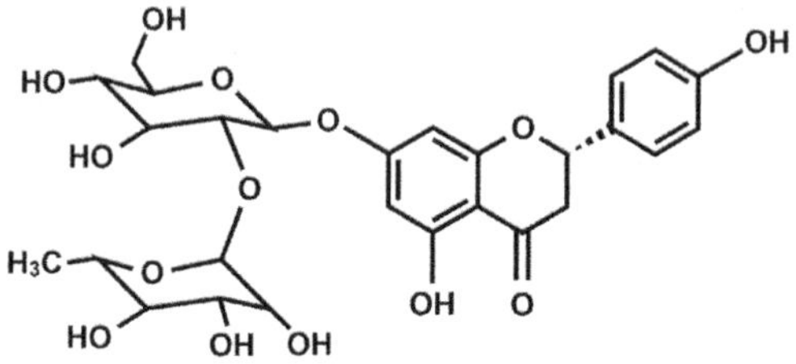

**NARINGIN**

With the molecular formula $C_{27}H_{32}O_{14}$ and a molecular weight of 580.4 g/mol, naringenin has the skeleton structure of a flavanone, a type of flavonoid, with three hydroxy groups at the 4', 5, and 7 carbons, thus its real name: 4',5,7-trihydroxy flavanone.

It is an aglycol, meaning without a glycol. With a glycol, or in its glycosidic form, naringin has an additional disaccharide neohesperidose attached via a glycosidic linkage at carbon 7, also called naringenin-7-rhamnoglucoside. [I know, drab science stuff]

Naringenin can also take the form narirutin (naringenin- 7-O-rutinoside) or naringenin-glucoside (naringenin-7-O-glucoside), depending on the sugar additive.

In foods, naringenin is generally present as naringin. Thereafter, it gets metabolized into various forms of naringenin, including a free form, naringenin glucuronide, and naringenin sulfate. In all, about twenty-three metabolites have been identified.

As a lipophilic molecule, thus insoluble in water, naringenin accumulates within membranes and eventually reaches greater concentrations there than in the plasma.[30]

This property also allows it to traverse the blood-brain barrier and to exert diverse neuronal effects.

The List...

## 2) MITOCHONDRIA: 3

There is good news in this category; both naringenin and naringin are free radical scavengers. Naringenin, however, is far more potent.[7]

Naringenin's antioxidant effect is due to its hydroxyl substituents (OH), demonstrating high reactivity against reactive oxygen and nitrogen species. In general, the antioxidant capacity of any given molecule increases with the number of OH radicals in the molecule, which, in the case of naringenin, has three.[14]

Naringin is less potent as the sugar or glycol add-on causes steric hindrance of the scavenging group.[1]

When compared to other flavonoids in test tube conditions, naringenin falls short.[14] Fortunately, it is highly lipophilic and accumulates in cell membranes, so it performs much better in actual cells.[14]

*"Naringenin protects against ROS in a model of neuronal damage, since it reduces their levels in neurons and decreases mitochondrial dysfunction and increases mitochondrial membrane potential."* [14]

The capacity of naringenin to enhance endogenous free radical scavengers far outpaces its primary scavenging ability, as it up-regulates the Nrf2 pathway and consequently increases the downstream antioxidants.[14,18,31]

In many, many models, it has been shown to enhance superoxide dismutase, catalase, glutathione transferase, glutathione reductase, and glutathione.[9,14,24,38]

The glutathione bump rings true in many studies, including one of my favorite examples: two hundred and sixteen, 54-wk-old Lohmann Pink-shell laying hens. Drynaria supplementation enhanced the total serum antioxidant capacity, glutathione peroxidase levels and had higher femur and tibial bone mineral density than the control group.[15]

As an additional bonus, naringenin is also a metal chelator.[14]

As a direct result of these qualities, naringenin strongly protects against lipid peroxidation. This has been validated in lungs, joints, retinas of streptozotocin-induced diabetic rats, cardiomyoblast cells, skin, testis, and livers in animal models.[14]

Naringenin also interacts with different types of channels within cell membranes. For example, it stimulates the mitochondrial calcium-dependent potassium channel, which causes an influx of potassium ions, a mild depolarization, and a decrease in the mitochondrial matrix calcium uptake, all of which contribute to stabilizing the mitochondria during cellular damage.[8, 16]

## 3) PATHWAYS: 2

In mouse models, naringenin enhances the expression of SIRT1. This is not surprising as its molecular structure is not far removed from that of resveratrol, another flavonoid, and prominent sirtuin activator.[25, 28]

There is also evidence, in rat models, that naringenin increases levels of SIRT3 within the mitochondria.[37]

In addition, naringenin up-regulates AMP Kinase.[19, 21, 37, 43]

## 4) QUALITY CONTROL: 1

In cell models, there is evidence that naringenin can increase DNA repair rates.

The molecule increases three principle enzymes in the DNA base excision repair (BER) pathway: 8-oxoguanine-DNA glycosylase 1 (hOGG1), apurinic/apyrimidinic endonuclease, and DNA polymerase β (DNA poly β).[12]

In human keratinocyte cells, treatment enhanced the removal of UVB-induced cyclobutane pyrimidine dimers (CPD) from the genome.[10]

## 5) IMMUNE SYSTEM: 2

In several rodent models, naringenin down-regulated the levels of NF-κβ,[4, 20] TNF-α and IL-6,[28] and IL-1β, and monocyte chemoattractant protein-1.[29]

In addition to suppressing cytokinin expression, naringenin has the unique property of promoting lysosome-dependent cytokine protein degradation, thus removing already produced cytokines.[39]

## 6) INDIVIDUAL CELL REQUIREMENTS: 3

One of the most salient features of naringenin is its ability to improve bone quality.[13, 27, 36, 41]

Put most simply: naringin "significantly inhibits bone loss, improves bone density, and enhances biomechanical anti-compression performance."[11]

Naringenin suppresses osteoclast activity[23, 32] and increases osteoblast activity to promote osteogenesis.[5, 23] It also inhibits the formation of fat cells or blocks adipogenesis in the bone marrow.[5, 23]

Naringenin does other innumerable, beneficial things as well:

It enhances bone marrow stem cell proliferation.[11]

It enhances the osteogenic differentiation of bone marrow stem cells.[17, 26]

It increases the mRNA expression levels of osteocalcin, alkaline phosphatase, and collagen Type I,[11] as well as osteopontin and bone morphogenetic protein-2 in neonatal rat skulls.[40] Lastly, it enhances the action of 1-α,25-dihydroxyvitamin D3 in promoting the secretion of osteoprotegerin by osteoblasts in vitro.[33]

Back to the pink-shell laying hens, treatment precipitated higher femur and tibial bone mineral density than the control group, with improved bone microstructure.[15]

In terms of human experience, there have been hundreds of years of confirmation. *Rhizoma Drynariae* is a well-recognized treatment for osteoporosis and bone nonunion in traditional Chinese medicine.[33]

## 7) WASTE MANAGEMENT: 2

There is decent evidence that naringenin can decrease glucose levels.

In cell cultures, specifically L6 myotubes from skeletal muscle, naringenin stimulated glucose uptake in a dose- and time-dependent manner.[43]

In mice, it increases the insulin concentration and lowers hepatic glucose-6-phosphatase activity, effectively reducing blood glucose levels.[1] In diabetic rats, treatment for twenty-one days decreased mean levels of fasting blood glucose and glycosylated hemoglobin and elevated serum insulin levels. The values obtained in the naringenin-treated animals actually approximated those observed in gliclazide (a prescription medication for Type II diabetics) - treated animals.[2]

## SYSTEMS KNOWN TO BENEFIT

### BONE
Enough already said.

### BRAIN
Naringenin crosses the blood-brain barrier and has beneficial effects in the brain.

In a male rat stroke model, i.e., middle cerebral artery occlusion followed by twenty-three hours of reperfusion, naringenin treatment successfully up-regulated the antioxidant status, decreased the infarct size, and lowered the levels of myeloperoxidase, nitric oxide, and cytokines. Most importantly, the functional recovery was vastly improved as compared to the control group.[20]

In a mouse model of depression, "data demonstrated that naringenin possessed potent antidepressant-like property(s) via the central serotonergic and noradrenergic systems."[34]

### ADIPOSE TISSUE
Naringenin is our friend here as well, as it suppresses fat accumulation.[21]

*"It has also been reported to have a great ability to modulate signaling pathways related to fatty acids metabolism, which favors fatty acids oxidation, impairs lipid accumulation in liver and thereby prevents fatty liver, besides efficiently impairing plasma lipids and lipoproteins accumulation."*[24]

In male-hooded rats, naringenin supplementation lowered adiposity as well as triglyceride levels.[1]

In human white adipocyte cultures (human fat cells in culture) and abdominal subcutaneous adipose tissue, treatment with naringenin for one to two weeks increased the expression of the genes associated with thermogenesis and fat oxidation, including uncoupling protein 1 and adipose triglyceride lipase.[21]

## DRUG INTERACTIONS

Naringin can alter the drug levels of some common medications; for example, it reduces the bioavailability of pitavastatin, imatinib, and some beta-2 agonists. It blocks the breakdown and thus extends the half-life of minoxidil. It also increases the bioavailability of some calcium channel blockers, such as verapamil.

## BIOAVAILABILITY

Most naringenin is consumed as naringin; luckily, gut microbiota converts naringin to naringenin within the intestine, which is then absorbed.[1] Unfortunately, the naringenin is poorly absorbed, with only 15% actually getting into the systemic circulation.[30]

The pharmacokinetics were examined in male endurance athletes, and oddly, the bioavailability in these athletes was lower than that in less trained individuals.[24]

## DOSE

Once again, determining a dose is challenging.

*"Ingestion of 150 to 900 mg doses of naringenin is safe in healthy adults, and serum concentrations are proportional to the dose administered. Since naringenin (8 µM) is effective in primary human adipocytes, ingestion of 300 mg naringenin twice/d will likely elicit a physiological effect."* [22]

My recommendation? 300 to 600 mg per day.

In terms of natural intake, grapefruits have 53 µg/mL, and orange juice has only 7 µg/mL.[1]

# CHAPTER 24

# POLYPODIUM LEUCOTOMOS

(0.2.0.3.3.0.0)

BENEFITS:

IMMUNITY

SKIN

*"P. leucotomos (a tropical fern plant) extract is rich in polyphenols with properties to counteract skin aging mechanisms from its antioxidant and anti-inflammatory properties: it inhibits oxidative stress, lipid peroxidation, dermal mast cell infiltration, inflammatory cytokines, DNA damage, and UV induced tumors."*

~ N. Philips 2010

**POLYPODIUM LEUCOTOMOS**

*Polypodium leucotomos* is a tropical fern native to Central and South America. Also known as Phlebodium aureum, golden polypody, golden serpent fern, cabbage palm fern, gold-foot fern, blue-star fern, hare-foot fern, the name *Polypodium leucotomos* is actually considered by some to be a derogatory term.

Meaning many feet, the word polypodium alludes to the foot-like appearance of the rhizome and its branches. This fern usually colonizes tropical rainforest canopies and subtropical forests. It's also found in cloud forests of the Caribbean and even reaches into the southern United States and into the swamps and hammocks of Florida.

The fern derives from Central American folklore, historically used to treat many ailments ranging from asthma to heart disease. More recently, studies have focused on skin conditions, including melasma, vitiligo, psoriasis, polymorphous light eruption, atopic dermatitis, post-inflammatory hyperpigmentation, photo-aging, and even skin cancer.

Clinical studies have shown that the extract provides protection against both UVB and UVA radiation, and, as a result, it is utilized as an oral sun protectant.

Protection from the sun is critical, so I'm going to emphasize this. If you spend a lot of time in the sun, really consider this as part of your protocol.

Names of the medicinal extracts include calaguala and anapsos, while commercial trademarked names include Heliocare, Fernblock, and Difur. These over-the-counter products have been marketed in Europe since 2000, both in topical and oral forms and, presently, are available in more than twenty-six countries. In the United States, it has been available as a dietary supplement since 2006. [Although it's really an adjuvant]

As per usual, the plant contains innumerable active ingredients. Of these, the phenolic components that are considered the most important are the cinnamic acids, including ferulic acid, p-coumaric acid, caffeic acid, vanillic acid, and chlorogenic acid (which is so important that it has its own chapter).[1,2,12]

CINNAMIC ACID

p-COUMARIC ACID

FERULIC ACID

CAFFEIC ACID

CHLOROGENIC ACID

*"...data indicate that extracts of this unique plant utilize multiple mechanisms for providing photoprotection and therapeutic activity that include reducing UV-induced cell damage, reducing oxidative stress and DNA damage, blocking UV radiation-induced immune suppression, and inhibiting the release of UV-induced levels of cyclooxygenase-2 and inflammatory cytokines."*[1]

The List...

## 2) MITOCHONDRIA: 2

To start this category, P. leucotomos is an effective scavenger of several reactive oxygen species, specifically the superoxide anion, hydroxyl radicals, singlet oxygen, and hydrogen peroxide.[4, 10, 12]

Of the components, the most robust antioxidants are ferulic and caffeic acids.[12] In addition to direct scavenging, polypodium enhances endogenous antioxidants.[10] In a study of irradiated animals, untreated rodents had decreased superoxide dismutase activity. In the treatment arm, the expression of superoxide dismutase did not increase, but the enzyme's activity did.[12]

In a rat study where the animals were first injected with beta-amyloid protein in the hippocampus and then treated with polypodium, results were mixed. Lower doses decreased superoxide dismutase activity in the hypothalamus, hippocampus, liver, and spleen, while in the cerebral cortex, there was a significant dose-dependent increase in SOD activity.[5]

## 4) QUALITY CONTROL: 3

This is the pivotal category for polypodium. The agent increases the ability to repair DNA damage, especially from UV-induced damage, most importantly cyclobutane pyrimidine dimmers or CPDs.[2, 12, 21]

In a UV-exposed mouse study, polypodium reduced DNA damage, as measured by 8-oxo-dG positive cells, by 59% at twenty-four hours and 79% at forty-eight hours. "Two weeks after UV exposure, mutations in P. leucotomos-fed-mice were approximately 25% less than those from mice treated with UV alone."[1]

In a hairless mice model, ten days of treatment with PL was followed by UV radiation. At seventy-two hours in the control group, $54 \pm 5\%$ of the cyclobutane pyrimidine dimers remained in the placebo group, versus only $31 \pm 5\%$ in the treatment contingency.[21]

This is thought to occur via the up-regulation of p53; "p53 up-regulation and activation has been reported to play (a) direct role in modulation of key regulatory genes in DNA damage repair and inflammation."[21]

The extremely good news is that we have human evidence in this category.

In people exposed to UV radiation after taking two doses of 7.5 mg/kg of P. leucotomos, there was a significant decrease in the mean erythema response at twenty-four hours, remarkably fewer sunburn cell numbers, and most importantly, fewer cyclobutane pyrimidine dimers.[1]

In a second study, ten healthy adult volunteers were exposed to doses of UVA radiation after having taken 240 mg, eight and two hours prior to the exposure. Skin biopsies taken twenty-four hours after irradiation showed that the mean common deletion value, a measure of mitochondrial DNA damage, in the untreated group increased by 160% over baseline while the mean common deletion values in the P. leucotomos-treated group decreased by 42%.[1]

There is also encouraging evidence in this category that polypodium, tested under the brand Fernblock, increases autophagy.[3]

All of this is especially important since skin cancer accounts for at least 40% of all human malignancies, and any reduction in relative risk can only be a good thing.

## 5) IMMUNE SYSTEM: 3

There is tremendous evidence that polypodium inhibits pro-inflammatory cytokines.[6, 10, 12, 17, 21]

Put quite simply... "*Polypodium leucotomos* extract, has a modulating effect on the in vitro production and release of cytokines by peripheral blood mononuclear cells of healthy subjects. At doses effective in vivo, Anapsos can stimulate PBMNc proliferation, delay IL-1β secretion and at the same time increase that of IL-2, IL-10, and INF-γ."[17]

A significant reduction has also been noted for COX-2,[1, 12, 21] TNF-α, and IL-6.[6, 12]

Turning to the immune system, polypodium continues to demonstrate a fantastic track record.

It is known to activate natural killer cells.[9, 16, 17] "In addition, it increases IL-2 and IFN-γ levels by the activation of Th2 lymphocytes, and IL-10, by the Th1 lymphocytes. Anapsos also increases immunoglobulins IgM, IgD and IgG[4] by B-lymphocyte activation in tonsils tissue."[9]

Many studies have, in fact, discovered that it tends to correct T lymphocyte imbalances, "specifically through the increase of the initially low T8 cells levels and subsequent normalization of the mean T4/T8 index."[8]

In addition, it also has been noted to reverse any UV radiation-induced immune suppression.[1]

Turning to humans, healthy athletes over the age of eighteen who took 480 mg *Polypodium leucotomos* extract twice daily for three months had fewer respiratory illnesses. In addition, relapses in the control group were higher at 37%, while only one person in the control group (14.2%) relapsed.[18]

Lastly, long-term treatment with oral polypodium reduces the signs and symptoms of atopic dermatitis in children and adolescents, including a reduction in inflammation and itching.[10]

## SYSTEMS KNOWN TO BENEFIT

### SKIN
Without question, skin is the chief beneficiary.

In addition to preventing damage from UV radiation, polypodium is useful for the treatment of atopic dermatitis, psoriasis,[1] vitiligo, melasma and has the potential to help with postinflammatory hyperpigmentation.[10]

Why is polypodium so amazing for the skin?

In addition to the anti-inflammatory properties, immune improvement, and the augmented DNA repair, it also decreases structural damage and promotes structural improvements.

For example, one of the challenges with UV radiation is the activation of matrix metalloproteinases, which dissolve collagen, elastin, and hyaluronic acid in the skin. Luckily, polypodium inhibits several of these.[12] In fibroblasts and keratinocytes especially, polypodium inhibits MMP-1[13] and MMP-2.[12]

In addition, it activates the TIMPs (TIMP-1 and TIMP-2 in fibroblasts), the tissue inhibitor of MMPs, a.k.a the antidote to MMP's.[12]

Polypodium promotes collagen production as well. Specifically, in fibroblasts, PL stimulates collagen Types I and V deposition in UV-irradiated fibroblasts and Types I, III, and V in non-irradiated fibroblasts.[12]

Elastin, the component that bestows stretchiness to the skin, gets increased as well.[13, 20]

PL also prevents membrane damage and lipid peroxidation induced by UV radiation. It blocks the resulting disarray of the actin cytoskeleton with the loss of cell-cell adhesion and cell-ECM adhesion after UV exposure.[12]

## POTENTIAL SIDE EFFECTS

No serious adverse events have been associated with the use of *Polypodium leucotomos*. The few reported side effects include gastrointestinal complaints and pruritus, occurring at a 2% rate.[19]

## DOSE

As per usual, the suggested dose range is relatively broad. One paper recommends 7.5 mg/kg, about 480 to 500 mg in an average-sized person.[11] The recommended dose of Difur is 120 mg, three times a day, for a total of 360 mg, however, doses up to 1,200 mg have also been used in studies.

My thoughts…400 to 500 mg per day.

# CHAPTER 25

# PYRROLOQUINOLINE QUINONE

(0.3.2.0.2.0.0)

BENEFITS:

BONE  BRAIN  SKIN

*"Through nutritive and environmental exposures, PQQ affects essential biological processes, influencing mitochondriogenesis, reproduction, growth and aging."*

~ K.R. Jonscher 2021

Pyrroloquinoline quinone is unfortunately graced with an impossible name to say quickly, so we are simply going to use its initials, PQQ.

Regardless, it is water-soluble, anionic, and an aromatic tricyclic o-quinone.

Whereas PQQ is essential for humans, it is not synthesized by mammals, and thus we acquire it through the consumption of plants and animals. The primary production, however, is most likely from microorganisms. Luckily, it is relatively ubiquitous and found in innumerable dietary sources, including fermented soybeans (natto), tea, green peppers, parsley, and kiwi fruit. We humans only have trace amounts, ranging from picomolar to nanomolar levels, although higher quantities have been identified in breast milk.

Therefore, of course, depending on one's diet, it is estimated that humans consume 0.1 to 1.0 mg of PQQ and its derivatives per day.

Moving along, PQQ is a very complex molecule, a.k.a. 4, 5-dihydro-4, 5-dioxo-1H-pyrrolo[2, 3-f]quinoline-2, 7, 9-tricarboxylic acid, which gets reversibly reduced to pyrroloquinoline quinol (4, 5-dihydroxy-1H-pyrrolo[2, 3-f]quinoline-2, 7, 9-tricarboxylic acid) or $PQQH_2$. [Thank goodness for abbreviations]

The process of converting from PQQ to $PQQH_2$ requires ascorbate, NAD(P)H, and thiol compounds such as glutathione, precipitating a two-electron reduction. Going the other way, PQQH$_2$ to PQQ, O$_2$ converts to a superoxide anion, which later breaks down into hydrogen peroxide. This is only important in that it explains its antioxidant capabilities.

**PQQ**  +2e$^-$ + 2H$^+$ ⇌  **PQQH$_2$**

Why are we talking about PQQ? Because it extends the lifespan of the worm *Caenorhabditis elegans*.[19]

It also has a reputation for two paramount properties: increasing mitochondrial biogenesis and as a nootropic.

We also know that a deficiency can precipitate negative outcomes, including growth impairment, immune dysfunction, and abnormal reproductive performance.[1]

The List...

## 2) MITOCHONDRIA: 3

*"PQQ was an effective antioxidant protecting mitochondria against oxidative stress-induced lipid peroxidation, protein carbonyl formation and inactivation of the mitochondrial respiratory chain."*[5]

The reduced form of PQQ, i.e., PQQH$_2$, exhibits potent anti-oxidative capacity. Overall, it is between 6 to 7-fold higher than that of vitamin C.[1]

Acting through the Nrf2-mediated signaling pathway, PQQ increases the levels of endogenous antioxidants, including superoxide dismutase, glutathione peroxidase, and catalase in the liver of rodents.[12] In healthy piglets, the dietary addition of PQQ increases mRNA levels of antioxidant genes (NQO1, UGT1A1, and EPHX1).[23]

In a small human study, ten subjects (five females, five males) were given a single dose (0.2 mg/kg) of PQQ. Based on malonaldehyde-related TBAR assessments, this improved the antioxidant potential.[4]

Now to the subcategory where PQQ really stands out - mitochondrial biogenesis. Don't worry about all of the studies; just know that there is an abundance of evidence in this realm. PQQ stimulates mitochondrial biogenesis. It increases mitochondrial DNA content, as well as the mitochondrially-encoded cytochrome c oxidase subunit 1 protein.[18]

It works the other way as well. PQQ deficiency in young mice reduces hepatic mitochondrial content by 20 to 30% and suppresses mitochondrial respiration. Luckily, PQQ supplementation in these deficient mice reverses the mitochondrial alterations and metabolic impairments.[1]

In out-of-shape humans, PQQ did not appear to elicit any improvements in aerobic performance or body composition. Still, it did impact mitochondrial biogenesis by way of significant elevations in PGC-1α protein content.[9]

## 3) PATHWAYS: 2
PQQ increases the expression of SIRT1 and SIRT3 in liver cell cultures. The effect on SIRT3 may be directly linked to the mitochondrial impact by way of peroxisome proliferator-activated receptor γ coactivator 1α, nuclear respiratory factor 1 and 2, and mitochondrial transcription factor A.[24]

In healthy piglets, PQQ activates the SIRT1 system.[23]

In a mouse model of Parkinson's disease, PQQ treatment not only dose-dependently alleviated the locomotor deficits and nigral dopaminergic neuron loss but activated AMP Kinase.[2]

And to round out this category, PQQ has also been found to be a potent mTOR inhibitor. "Moreover, it was found to be a dual mTORC1 and mTORC2 inhibitor that inhibits the entire mTOR kinase-dependent function and feedback commencement of the PI3K/Akt pathway."[14]

## 5) IMMUNE SYSTEM: 2
In microglia cell culture, PQQ significantly suppresses proinflammatory mediators such as iNOS, COX-2, TNF-a, IL-1β, IL-6, MCP-1, and NF-κβ.[22]

In mouse brains, PQQ "demonstrated marked attenuation of neuro-inflammation,"[22] and in mouse muscle, the proinflammatory cytokines IL-6, IL-1β, and TNF-α were reduced after treatment.[15]

In piglets with gut inflammation, "dietary supplementation with PQQ can effectively alleviate jejunal mucosal inflammatory injury by inhibiting NF-κβ pathways and regulating the imbalance of colonic microbiota in piglets challenged with *E. coli* K88."[6] Meanwhile, healthy piglets also benefit, with a reduction in inflammation-related genes (IL-2, IL-6, TNF-α, and COX-2) via the SIRT1/NF-κβ deacetylation signaling.[23]

## SYSTEMS KNOWN TO BENEFIT

### BRAIN

Here we go...does it actually help your brain?

In rats on a PQQ supplemented diet, treatment enhanced learning ability. In addition, after experiencing hyperoxia-induced oxidative stress for forty-eight hours, rats on PQQ had better memory retention than the controls.[1]

Turning to people, in a randomized, placebo-controlled, double-blinded study, forty-one elderly, otherwise healthy people consumed 20 mg of BioPQQ™ per day for twelve weeks. The results strongly suggest that PQQ can prevent reductions in brain function, especially attention and working memory.[11]

BioPQQ® also has been shown to improve language skills in humans.[21]

As a modus operandi, PQQ enhances nerve growth factor production (NGF).[17] A protein composed of 118 amino acids, NGF is well known as a neurotrophic factor required for the development and maintenance of peripheral sympathetic and sensory neurons. PQQ has a stimulatory effect on its synthesis and secretion in astroglial and fibroblast cells.[1]

Lastly, seventeen adult, female subjects participated in an open-label trial to evaluate the effectiveness of PQQ on stress, fatigue, quality of life, and sleep. They ingested 20 mg of PQQ daily for eight weeks. In the end, all six measures of vigor, fatigue, tension-anxiety, depression, anger-hostility, and confusion significantly improved.[1]

## BONE

Studies have shown "that supplemental PQQ played a role in anti-osteoporosis by up-regulating antioxidant capacity, inhibiting oxidative stress and reducing DNA damage, down-regulating CDKI proteins levels, and decreasing cell apoptosis."[7]

In orchidectomized male mice, PQQ supplementation prevented testosterone deficiency-induced osteoporosis by inhibiting oxidative stress and DNA damage. It stimulated osteoblastic bone formation and inhibited osteoclastic bone resorption.[20]

In the female counterpart, ovariectomy-induced osteoporotic mice, PQQ prevented bone loss and improved bone strength. It prohibited osteocyte senescence and the senescence-associated secretory phenotype (SASP), subsequently promoting osteoblastic bone formation and inhibiting osteoclastic bone resorption, which was comparable to treatment with exogenous estrogen.[3]

## SKIN

PQQ does several beneficial things for the skin. It is useful in the prevention and treatment of hyperpigmentation disorders. It reduces epidermal water loss, improves skin moisture, texture, viscoelasticity, and reduces mast cells in the dermis and epidermis.[17]

## DOSE

PQQ dietary supplements have been available in the US since 2009, after the official acceptance of notification by the Food and Drug Administration, and no adverse effects have been reported.

Studies have used 20 mg/day [1,16] up to 60 mg/ day.[1]

# CHAPTER 26

# SALIDROSIDE

(0.3.2.2.2.2.2)

BENEFITS:

ACCLIMATION  BONE  BRAIN

*"In addition to its multiplex stress-protective activity, Rhodiola rosea extracts have recently demonstrated its anti-aging, anti-inflammation, immunostimulating, DNA repair, and anti-cancer effects in different model systems."*

~ Y. Liu 2017

Salidroside is one of my favorite agents, hidden away on the slopes of breathtaking mountains. Despite being present in many different plants, it most famously originates from *Rhodiola rosea*, the extracts of which contain about 1% salidroside. [Cistanche also contains Salidroside]

Rhodiola, a genus in the *Crassulaceae* family, grows in icy, high-altitude locations such as the Himalayas and mountainous regions throughout Europe and Asia. Of the more than two hundred species, over seventy are in China, in Tibet, Qinghai, Yunnan, Sichuan, and other alpine provinces.

Also known as Roseroot, Arctic Root, Golden Root, and Tibetan Ginseng, the plant is renowned in ancient Chinese medicine for replenishing qi (vital energy), activating blood circulation, unblocking blood vessels, and enhancing mental function. In European folk medicine it was used to combat fatigue. In the present day and all over the world, it is celebrated for alleviating fatigue, depression, and anxiety and is considered a brain tonic.

Of note, and making it one of my favorite go-to's when I travel, salidroside not only decreases fatigue but enhances physical performance and prevents high altitude sickness.

Regardless, we are examining rhodiola, known as *Hongjingtian* in Chinese, because many consider it to possess anti-aging activity.

Chemically known as 2-(4-hydroxyphenyl) ethyl-β-D-glucopyranoside, salidroside is a glucoside of tyrosol and is water-soluble.

**SALIDROSIDE**

The List...

## 2) MITOCHONDRIA: 3

There is significant evidence that salidroside scavenges free radicals in many settings, from human umbilical vein endothelial cells[9] to the hippocampal neurons in a rat model of Alzheimer's disease.[33]

Salidroside also activates nuclear factor-erythroid factor 2-related factor 2 (Nrf2), a transcriptional factor that regulates the production of the endogenously produced antioxidants.[4, 14, 22, 34, 50]

Turning to rodent examples, diabetic mice given salidroside for ten weeks had reduced reactive oxygen species and malondialdehyde levels with increases in superoxide dismutase, catalase, and glutathione in testicular tissue (probably in other tissues as well).[15]

In exhausted swimming rats, salidroside elevated not only their exercise tolerance but also enhanced antioxidant liver enzymes (CAT, SOD, and GSH-Px).[39]

Lastly, in the mitochondrial category, salidroside stimulates mitochondrial biogenesis.[48]

## 3) PATHWAYS: 2

There is limited evidence that salidroside activates AMP Kinase.[17, 46] Specifically, in human umbilical vein endothelial cells (HUVECs), salidroside up-regulated AMP Kinase phosphorylation while down-regulating mTOR phosphorylation under oxidative stress.[47]

As mentioned in the mitochondrial section, salidroside increases mitochondrial biogenesis, and this is thought to occur through the activation of SIRT1.[29]

In fact, SIRT1 activation occurs in human umbilical vein endothelial cells,[44] regular rat brains,[10] hypoxic rat brains,[48] in rats with Alzheimer's,[48] and in obese mice.[41]

Whereas human evidence is lacking, all indications point to salidroside as a SIRT1 activator.

As well, in a hyperhomocysteinemia mouse model, there is isolated evidence that salidroside up-regulates the expression of SIRT3.[38]

## 4) QUALITY CONTROL: 2

Salidroside increases DNA repair capabilities, stimulating the activity of poly (ADP-ribose) polymerase-1 (PARP-1), a component of the DNA base excision repair pathway. Biochemically, this is feasible as PARP1 has a tryptophan-glycine–arginine-rich domain to which salidroside binds.[23]

In cell cultures and mice, salidroside-activated PARP1 enhances the repopulating capacity of cells by accelerating oxidative DNA damage repair.[24]

Additionally, salidroside increases autophagy in human umbilical vein endothelial cells under oxidative stress.[46]

## 5) IMMUNE SYSTEM: 2

Whereas there are no human studies to confirm the potency of salidroside as an anti-inflammatory, there is plenty of rodent proof.

Evidence confirms the inhibition of nuclear factor-κβ-mediated inflammation in rats[10, 30, 48] and TNF-α, interleukin-1β, and interleukin-6 in arthritic rats. Of note, in these arthritic rats, researchers noted a decrease not only in the systemic circulation but also in the hippocampus, confirming that salidroside can pass through the blood-brain barrier.[49]

In fact, in rats undergoing cerebral perfusion challenges, salidroside significantly reduced brain infarct size and cerebral edema, in addition to decreasing the levels of interleukin-6, interleukin-1β, and tumor necrosis factor-α in the serum.[6]

My favorite example, however, is in a cohort of smoking mice. While these rodents puffed away, salidroside inhibited multiple pro-inflammatory cytokines, including tumor necrosis factor-α, interleukin-6, and interleukin-1β in the serum and lungs.[27]

Salidroside also is capable of binding COX-2 and inhibiting the production of prostaglandin E2.[35]

## 6) INDIVIDUAL CELL REQUIREMENTS: 2

Stem cells in culture respond favorably to salidroside, promoting the proliferation of bone marrow mesenchymal stem cells and increasing the expression and secretion of stem cell factors.[3]

In addition, salidroside induces rat mesenchymal stem cells to differentiate into dopaminergic neurons and induces mouse mesenchymal stem cells to differentiate into neuronal cells.[7, 26, 45]

## 7) WASTE MANAGEMENT: 2
In diabetic mice, salidroside precipitates a hypoglycemic effect,[20] and in a mouse model of galactose-induced aging, salidroside decreased the production of AGEs.[11]

This is due, in part, to an increase in pancreatic β-cell mass and β-cell replication, as seen in diabetic and obese mice. The increase in pancreatic beta cells boosted insulin output and lowered glucose levels.[17]

Before we leave this section as a whole, I want to point out that this agent would probably have higher scores if there were human studies. The lack of which can only leave us guessing and extrapolating.

## SYSTEMS KNOWN TO BENEFIT

### BONE
Salidroside both increases the activity of osteoblasts and decreases that of osteoclasts, resulting in better bone.

In cell culture, salidroside stimulates osteoblast differentiation and mineralization. It can also promote angiogenesis within a bone callus and accelerate fracture healing.[13]

In ovariectomized mice given salidroside supplementation for three months, these ladies showed improvements in trabecular bone microarchitecture and bone mineral density in the fourth lumbar vertebra and distal femur.[42] This holds for ovariectomized rats as well.[5]

Thus, salidroside acts by "coordinating the coupling of angiogenesis-osteogenesis in the bone environment. Therefore, we have discovered an ideal molecule that simultaneously enhances angiogenesis and osteogenesis and thereby accelerates bone healing."[12]

### BRAIN
*"Rhodiola rosea L. is widely used to stimulate the nervous system, extenuate anxiety, enhance work performance, relieve fatigue, and prevent high altitude sickness."*[28]

There is plenty of evidence that salidroside improves wellbeing and overall happiness. A single oral administration of R. rosea induced antidepressant, adaptogenic, anti-anxiety, and stimulating effects in mice.[32] In rats with Alzheimer's disease, salidroside for twenty-one days improved learning and memory.[48] In olfactory bulbectomized rats, chronic salidroside treatment exhibited an antidepressant-like activity comparable to amitriptyline, a common antidepressant.[48]

As a potential etiology, salidroside restored the capacity of the dentate gyrus to generate new neurons, preventing learning and memory deficits during aging in mice.[16] And lastly, in senescence-accelerated mice, salidroside alleviated hippocampus-dependent memory impairment.[36]

Turning to people...
In this category, there is a plethora of human evidence that salidroside improves attention, cognitive function, and mental performance.[28,31]

As an example study, eighty mildly anxious participants took 400 mg of *Rhodiola rosea* daily for fourteen days. The experimental group reported a significant reduction in anxiety, stress, anger, confusion, and depression relate to controls.[8]

On a personal note in this category, I recommended salidroside to a fellow mountain climber. Not only did he report an easier time with the high altitude, he commented on how happy he had become.

## FATIGUE

Remarkably, treatment with R. rosea in rats prolonged the duration of exhaustive swimming by almost 25%.[1] In mice as well, salidroside improves fatigue. It lengthened swimming time, increased hemoglobin, augmented muscle and liver glycogen, and diminished blood lactate levels.[21]

Turning to people, eleven physically active college females took salidroside for three days with a bonus dose immediately prior to exercise. Supplementation enhanced anaerobic performance in repeated bouts of exercise.[2]

## HYPOXIA

Time for another side trip...
In response to low oxygen levels, cells produce something called the hypoxic

inducible factor (HIF-1a), which accumulates at the cellular level. Actually, cells make HIF continuously, but it disappears very quickly under normal oxygen conditions. When oxygen levels are low, HIF-1a accumulates, activates Hypoxia response elements (HREs), which in turn trigger metabolic alterations, helping cells function with less available oxygen.

The first of these changes is in the mitochondria. HIF-1 mediates a subunit switch in the cytochrome c oxidase complex within the electron transport chain, improving electron transfer efficiency. It also affects target genes such as heme oxygenase-1, vascular endothelial growth factor (VEGF), and glucose transporters, with the ultimate result of cells being more dependent on glycolysis for energy.[19]

Back to salidroside... Treatment of multiple cell types with salidroside increases the hypoxic inducible factor (HIF-1a) when exposed to hypoxic conditions, including cardiomyocytes[43] and human umbilical vein endothelial cells.[40] In cardiomyocytes, this decreased the number of necrotic and apoptotic cells in a dose-dependent manner.[43]

What does this mean for us? It means increased survival in low oxygen conditions, either high altitude hiking, during a stroke or cardiac arrest or simply with anaerobic exercise. [Thus the previous recommendation]

## DOSE

Unlike many of the agents we talk about, the bioavailability of salidroside is quite high.[25]

The suggested dose ranges from 50 to 200 mg two times daily, with the most common being 200 mg twice daily.[8,18]

The toxicity is also very low, with doses as high as 600 mg/kg/day in some clinical trials reporting no side effects.[48]

# CHAPTER 27

# SHILAJIT

(0.2.0.0.2.1.0)

BENEFITS:

ACCLIMATION  BRAIN  VIRILITY

*"This substance is known as Çilájatu and cures all distempers of the body."*

~ S. Bhattacharya 1995

Shilajit is a substance that generally warrants the reaction: "Really? I'm supposed to eat that?" This was undoubtedly my response when I first heard about shilajit. When I first consumed it, my response was worse; it is not a joyful culinary experience.

What is it? And how is it pronounced? [It's pronounced MC Hammer style... shilajit to quit. Hey, hey!]

It's a pale-brown, blackish-brown goo, ranging from ooze to a mushy gum, that seeps out of mountain rocks in the summer. Officially called an exudate, it is found between the elevations of 1,000 to 5,000 meters and in almost every high mountain range across the globe. Most commonly harvested in the Himalayas, it is also in the countries of Kashmir, Afghanistan, Nepal, Bhutan, Pakistan, China, Tibet, and Russia. There is even shilajit in the South American Andes, described as a dark-red, gummy matrix.

As if consuming rock ooze isn't tempting enough, the aroma and taste aren't any better. Descriptions include smelling like stale cow urine and having a bitter, pungent and astringent taste. Despite this unpleasantness, it still remains on my personal protocol - but I make sure I take it with something to mask the taste.

And yet, people consume this stuff worldwide. Children in Nepal and North India drink it in their morning with milk, and Sherpas claim it gives their already strong population even better health and energy levels.

What is this stuff, exactly? There are many descriptions of this material, none of which sound very appetizing. It is made up of decomposed organic plant matter and microbial metabolites mixed with minerals that have leached out of the rocks. The polite word for this is *humus*, i.e., organic components of soil formed by the decomposition of leaves and other plant material. [not to be confused with Hummus...Reminds me of desert vs dessert]

*"Shilajit is a compact mass of vegetable organic matter, which is composed of a gummy matrix interspersed with vegetable fibers and minerals."* [26]

As this is not factory-made, there are dramatic variations in the final composition depending on its source. As could be expected, shilajit samples collected from different places exhibit variations in chemical characteristics and bioactivities.[25]

Why are we even talking about this bizarre substance? Because for over 3,000 years, it has been considered a rejuvenator and an anti-aging compound. In Ayurvedic medicine, it is a rasayana, an early term concerning the means by which we can invigorate the body and lengthen lifespans.[6]

Despite its drawbacks, it's a popular agent in a vast number of cultures and thus carries many titles.

Shilajit, the original Sanskrit nomenclature, translates as the *Conqueror of mountains* and *Destroyer of Weakness* and/ or the *Winner of rock*.

Other Sanskrit names are *Silajit* and *Silaras*.

Across the globe: [14]
Bengali: *Silajatu*
Tamil: *Uerangyum*
Arabic: *Hajar-ul-musa*
Russian: *Mummio or Mumie*
Persian: *Momio*
German: *Mumie*

In western culture, it's Mineral pitch, Jew's pitch, Mineral wax, or my personal favorite... Asphalt. [Doesn't that sound delicious?]

Because the ooze origins are widespread and the rocks are quite varied, it has also been categorized historically by its predominant metal type. A prevalence of gold, for example, is called suvarna, while that of silver is rajat. Similarly, tamra is copper laden, loha is iron, naag is lead, and vanga is tin.[14]

The color variations, as well, reflect these components. Savrana, made of gold, is, of course, gold-ish red, while rajat has a silver-white tone. Tamra is a bit blue, while lauha leans toward blackish-brown. This last one, lauha, the most common of the shilajits, is found in the Himalayan ranges and is supposed to be the most effective therapeutically.[14]

Shilajit has been used extensively over the centuries for many illnesses, and like most medicinal cures from long ago, the list of treatable diseases is seemingly limitless. This includes diabetes, jaundice, gall bladder and renal calculi, an enlarged spleen, digestive troubles, fermentative dyspepsia, adiposity, hysteria, neurological diseases, amenorrhoea, dysmenorrhoea and menorrhagia, urinary diseases, tuberculosis, leprosy, eczema, anemia, anorexia, chronic bronchitis, asthma, fracture of bones and many other clinical conditions.[20]

Shockingly, the composition of shilajit is remarkably consistent. It is mainly humic substances, including fulvic acid, that account for around 60% to 80% of the total nutraceutical compounds, plus some vitamins and minerals (20 to 40%) and up to 5% trace elements (including Fe, Ca, Ag, Cu, Zn, Mg, Mn, Mo, Pb, P).[6] Because of the huge variety of vitamins and minerals, it may, in fact, serve as a de facto *Fred Flintstone multivitamin.*

In addition, it contains dibenzo-alpha pyrones, small peptides and lipids, uronic acids, benzoic acid, hippuric acid, phenolic glucosides, and some amino acids. There are a zillion, plus or minus, other agents thrown in there as well, but these are relatively minute and inconsistent.

Of all the compounds, the major physiological action is thought to be due to fulvic acid and the dibenzo-alpha pyrones. Notably, fulvic acid is soluble in water, has a low molecular weight (around 2 kDa), and is well absorbed in the intestinal tract.[6]

The List...

## 2) MITOCHONDRIA: 2

The free radical scavenging activity and antioxidant capacity are well documented both in test tubes and cultures,[9,17,19] and attributed primarily to the dibenzo-alpha pyrones and fulvic acid.[21]

*"Fulvic acid plays as a bi-directional super antioxidant i.e. as electron donors and acceptors, depending upon the need for balance in the situation. If it encounters free radicals with unpaired positive electron it supplies an equal and opposite negative charge to neutralize it, likewise, if free radicals carry a negative charge, the fulvic acid molecule can supply positive unpaired electron to nullify that charge."* [20]

In addition, in lab animals and humans, shilajit increases endogenous antioxidants. In a study of rats given an intraperitoneal dose, scientists demonstrated an increase of superoxide dismutase, catalase, and glutathione peroxidase.[21]

Examining rat brains from animals fed various amounts of shilajit, scientists reported a dose-related increase in superoxide dismutase, catalase, and glutathione peroxidase in the frontal cortex and striatum.[2]

Moving to people, healthy young humans ranging in age from 16 to 30 were given two grams of shilajit daily, which significantly increased blood levels of superoxide dismutase, vitamin E, and vitamin C.[20] This is key as most studies are done on older folks, and it proves that even healthy youngsters can benefit.

There is also some evidence that shilajit improves mitochondrial function.[24,28] Details however, are still vague.

## 5) IMMUNE SYSTEM: 2

Innumerable mouse and rat studies confirm the anti-inflammatory activity of shilajit. It appears to work in several models of researcher-induced inflammation, including rat arthritis, paw edema, and ear edema. Only one rat study to date confirmed this finding on a cytokine level, however, citing diminished levels of IL-1, IL-6, and TNF-α.[21]

Shilajit improves immunity as well, activating macrophages and splenocytes in rodent models.[13]

## 6) INDIVIDUAL CELL REQUIREMENTS: 1

In culture, shilajit is a "potent stimulator of osteoblastic differentiation of mesenchymal stem cells and an inhibitor of osteoclastogenesis."[21]

This was demonstrated clinically in rat models where the mechanical strength of bone tissue (femur and tibia) increased after treatment.[1]

## 7) WASTE MANAGEMENT: 0
Despite historic references suggesting the possibility of lowering blood glucose, no confirmation was found experimentally.[3]

## SYSTEMS KNOWN TO BENEFIT
There are a plethora of isolated examples demonstrating the benefits of shilajit. One study proved that oral intake for 14 weeks (125 or 250 mg bid) stimulated genes necessary for muscle development in obese women.[7]

In another study, twenty healthy young volunteers consumed two grams of shilajit for 45 days. Researchers found a significant reduction in serum triglycerides and cholesterol with a simultaneous improvement in HDL cholesterol.[20]

### BRAIN
Shilajit is clearly advantageous for rodent brains. In normal rats, oral intake "augmented learning acquisition and memory retrieval" and well as making them less anxious.[8]

This may be due to an observed increase in brain dopaminergic activity, which is associated with the nootropic and anxiolytic effects of the agent.[11]

In addition, rats suffering from researcher-induced head trauma fared far better on shilajit than controls. They had reduced brain edema, less disruption of the blood-brain barrier, and a more normal intracranial pressure after the injury.[12]

No evidence yet exists for people.

### ENERGY
*"Animal and human studies support its use as a "revitalizer," promoting physical and mental energy, enhancing physical performance, and relieving fatigue in association with enhanced adenosine triphosphate production."*[23]

### ALTITUDE
Native, high-altitude dwellers, as well as mountain climbers, take advantage of shilajit. It augments energy levels, reduces recovery time after injury, increases

the strength of muscles and bones, and alleviates mental stress. As a diuretic, it treats both cerebral and pulmonary edema, and as an anti-inflammatory, it reduces overall aches and pains.[13] Additionally, shilajit protects mountaineers from sunburn and eye problems precipitated by high-intensity UV radiation.[17]

Fulvic acid which "stimulates blood formation, energy production, and prevents cold exposure and hypoxia"[13] is thought to be responsible for many of these benefits.

Take this with salidroside for high altitude trips, and you should be good to go!

## MALE VIRILITY

Shilajit researchers are very interested in its effects on men and their ability to reproduce. Thus we are going to skip the countless rodent studies[15,18] and move on. In men, both healthy between the ages of 45 and 55[16] and infertile men,[5] multiple measures of virility improved with treatment, including testosterone levels, sperm count, and sperm motility.[16]

## POTENTIAL CONCERNS

Before you elect to go licking the Himalayan cliffs, the biggest issue with *fresh*, native shilajit is that it's contaminated with polymeric quinines, microbial debris, fungal toxins, and heavy metals.[6,22,27]

Thus, prior to ingestion, most commercially available products are purified; and conveniently, in many studies, the refined product seems to be even more efficacious.

*"These findings reinforce our earlier postulate that purification of shilajit is an imperative necessity to ensure its optimum therapeutic effect. This would also safeguard from potential health risks associated with prolonged ingestion of raw shilajit containing free radicals and fungal toxins."*[8]

On the other hand, if you choose to evaluate your own native ooze, the historic recommendations are as follows:

*"It should puff upon subjected to fire and burn without smoke. On adding to water it should not dissolve completely but leave a trail and disintegrates as it travels from the surface of the water to the bottom of the container."*[25]

## DOSE

The dose is, of course, a relative suggestion, but as a place to start, studies range anywhere from 150 mg to 250 mg, once or twice daily. Another source recommends 300 to 500 mg daily, with milk and in divided doses.[13]

Interestingly, and in opposition to most of the agents in this book, shilajit is slowly metabolized and reaches a maximum level in the blood after twelve to fourteen hours of consumption.[13]

Thus, shilajit is considered safe, assuming it is purified and not taken in excessive doses. The one-time oral lethal dose is thought to be greater than 2000 mg/kg,[16] which is probably equivalent to consuming an entire mountain range. Taken chronically, 0.2–1.0 grams/kg is considered safe.[16]

Therefore, aim for 200 to 300 mg per day.

Of note, one interesting warning suggests not consuming it with *meat of pigeon*.[13]

# CHAPTER 27

# SPERMIDINE

(3.2.2.3.2.1.2)

BENEFITS:

BRAIN     FAT     HEART

*Polyamines, spermidine and spermine, are synthesized in every living cell and are therefore contained in foods, especially in those that are thought to contribute to health and longevity. They have many physiological activities similar to those of antioxidant and anti-inflammatory substances such as polyphenols. These include antioxidant and anti-inflammatory properties, cell and gene protection, and autophagy activation... There is a close relationship between polyamine metabolism and DNA methylation.*

*~ K. Soda 2022*

To start with the obvious question, yes, spermidine is related to sperm. It was discovered by Antoni van Leeuwenhoek in human semen in 1678. Using self-made microscopes, he technically is known for the identification of spermatozoa in 1677, but the source of this specimen has never really been discussed. [I think we can all guess where it came from]

After being mesmerized by these aquatic athletes, he also noticed crystals in the same pool; these he took note of, but never named. Two hundred years later, in 1888, two very unimaginative scientists, A. Landenburg and J. Abel, bequeathed the title of spermidine on these crystals.

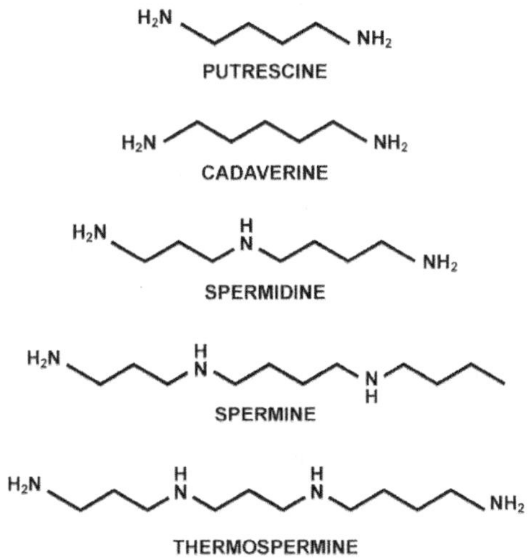

```
H₂N ~~~ NH₂
PUTRESCINE

H₂N ~~~~ NH₂
CADAVERINE

H₂N ~~ NH ~~~ NH₂
SPERMIDINE

H₂N ~~ NH ~~~ NH ~~ 
SPERMINE

H₂N ~~ NH ~~ NH ~~~ NH₂
THERMOSPERMINE
```

Spermidine, all jokes aside, is an aliphatic polyamine that is present in all living organisms. It is also a polycation, being positively charged at physiological pH. Other prominent and related polyamines include spermine, and the precursors putrescine and cadaverine.

Looking at these molecules, you will notice that they are all linear carbon chains, just getting progressively longer, with nitrogens at both ends, and in the middle of spermidine and spermine. [Not that this is necessary to know, but spermidine is volatile and confers the characteristic smell and taste to sperm.]

Of note, putrescine and cadaverine are both colorless liquids produced by the breakdown of amino acids. The two compounds are largely responsible for the foul odor of putrefying flesh, but also contribute to the odor of such processes as halitosis and bacterial vaginosis. [The naming of these substances is unfortunate, as they do some amazing things, yet seem a bit repulsive on the surface.]

Polyamines, in general, are essential for cell growth and proliferation, tissue regeneration, they regulate ion channels, provide DNA and RNA stability, inhibit inflammation, regulate DNA methylases, control protein acetylation, and assist with stress resistance.

Spermidine, unfortunately, despite being ubiquitous and doing these amazing things, has been underappreciated for many years.

We now know, however, that polyamines, and especially the tissue concentrations of spermidine, decline in an age-dependent manner in both model organisms and humans, and that supplementation can prolong lifespan and reduced age-related pathology.[31]

Where can we find spermidine naturally? Conveniently, as I mentioned, spermidine is ubiquitous in living things; it is actually found in all food items that contain nuclei, including both animal and plant parts. Of note, it is scarcely present in milk or in hyperprocessed food items. But it's prevalent in foods that are generated by bacterial or fungal fermentation, including smelly cheese, fermented soy, and specific fruits, such as the durian fruit that has a sperm-like odor. Additionally, the content is high in wheat germ.

[Another tangent into strange fruits...]
*The durian fruit, also called the "king of fruits," comes to us from Thailand and Malaysia, and is notable for its large size, strong odour, and thorn-covered rind. While the famous British naturalist Alfred Russel Wallace described the fruit's flesh as "a rich custard highly flavored with almonds," it has probably more accurately been described as having the aroma of rotten onions, turpentine, and raw sewage.*

So, back to the basic question, why are we talking about spermidine? Because polyamines, and spermidine in particular, increase the lifespan of multicellular organisms, including yeast, nematodes, fruit flies, mice, and human peripheral blood mononuclear cells. In humans, there aren't any prospective studies, but there seems to be an association between high dietary spermidine intake and reduced mortality.[18,19]

In fact, in an all-cause mortality study correlated to spermidine intake, there was a significant drop in mortality across the three tiers of intake. This difference equated to a gain in 5.7 years between the lowest and highest intake groups.[14]

*"Our findings lend epidemiological support to the concept that nutrition rich in spermidine is linked to increased survival in humans."* [14]

On an interesting note, spermidine levels are not depressed in healthy nonagenarians and centenarians, who retain whole-blood concentrations comparable to younger (middle-aged) individuals. This probably contributes to their longevity.

More specifically, spermidine levels in humans aged between sixty and eighty were lower than in humans below fifty, but humans older than ninety have levels similar to people below fifty.

The addition of oral spermidine also seems to be beneficial to organisms. Given to aging fruit flies, their climbing activity vastly improves.[24] [As an aging climber myself, I hope this holds true for people]

As well, in an aging mouse study, a high polyamine diet allowed rodents to maintain a thicker coat and have higher activity levels over time.

The List...

## 1) DNA ALTERATIONS: 3

Spermidine performs incredibly well in this category, both as an epigenetic modifier and as a protector of DNA.

To start, in human dermal fibroblasts, spermidine reduces the expression of histone acetyltransferase (HAT)[27], and induces protein deacetylation.[19]

Interestingly, spermidine triggers deacetylation in the cytosol and acetylation in the nucleus.[24]

This becomes particularly important because spermidine diminishes cytosolic protein acetylation by inhibiting the activity of several acetyltransferases. Specifically, this includes the E1A-associated protein p300 (EP300), and its inhibition is sufficient to acutely induce autophagy, which is a key mechanism for many of the actions of spermidine.

"EP300 directly inhibits acetylation of several autophagy-essential autophagy-related (ATG) proteins and indirectly stimulates deacetylation of tubulin by inhibiting α-tubulin acetyltransferase 1 (αTAT1). Therefore, the inhibition of EP300 by spermidine causes deacetylation of ATG proteins and increases acetylation of tubulin, thus stimulating autophagic flux."[20]

The effects of autophagy will be covered more down below.

At the same time, low spermidine levels can lead to reduced DNA methylation levels by inducing the accumulation of dcAdoMet, which inhibits DNA methyltransferase (DNMT). And in fact, abnormalities in global DNA methylation associated with aging can be partially reversed by exogenous polyamines, supporting the idea that the reduction of tissue polyamines during aging is closely related to alterations in DNA methylation.[28]

In addition to the epigenetic modifications, spermidine has an even greater influence on DNA. One might even call it an Epi- Epigenetic effect. I also like to think of it as the bubble wrap phenomenon.

Because spermidine is a cation (+3 at neutral pH), it likes to cozy up to negatively charged molecules, and DNA is a favored choice. Spermine (+4) does this as well. The preferred binding site is the major groove on DNA for all of the polyamines, but spermidine and spermine can occupy more varied sites, including binding along the backbone and bridging of both the major and minor grooves.[4]

This arrangement is beneficial for DNA, as the DNA and even RNA, become more stable once wrapped with spermidine. This bonding is non-specific electrostatic binding and does not alter the secondary structure of DNA. But it does allow the DNA to become more condensed and thus protected from harmful attacks.[7]

Put another way, the condensed structure reduces the accessibility of attack sites.

In fact, a very high percentage of both spermidine and spermine are bound by ionic interactions to nucleic acids, proteins, and other negatively-charged molecules in the cell. This condensation of the DNA makes it more stable, especially from thermal denaturation, or melting in high heat.[17] It has also been shown to have protection from radiation.[25,30]

## 2) MITOCHONDRIA: 2

The effect of spermidine on radicals does not seem to be straightforward.

On one hand, there is clear evidence that it is effective against certain free radicals. For example, both spermidine and spermine are potent against singlet molecular oxygen.[13] In fact, spermine is thought to be the polyamine with the strongest antioxidant properties, secondary to the higher number of positive charges. Metal chelation is also important here, which prevents the formation of hydroperoxides and delays the generation of secondary oxidation compounds.[26] On the other hand, polyamine catabolism can generate potentially toxic products such as $H_2O_2$ and polyamine-derived aldehydes.[28]

Spermidine, does, however, improve the status of endogenous antioxidants. In aging rodent heart cells, supplementation reversed the decrease in superoxide dismutase and catalase.[31] In prematurely-aged mice (SAMP8), spermidine and spermine robustly decreased MDA levels and improved the activity of superoxide dismutase in the aging brain.[33]

Other cool mitochondrial tricks:

### MITOCHONDRIAL BIOGENESIS
Data demonstrate that spermidine stimulates mitochondrial biogenesis in senescent cardiomyocytes.[31]

### IMPROVES MITOCHONDRIAL FUNCTION
Spermidine improved the structure and function of mitochondria from aged cardiac muscles and skeletal muscle (stem) cells.

### INHIBITS MITOCHONDRIAL PERMEABILITY TRANSITION PORE (MPTP) OPENING
This topic was brought up in the CoQ10 chapter. (feel free to glance back for a quick refresher)

Spermidine and spermine both inhibit pore opening, but with no definitive etiology.

## 3) PATHWAYS: 2
In this category, spermidine has positive effects on AMP Kinase and SIRT1.

This increase in AMP Kinase has been demonstrated in the brain of prematurely-aged mice (SAMP8) after treatment with both spermidine and spermine.[33]

In obese mice, spermidine increased the phosphorylation of hepatic AMP-activated protein kinase. To further prove the point, the spermidine-mediated molecular effects were abolished by compound C, an inhibitor of AMP Kinase, in primary hepatocytes.[10]

Because spermidine activates the phosphorylation of AMP Kinase, this, in turn, antagonizes mTOR.[18]

Thus, spermidine, thru the AMP kInase pathway, has been classified as a caloric restriction mimetic, recognizing that its actions are similar to those that are precipitated by caloric restriction.[18]

Turning to the sirtuins, a six-week spermidine supplementation in old rats increased SIRT1, as well as PGC-1α, and nuclear respiratory factors 1 and 2.[31]

Lastly, SIRT1 was increased in human dermal fibroblasts after spermidine.[27]

## 4) QUALITY CONTROL: 3
Yet again, spermidine rocks in this category.

First, spermidine stimulates DNA polymerase B. What does DNA polymerase B do you ask? It plays a central role in the base excision DNA repair pathway. Thus, spermidine promotes improved DNA repair mechanisms. This effect is concentration-dependent however, and less is actually better than more.

*"The stimulation of DNA polymerase B activity by spermidine is much greater at lower concentrations of enzyme than at higher concentrations and so the enzyme concentration curve for DNA polymerase B is more linear in the presence of spermidine than in the absence."*[6]

And now for the main event...

*"Accumulating evidence indicates that SPD's beneficial effects on aging are mainly due to the induction of autophagy."*[31]

This is the category where our protagonist really shines. So, to fully appreciate it, let's review autophagy for a moment.

Autophagy, a.k.a. self-eating, is essentially cellular recycling. It takes suboptimal cellular pieces and parts, and reuses the old elements to make new, bright, and shiny cell components. This process ensures general cell homeostasis, makes a cell more efficient, and removes toxic cytoplasmic material that otherwise would accumulate during aging.[20]

Autophagy is extremely important in long-lived cells and is known to increase the life of cells and organisms. It is promoted in times of energy deficiency and reduced in times of excess nutrient availability.

Back to our protagonist... Spermidine has been shown to quickly induce autophagy in yeast, worms, flies, mice, and cultured mammalian cells. As well, there are myriad mouse models where this is true. For example, short-term administration of spermidine to old mice reversed the age-associated defect of autophagy and mitophagy in muscle stem cells.

In fact, intra-peritoneal administration of spermidine in mice will induce autophagy in vivo in multiple tissues such as the heart, liver, and muscle within four to twenty-four hours.

Oral supplementation of spermidine in drinking water triggers autophagy after two to four weeks in mouse cardiac tissue and also reverses the age-dependent decline of autophagy in the aorta.

How does this work? [This was partly addressed in the DNA section, but that was several paragraphs ago. It's also going to come up in other sections, so it's worth repeating]

Spermidine inhibits the activity of several acetyltransferases, including that of E1A- (adenovirus early region 1A) associated protein p300 (EP300). EP300 is key to inhibiting autophagy, thus its inhibition is sufficient to induce autophagy.[20]

Spermidine also stabilizes the microtubule-associated protein 1S (MAP1S), which is necessary for autophagy.[20]

More details.... Spermidine is an inhibitor of the acetyltransferase EP300, an enzyme that transfers acetyl groups from acetyl coenzyme A (CoA) on lysine residues of cytoplasmic and nuclear proteins.

There is a competition between spermidine and acetyl CoA for binding to the EP300 catalytic site. Acetyl CoA, which increases with nutrient excess, functions as an endogenous inhibitor of autophagy. Excess acetyl CoA causes the hyperacetylation of multiple proteins involved in the regulation or execution of autophagy, thus stalling the process. Fasting or caloric restriction reduces acetyl CoA levels, triggering autophagy.

Thus, spermidine competitively blocks acetyl CoA from binding to the EP300, making the cell think it is nutrient-deprived, therefore stimulating autophagy.[20]

In cell cultures, spermidine prevents the aging- and Osteoarthritic-related decrease in autophagy and may protect against osteoarthritis.[29]

As a side note, aspirin, which, like spermidine, inhibits EP300 by competing with acetyl CoA, is well known for its role in cancer prevention.

In addition, the combination of spermidine and resveratrol shows synergistic effects on the induction of autophagy.

## 5) IMMUNE SYSTEM: 2

In this category, spermidine promotes M2 polarization of macrophages, meaning it increases the macrophages that are responsible for the resolution phase of inflammation and the repair of damaged tissues. Thus, it also suppresses NFκβ-dependent proinflammatory cytokines.

These inhibitory macrophages also suppress autoimmune-reactive T cells. In neurodegenerative models, this inhibitory effect of spermidine on autoimmune-reactive T cells translates into the prevention of demyelination.

Spermidine also favors the formation of CD8+ memory T cells. In old mice, spermidine restores the CD8+ T cell response to vaccines, allowing a more vigorous, youthful, and appropriate antibody response.[5]

Interestingly, CD8+ T cells from aged mice and humans are known to have reduced autophagy, and the up-regulation of autophagy in these cells seems to restore their immune responses back to youthful levels. Thus, the improvement in the immune system may be secondary to the induction of autophagy.[23]

Spermidine also suppresses cytokines, mostly through the suppression of TNF-α.[18,24]

As well, polyamines trigger the production of anti-inflammatory cytokines while decreasing the production of pro-inflammatory ones. Unfortunately, polyamine metabolism does generate some cytotoxic products that can cause inflammation, but the overall balance is in favor of anti-inflammatory effects.[24]

In a human study of patients with cognitive impairment, spermine down-regulated all investigated cytokines in a dose-dependent manner. Oddly, spermidine led to an up-regulation of some cytokines for lower dosages, while high dosages down-regulated all cytokines apart from up-regulated IL-17A.[9]

Thus, supplementing spermidine in aging human cohorts seems to recover autophagy and T cell function.[2]

## 6) INDIVIDUAL CELL REQUIREMENTS: 1

Once again, autophagy comes to the rescue. In the hearts of aging rats, spermidine decreased the number of senescent cells.[31]

This effect also extends to epithelial stem cells where the stemness-enhancing effects promote hair follicle regeneration in human hair follicle cultures.

With any hope, this attribute will be explored in more detail as many, many people ask about hair growth.

## 7) WASTE MANAGEMENT: 2

This particular aspect of spermidine was noted twenty years ago but seems to have gotten lost in the shuffle.

Spermidine, on top of everything else, has potent anti-glycation effects, comparable to aminoguanine and carnosine. Unfortunately, there has been one, and only one, study here.[12]

However, in a mouse study examining aortic stiffness, spermidine supplementation reduced the AGEs and collagen in the treated, older groups.[15] As well, the exogenous administration of spermidine in diabetic rats resulted in an improvement of blood glucose and a concomitant reduction of HbA1c levels.[28]

## SYSTEMS KNOWN TO BENEFIT

Among other things, in mice, spermidine helps counteract the age-associated disruption of circadian rhythms, and there is evidence that spermidine promotes hair growth and epithelial stem cell function in human hair follicle organ cultures.

In addition....

## CARDIOVASCULAR DISEASE
There are strong indications that spermidine can help prevent cardiac aging.

In old mice, dietary spermidine improves diastolic function and left ventricular elasticity. It reduces cardiac hypertrophy and improves the elastic properties of cardiomyocytes.[8]

In obese mice, twenty weeks of treatment reversed age-induced arterial stiffness and alleviated the formation of atherosclerotic plaques.[15] Spermidine supplementation in rodents limited Abdominal Aortic Aneurysm formation, which was associated with the preservation of aortic structural integrity. Treatment attenuated aortic inflammatory infiltration, it reduced circulating inflammatory monocytes, and it increased autophagy-related proteins.[16] Lastly, in rats that had had heart attacks, spermidine reduced the infarct size, improved cardiac function, and attenuated myocardial hypertrophy.[34]

Moving to people, we only have association studies. However, the intake of dietary spermidine was found to inversely correlate with the incidence of cardiovascular disease and death in the Bruneck cohort, which was a random sample of 1,000 men and women that ran from 1990 to 2018. As well, a cross-sectional regression meta analysis of nutritional polyamine content with CVD-caused mortality rates from forty-eight Western countries identified negative associations of spermidine and spermine with cardiovascular disease.[18]

## ADIPOSE TISSUE
On a cellular level, polyamines seem to be critical for the formation of adipose cells, which would suggest that limiting spermidine might be beneficial for fat control.[28] More specifically, both spermidine and spermine are essential factors during the early stages of adipocyte differentiation for preadipocytes.[24]

On the other hand, spermidine inhibits the expression of lipogenic genes in vivo and in vitro.[10] And in fact, high-fat diet-induced, obese mice treated with spermidine demonstrated decreased body weight and decreased subcutaneous and visceral fat content. As well, the hepatic intracellular and serum triglyceride and total cholesterol concentrations were reduced.[10]

Evidence also suggests that polyamines in white adipose tissue, liver, and skeletal muscle stimulate energy expenditure and can help burn off the fat surplus.

Finally, spermidine administration was able to prevent lipid accumulation and necrotic core formation in vascular smooth muscle cells through the induction of cholesterol efflux in an experimental model of atherosclerosis.[24]

## BRAIN

To start with, polyamines interact with several neurological systems, including the opioid pathway, glutamine signaling, and it acts to limit neuroinflammation.[11]

Among the polyamines, spermidine is known to be the most prevalent in the human brain.[11] Importantly, it acts in the hippocampus, specifically on the mossy fiber-CA3 synapses, formed by dentate granule cell axons connecting with CA3 pyramidal neurons. It rescues the age-dependent decreases in synaptic vesicle density and restores the defective presynaptic MF-CA3 long-term potentiation. It does not, however, affect the more famous CA3-CA1 synapses.[21]

In addition, spermidine increases the level of nerve growth factor and Brain-derived neurotrophic factor in mice. Brain-derived neurotrophic factor is associated with synaptic plasticity, neurogenesis, learning, memory, and cognition.[33]

Therefore, overall, long-time administration of polyamines, spermidine and spermine, have been shown to delay brain aging and improve cognitive dysfunction (at least in mice). [Anyone remember the Rats of Nimh?]

This has been explored in humans in a three-month, randomized, placebo-controlled double-blind phase IIa pilot trial. In cognitively intact humans with subjective or subtle cognitive decline aged 60 to 80, treatment with spermidine "moderately enhanced" memory and mnemonic discrimination ability improved as well.[32]

## AUTOIMMUNE DISEASE

This is going to seem a little circuitous, but give it a minute.

Some autoimmune diseases result when there is chromatin disruption leading to the exposure of previously sequestered, or protected DNA.

The exposure of these otherwise sequestered sequences can lead to the reverse transcription of a gene, yielding auto-antigenic, hypo-methylated DNA fragments, which are known to elicit autoimmune responses. Thus, altered levels of polyamines leading to chromatin disruption can precipitate autoimmune diseases.[3]

Increased doses should therefore help to prevent autoimmune disease by protecting DNA. Spermidine may also reduce the risk of these diseases through inhibitory macrophages that suppress autoimmune-reactive T cells.

## MALE VIRILITY

There is more spermine than spermidine in human semen, and one of the things it does is to increase glucose and decrease fructose utilization by sperm. In fact, there is a positive correlation between concentrations of spermidine and spermine and the motility of ejaculated ram spermatozoa.

Not shockingly, the addition of polyamines to human sperm cultures can increase motility in those with reduced or absent motility. In studies, both spermidine and spermine have been associated with improvement in in vitro fertilization and pregnancy in mice and humans.[22]

## METABOLISM

Spermidine starts with the amino acid arginine, which is converted to ornithine, then to putrescene, and finally to spermidine. Spermidine is then interchangeable with spermine, an even longer nitrogen-containing chain. In mammalian cells, the polyamine biosynthetic pathway is four steps, catalyzed by enzymes mainly located in the cytosol. In this process, ornithine, and S-adenosyl methionine are used as substrates.[28]

In addition to cellular biosynthesis, other sources are equally important. In fact, it is generally believed that oral consumption accounts for the vast majority of body polyamines. Because they are relatively stable compounds, capable of resisting both acidic and alkaline conditions, these molecules make it unharmed through the stomach and into the intestine where they are absorbed.

In addition to oral intake, the gut microbiome is also capable of producing polyamines. In general, spermidine in the upper part of the GI tract is most likely derived from the diet, whereas polyamines present in the lower parts are mostly synthesized by intestinal microbiota.

Once in the gut, the absorption of these molecules appears to be rapid, but they are partly metabolized in the intestinal wall before reaching the systemic circulation. Thus, the highest levels of polyamines are found in the intestine, followed by the thymus and liver.

## GASTROINTESTINAL TRACT

This high concentration of polyamines in the gut turns out to be pretty beneficial. Because these molecules are necessary for supporting rapid cell turnover, they are key to intestinal epithelial cells which have a very short half-life. Polyamines are also crucial for enhancing the integrity of the intestinal barrier. They stimulate the production of intercellular junction proteins, such as occludin, zonula occludens 1, and E-cadherin, which are essential for regulating the paracellular permeability and reinforcing epithelial barrier function.[28]

In addition, gut polyamines serve a central role in the regulation of mucosal adaptive immunity.[26] Insufficient polyamine intake, in fact, is associated with reduced tolerance to dietary allergens and a high intake of spermine is associated with a decreased risk of food allergy in children.

## POTENTIAL RISKS

The association of spermidine and cancer risk is much like many molecular agents that are beneficial to cells. Sometimes substances are too helpful, meaning they benefit both normal cells and abnormal cells.

There are increased polyamine concentrations caused by enhanced biosynthesis in skin, breast, colon, lung, and prostate cancers. As well, polyamines have been implicated as potential enhancers of tumorigenesis.[19]

On the other hand, in animal models that do not have cancer, lifelong spermidine treatment does not seem to increase the risk of malignancy, and in fact has been shown to decrease the risk. Thus, the going theory is that polyamines can inhibit the emergence of tumors, but may promote the growth of existing ones. In human studies, as of yet, there has been no evidence of adverse events. Renal failure may also be of concern as putrescine and acrolein, somewhat toxic byproducts of spermidine and spermine catabolism, are eliminated by the kidneys. Therefore, these metabolites can potentially accumulate in patients with chronic renal failure. Thus, based on preclinical studies and common sense, possible contraindications of spermidine administration are cancer and renal failure.[19]

## DOSE

First off, there are no recommended doses of any of the polyamines. You need more as an infant, in times of physical repair or rapid cell growth, and when you are getting older. The getting older problem has to do with the age-related decline in de novo synthesis. But the question still remains...what's the dose?

In food studies, it has been determined that the mean polyamine intake in the European adult population is roughly 354 μmol/day. It's the lowest in the UK, and the highest in Italy and Spain, of course depending on the standard diets. In these Mediterranean regions, it reaches 700 μmol/d. Meanwhile, Japan (200 μmol) and the US (250 μmol) fall way below the standard.[1]

One author proposed an intake of around 540 μmol/day, taking into consideration the recommendations that promote a high consumption of fruits, vegetables, and cereals.[26]

Interestingly, alcoholic beverages like beer have been previously reported to contain fair amounts of polyamines, particularly putrescine.

In terms of supplements, spermidine is generally consumed as wheat germ extract, with recommendations for 800 to 900 mg of wheat germ daily. This works out to be about one mg (or a little less) of spermidine daily.

While the US FDA has not taken a position on spermidine supplementation, the EU's European Food Safety Authority has set an upper daily limit at 6 mg of supplemental spermidine. No human clinical trials have been conducted with amounts higher than this.

Naturally occurring spermidine and the other polyamines involved in the spermidine and spermine recycling loop are in the food supply, but synthetic spermidine has not been tested in humans.

In terms of actual recommendations for food-derived spermidine, the dose range is 1-6 mg, which are the amounts that have been or are currently being used in clinical trials.

The best bet for now? 1 to 2 mgs daily.

# CHAPTER 29

# URSOLIC ACID

(1.2.2.2.2.0.2)

BENEFITS:

BRAIN

FAT

MUSCLE

*"Amongst other pharmacological properties of UA, one can mention protective effect on lungs, kidneys, liver and brain, anti-inflammatory properties, anabolic effects on skeletal muscles and the ability to suppress bone density loss leading to osteoporosis. Ursolic acid also exhibits antimicrobial features against numerous strains of bacteria, HIV and HCV viruses and Plasmodium protozoa causing malaria."*

*~ L. Wozniak 2015*

Ursolic acid, also known as urson, prunol, micromerol, and malol, is a pentacyclic triterpenoid identified in the waxy coating of apples in the 1920s. It naturally occurs in a vast number of herbs such as basil, rosemary, sage, peppermint, oregano, and thyme, in addition to common fruits such as pears, cranberries, and prunes. But above all, apple peels are the best source.

*The peel is great for you; the seeds are not. Apple seeds contain amygdalin, a substance that releases cyanide into the bloodstream. How many can kill you? Somewhere over 150 - luckily, an average apple has only 5 to 8 seeds.*

Ursolic acid (UA) is formally known as 3b-hydroxy-12-urs-12-en-28-oic acid. More importantly, it is a large, lipophilic molecule, having a molecular weight of 456.7g/mol with thirty carbons ($C_{30}H_{48}O_3$), constituting a five-membered ring.

These rings are built from isoprenoid units, putting it in a family with vitamin E and CoEnzyme Q.

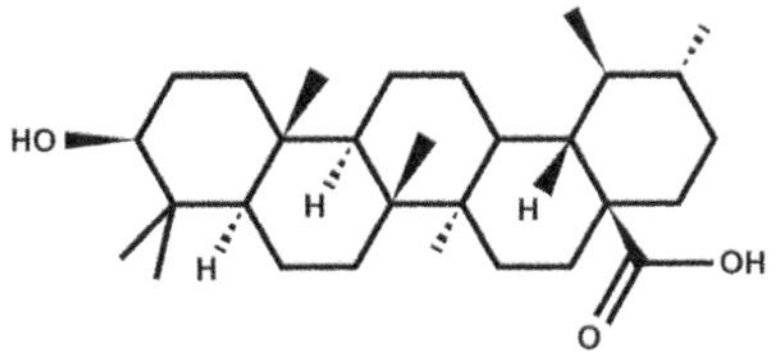

**URSOLIC ACID**

**OLYMPIC RINGS**

Of the many properties of ursolic acid, it has anabolic effects on skeletal muscle and increases endurance, which has made it quite popular with athletes. Thus, I am entertained by its structural similarity to the Olympic Rings.

The List...

## 1) DNA ALTERATIONS: 1

Scientists have just started investigating the epigenetic effects of ursolic acid. We have a few details but no big picture as of yet.

For example, there is evidence that UA demethylates the first fifteen CpG sites of the Nrf2 promoter region, correlating with an increased expression of Nrf2 in cell cultures. This should translate into an increased expression of endogenous antioxidants.[15]

Ursolic acid also has been noted to reduce the expression of epigenetic modifying enzymes, such as the DNA methyltransferases DNMT1 and DNMT3a, and the histone deacetylases HDAC1, HDAC2, HDAC3, HDAC8 (Class I) and HDAC6 and HDAC7 (Class II), and HDAC activity.[15]

This pattern is evident in a male rat model where UA inhibits DNA methyltransferases and histone deacetylases in leukocytes.[34]

## 2) MITOCHONDRIA: 2

Generally, in this category, most agents have free radical scavenging activity. Ursolic acid does not. It does, however, up-regulate antioxidant defenses.[9] In mice with induced strokes, ursolic acid promoted the activation of the Nrf2 pathway.[19] [As predicted above]

There is also limited evidence that UA induces mitochondrial biogenesis in skeletal muscle.[5]

## 3) PATHWAYS: 2

In preadipocytes, ursolic acid increases the activity of AMP Kinase and the protein expression of SIRT1,[31] and in isolated satellite cells, UA elevates the expression of SIRT1 about 35 fold.[2] In mice given ursolic acid twice a day for seven days, SIRT1 increased about five fold, and SIRT6 increased about six times in the liver.[8] In another mice model, UA for seven days increased SIRT1 3.5 fold and SIRT6 about 1.5 times in the hypothalamus.[1]

Moving up the animal hierarchy, obese quails treated with UA also demonstrated enhanced SIRT1 expression.[13]

There is also good evidence that UA increases AMP Kinase, as shown in preadipocytes[11] and myotubes.[5]

Meanwhile, obese rats after six weeks of supplementation demonstrated a markedly reduced body weight, increased energy expenditure, decreased free fatty acids in serum and skeletal muscle, and decreased triglyceride content in skeletal muscle in addition to activated AMP Kinase.[6]

In a more controversial category, ursolic acid interacts with the mTOR pathway, although there is evidence that it both activates it,[16] as well as blocks it.[22]

## 4) QUALITY CONTROL: 2

In the same chunky quails from above, ursolic acid treatment induced autophagy.[13]

In treated melanocytes, UA promoted melanosomal autophagy, or melanophagy, potentially inhibiting age-related skin pigmentation.[23]

The increase in autophagy also holds in atherosclerotic mice fed a western diet, increasing macrophage autophagy and reducing atherosclerotic lesion size.[18]

Eight weeks of treatment in yet another group of fat mice reduced liver and adipose tissue mass, adipocyte size, plasma leptin concentrations, plasma triglycerides, and low-density lipoprotein cholesterol concentrations. Meanwhile, it elevated the high-density lipoprotein cholesterol and adiponectin concentrations compared with controls. These effects were attributed to the activation of hepatic autophagy.[12]

In terms of DNA repair, in a single 2010 study, DNA damage created by $H_2O_2$ in cell cultures was not only decreased by ursolic acid, but the repair activity of the base excision repair pathway increased.[26]

## 5) IMMUNE SYSTEM: 2

Ursolic acid has known anti-inflammatory properties. In mice with elevated blood sugar, UA down-regulated the expression of iNOS and COX-2, decreased interleukins IL-1β and IL-6, and decreased tumor necrosis factor-α levels in the prefrontal cortex.[20]

It reduced pro-inflammatory cytokines induced by neuropathic pain in rats[4] and whole-body radiation-induced inflammatory responses in mice, as indicated by the decreased production of TNF-α, IL-6, and IL-1β.[29]

## 7) WASTE MANAGEMENT: 2

In diabetic mice, ursolic acid significantly decreased AGE production and reduced the expression of AGE receptors in the prefrontal cortex.[20, 33]

The glucose reduction occurs as UA inhibits aldose reductase and sorbitol dehydrogenase, two major enzymes in the polyol pathway, consequently lowering AGE levels.[33]

In fact, in a head to head test-tube study, UA reigned supreme over boswellic acid, corosolic acid, and ellagic acid, in terms of inhibitory activity against aldose reductase, and the subsequent generation of AGEs.[28]

*"One of the best-characterised pharmacological effects of UA is related to*

amelioration of insulin resistance and antidiabetic properties through a variety of mechanisms. In this regard, UA (10 mg/kg) by its own but even better in combination of metformin (150 mg/kg) has been shown, not only to enhance insulin sensitivity but also to improve cognitive impairment."[9]

## SYSTEMS KNOWN TO BENEFIT

### MUSCLE
One of the most standout features of ursolic acid is its anabolic effect on skeletal muscles.

At the cellular level, UA boosts muscle development by enhancing the number of muscle satellite cells. In addition, it doubles myoglobin expression in individual muscle cells.[2]

UA also improves skeletal muscle mass by the activation of growth hormone, insulin-like growth factor-1, and the mammalian target of rapamycin.[14]

In addition, it reduces the loss of muscle or muscle atrophy after denervation.[32]

In a mouse study, skeletal muscle mass, fast and slow muscle fiber size, grip strength, and exercise capacity were all increased after treatment.[17]

### ENDURANCE
UA improves endurance exercise capacity, at least in traumatized mice (as assessed by a weight-loaded swimming test).[5] [Akin to the Salem witch swimming tests]

### LIPID PROFILE & ADIPOSE TISSUE
New Zealand rabbits with food-induced hyperlipidemia had decreased plasma cholesterol and triglycerides with attenuated liver steatosis after treatment.[30]

In a parallel fashion, treated hypercholesterolemic hamsters demonstrated a 15% reduction in plasma cholesterol and a 9% reduction in cholesterol absorption from the intestine.[10]

These improvements occur by several mechanisms, including inhibition of preadipocyte differentiation and adipogenesis,[11] the inhibition of pancreatic lipase activity, an increase in brown fat and consequent thermogenesis, and lipolysis.[31]

In people, a randomized, double-blinded, placebo-controlled clinical trial in twenty-four patients (30 to 60 years) with a diagnosis of metabolic syndrome received either 150 mg of ursolic acid or placebo for twelve weeks. Treatment led to a transient remission of metabolic syndrome, reducing body weight, BMI, waist circumference, fasting glucose, and increasing insulin sensitivity.[25]

## BRAIN

Ursolic has many qualities that lend it to being beneficial to the brain. It increases endogenous antioxidants, it's an anti-inflammatory, and it can cross the blood-brain barrier.

It also modulates the monoaminergic system, affecting mood and cognitive dysfunction.[27] Specifically, ursolic acid interacts with the serotonin and noradrenergic systems, but not the glutamate or opioid systems.[7] There is also evidence that the anti-anxiolytic activity can be reversed by flumazenil, a standard antidote for chemically reversing benzodiazepines, suggesting that it acts in part through the benzodiazepine-binding site of GABA receptors.[9]

In a mouse model of Parkinson's disease, characterized by a degeneration of dopaminergic neurons in the substantia nigra, UA restored altered dopamine levels and protected dopaminergic neurons. As a result, it improved behavioral deficits.[24]

In addition, ursolic acid displays anticonvulsant, antidepressant, and anxiolytic activities in mice.[9]

## POTENTIAL SIDE EFFECTS

If you are a man and want to conceive, this might be a little concerning.... ursolic acid reduces spermagenesis.[21]

In addition, if you have a propensity to be hyper-coagulable or clot excessively, there is some concern that this triterpene can make platelets more susceptible to clumping and could potentially aggravate any cardiovascular conditions.

## DOSE

As per usual, there are not enough studies to determine an optimal dose. Options range from 150 to 450 mg a day.[25]

# CHAPTER 30
# VITAMIN C
## (2.3.0.1.3.2.1)

BENEFITS:

BRAIN   IMMUNITY   SKIN

*"Vitamin C is an essential micronutrient for humans, with pleiotropic functions related to its ability to donate electrons. It is a potent antioxidant and a cofactor for a family of biosynthetic and gene regulatory enzymes."*

~ A. Carr 2017

The deficiency of vitamin C has made history for hundreds, if not thousands of years. Scurvy, the associated clinical disease, was ever-present in medieval siege warfare, but it was catapulted to notoriety during the age of sea exploration. In essence, anytime fresh food was unavailable, scurvy would set in.

In the 1740s, thanks to Lord George Anson, who circumnavigated the globe, the world became more aware of the problem. In addition to losing five of his six ships, he lost 1,400 out of 1,900 men, most thought to be from scurvy. That being said, he himself did not suffer and returned to London in 1747 a wealthy man, having captured a galleon filled with pieces of eight.

In that same year, i.e., 1747, and in one of the first reported, controlled, clinical experiments in the history of medicine, James Lind set out to conquer the disease.

Despite the fact that he, and many others, thought scurvy was a putrefaction of the body which could be treated by acids, he trialed six different treatments in scurvy-laden sailors. Groups were given either cider, vinegar, sulfuric acid, seawater, a spicy paste, or the lucky group, which received oranges and lemons. Only one sailor improved, and that one had received an orange.

Lind published *A Treatise of the Scurvy* in 1753, but unfortunately, no one took note. The English navy was finally proactive in 1794 when lemon juice was provided to sailors on board the Suffolk for a twenty-three-week, non-stop voyage to India. Equivalent to roughly 10 mg of vitamin C, no serious outbreak of scurvy occurred. This astonishing outcome eventually precipitated the addition of lemon juice to the entire English fleet.

Despite being aware of its importance, vitamin C wasn't discovered until 1912. It was chemically isolated in 1928, and in 1933, it was the first vitamin to be chemically produced.

Despite the sea voyage lessons, scientists persisted in their inquiries. In 1968, six healthy men imprisoned in the Iowa State Penitentiary were given a diet completely devoid of vitamin C. While two prisoners managed to escape, the remaining four developed the disease.

## WHAT IS SCURVY EXACTLY?

Scurvy is characterized by weakening of collagenous structures, poor wound healing, impaired immunity, and much more. Of interest, one of the first signs noted by Lind in the 1700s was lassitude. He realized that sailors would lose the initiative and the will to work a month or two at sea but could function normally when compelled to do so. Another early sign was loose teeth.

The complete disease spectrum includes hypochondriasis and depression, perifollicular hyperkeratosis with coiled hairs, swollen and friable gums, anemia, petechial hemorrhage, erythema, arthralgia and/or joint effusions, breakdown of old wounds, bleeding into various tissues, fever, shortness of breath, infections, and confusion.

The other thing that has been forgotten over time was the high incidence of lobar pneumonia with scurvy, as noted by Alfred Hess in 1920.[7]

I think it's vital in today's world to be aware that scurvy outbreaks occur during infectious epidemics, especially in the malnourished.[3] This is because fighting infection requires increased amounts of vitamin C.

Now that I have obsessed about vitamin C deficiency, what exactly is it?

*What is a vitamin? Any of a group of organic compounds which are essential for normal growth and nutrition and are required in small quantities in the diet because they cannot be synthesized by the body.*

Vitamin C is the generic term for L-threo-hexo-2-enono-1,4-lactone, which constitutes a low molecular weight carbohydrate structurally related to glucose. It is a gluconic acid lactone derived from glucuronic acid and water-soluble ketolactone with two ionizable hydroxyl groups.[15]

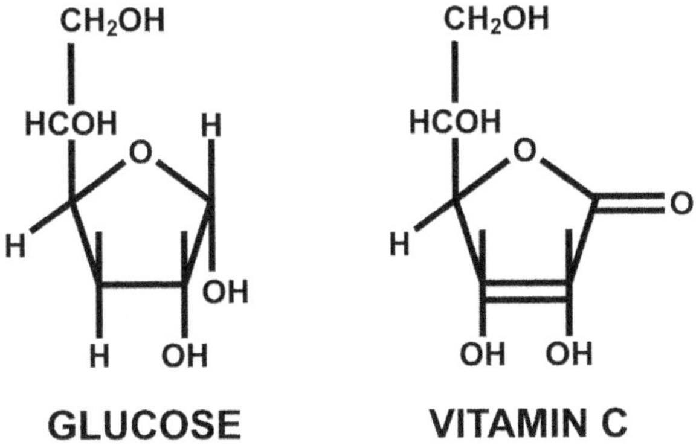

**GLUCOSE**

**VITAMIN C**

Called L-ascorbic acid, the l-enantiomer of ascorbic acid or ascorbate, it is one of the hydrophilic antioxidants that accumulate in the aqueous phase of the cell. Of note, the name ascorbic actually means "anti-scurvy."

Depending on the pH, there are different forms. Ascorbic acid exists only at low pH, but in solutions above a pH of 5, it is predominantly found in the ionized form, ascorbate.

Vitamin C is a vitamin as we cannot synthesize the compound ourselves. Thanks to some fluke of evolution, most mammals can, but primates, bats, guinea pigs, some fish, and we humans cannot. Somewhere along the line, we lost the enzyme gulonolactone oxidase (GLO), which is the terminal enzyme in the standard pathway.

## FUNCTIONS

As I have suggested, vitamin C is an extremely busy molecule. Ascorbate acts as an electron donor for fifteen mammalian enzymes as well as three fungal ones.

C is essential for the function of two monooxygenases and twelve dioxygenases.

The first monooxygenase, dopamine β-monooxygenase, a.k.a. dopamine β-hydroxylase, produces norepinephrine from dopamine in the adrenal gland.[11]

Peptidylglycine α-amidating monooxygenase, meanwhile, is an enzyme that catalyzes the conversion of glycine amides to glyoxylate and amides and is involved in the biosynthesis of many signaling peptides and some fatty acid amides.

Of the twelve dioxygenases, most are hydroxylases.

In the really important category, vitamin C is a cofactor for the lysyl and prolyl hydroxylases required to stabilize the tertiary structure of collagen.

It is also a cofactor for two hydroxylases involved in carnitine biosynthesis, a molecule required to transport fatty acids into mitochondria and generate metabolic energy.

Asparagyl and prolyl hydroylases are required for the down-regulation of the pleiotropic transcription factor hypoxia-inducible factor-1α (HIF-1α), which utilizes vitamin C as a cofactor.[11]

In the epigenetics category, C is essential for Ten-eleven translocation methylcytosine dioxygenase, which will be discussed below.

## RECEPTORS

Transportation of vitamin C and influx into cells is dependent on two sodium-dependent transporters that are specific for ascorbic acid and its oxidation product dehydroascorbic acid. This accounts for the differential accumulation in different tissues.

The sodium-dependent vitamin C transporters (SVCT) 1 and 2 are encoded by the SLC23A1 and SLC23A2 genes and belong to a family of nucleobase transporters that are highly conserved through evolution.

SVCT1s can be found in intestinal epithelial cells, renal tubules, and the liver. The SVCT2 isoform is expressed in more specialized cells such as neurons and in the placenta, characterized by high affinity at low transport efficiency. Mutations in the gene for this protein (SLC23A2) lead to a decrease in plasma ascorbate concentrations and are associated with an increased risk of acute coronary syndrome, gastric cancer, hematologic malignancies, and glaucoma.[10]

In contrast to the specific cell transport systems, non-specific intracellular transport of ascorbic acid occurs through glucose transporters (GLUT). This process occurs due to the similarity in structure between glucose and the oxidized form of vitamin C, dehydroascorbic acid, or DHA. Once inside the cell, DHA is reduced to ascorbic acid.[15]

Because of the competition with glucose at these receptors, there is reduced absorption of dehydroascorbic acid at high glucose concentration in some cells (e.g., muscle cells).[10]

## METABOLISM

There are significant conversions between the various forms of vitamin C. L-ascorbic acid is converted to an ascorbate and then to an ascorbate free radical with the loss of one electron. When a second electron is lost, this becomes dehydroascorbic acid or DHA, which exists in both hydrated and anhydrous forms.

ASCORBIC ACID — ASCORBATE — ASCORBYL RADICAL — DHA

2,3-DTG

Both the ascorbate radical and DHA molecules are reversibly reduced to ascorbate. The half-life of DHA is extremely short, in fact only minutes, as it's prone to hydrolytic ring rupture. Once this ring structure is lost, the product 2, 3-diketogulonic acid cannot revert and gets metabolized into oxalic acid.

As oxalate is one of the major end products of vitamin C breakdown in humans, there is potential for accumulating calcium oxalate and creating kidney stones.

The List...

## 1) DNA ALTERATIONS: 2

This is going to seem long-winded, but it makes sense in the end.

There is an unusual nucleotide in mammalian DNA, 5-hydroxymethylcytosine (5hmC), which is essentially a regular cytosine that got methylated along the way. Although this altered nucleotide represents less than 1% of total nucleotides, high levels have been observed in cerebellar Purkinje cells and granule neurons, suggesting a role in brain function.

This particular nucleotide is formed by the activity of a group of enzymes, Ten-eleven translocation methylcytosine dioxygenases (TET: TET1, TET2, TET3), that catalyze the transformation. Ascorbate, of course, serves as a cofactor for these dioxygenases. Thus, ascorbic acid increases 5hmC production in a TET-dependent manner. [2, 9]

Next on the list is the Jumonji C-domain-containing histone demethylases, which also require ascorbate as a cofactor for histone demethylation.[17] Vitamin C enhances the activity of both DNA and histone demethylases and, via this mechanism, improves somatic cell reprogramming and nudges embryonic stem cells towards the naive pluripotent state.[5]

*Thus, "the variation in ascorbate bioavailability can influence the demethylation of DNA and histones: in addition, ascorbate deficiency can present at different stages of aging and could be involved in the development of different age-related diseases."*[14]

### TELOMERES

It is also possible that vitamin C can help telomeres.

Bone marrow mesenchymal stem cells derived from senescence-accelerated (SAMP6) mice were treated with increasing concentrations. Not only did treatment significantly improve stem cell proliferation, it did so by increasing telomerase activity and TERT expression.[14]

In periodontal ligament stem cells, vitamin C induced telomerase activity, leading to the up-regulation of extracellular matrix Type I

collagen, fibronectin, and integrin β1, stem cell markers Oct4, Sox2, and Nanog as well as osteogenic markers RUNX2, ALP, OCN.[16]
[I love useless details!]

## 2) MITOCHONDRIA: 3

As we discussed earlier, vitamin C and its molecular relatives shuffle protons and electrons between ascorbic acid, ascorbate, the ascorbic radical, and dehydroascorbate.[3]

This is possible because hydroxyl groups at the double bond in the lactone rings are donors of these protons and electrons. As a result, these hydroxyl groups are quite reactive, protecting against singlet oxygen, hydrogen peroxide, peroxide radicals, and hydroxyl radicals. Unfortunately, once DHA gets hydrolyzed to 2,3-diketogulonic acid, the ring is irreversibly broken, and the game is over.[10]

Vitamin C is the most potent water-soluble antioxidant in the body, gets concentrated in the cytosol, and is the principal antioxidant that quenches aqueous peroxyl radicals and lipids in plasma ex vivo.[11]

Of note, glutathione is also water-soluble and hangs out in the cytosol, in contrast to vitamin E that lives within the lipid fraction of the cell.[13]

Moving to cell cultures, human bone marrow mesenchymal cells undergoing replicative senescence were treated with ascorbic acid. This eliminated excess free radicals and restored the endogenous antioxidant enzymatic activity.[14]

Vitamin C also serves as a chelator; it reduces ferric to ferrous ($Fe+3 \rightarrow Fe+2$) iron. By generating soluble iron complexes, it efficiently enhances the absorption of non-heme iron by the intestine.[5]

Too much vitamin C can be detrimental, however, it can actually promote free radical production by catalyzing the reduction of transition metal ions. Reduced iron ions react with hydrogen peroxide to form reactive hydroxyl radicals or peroxide ions (in the presence of oxygen).[10]

Finally, in healthy elderly subjects, the daily intake of star fruit juice with a reasonably high ascorbic acid content acted as a free radical scavenger, maintained low levels of lipoperoxidative stress, and restored GSH levels.

Treatment also reduced pro-inflammatory cytokines, especially TNF-alpha and interleukin-23 (IL-23).[14]

## 4) QUALITY CONTROL: 2

In cell cultures, the addition of vitamin C up-regulated the Activator protein 1 transcription factor, which in turn modulated the Nucleotide excision repair (NER) process. If you recall, NER is a DNA repair system that removes anomalous DNA adducts. This is hypothesized to occur by the induction of temporary oxidative stress by vitamin C, demonstrating that we require some, but not too much, oxidative stress.

*"The hypothesis that vitamin C stimulates DNA repair by temporarily inducing oxidative stress was supported by gene expression analysis. A change in expression of several dozen genes was observed after CCRF-HSB-2 cell exposure to 150 μmol/L vitamin C or ascorbic acid 2- phosphate (AA2P) and incubated up to 24 h. GdC (deoxycytidine glyoxal) levels in culture supernatant are increased after vitamin C treatment which indicates enhanced repair processes —NER removes adducts from DNA."* [10]

In one human study, hemodialysis patients with chronic kidney failure were given 300 mg of vitamin C. The treatment up-regulated 8-oxoguanine glycosylase, a key enzyme for Base excision repair (BER), another DNA repair mechanism. This paralleled a decrease in the 8- hydroxy-2-deoxyguanosine (8-OHdG), a marker of DNA damage, in peripheral blood lymphocytes.[10]

## 5) IMMUNE SYSTEM: 3

Vitamin C really stands out in this category, but that should come as no surprise - anyone with a runny nose or scratchy throat loves it.

*"Vitamin C orchestrates the function of the human immune system by supporting both the innate and adaptive immune system including epithelial barrier function, chemotaxis and antimicrobial activities of phagocyte cells, natural killer (NK) cell functions, and lymphocyte proliferation and differentiation."* [8]

To start with, we know that a deficiency in vitamin C can precipitate or aggravate infection. Scurvy is associated with severe lobar pneumonia, for example. However, even in the non-deficient category, there is evidence that vitamin C can prevent and treat symptoms of the common cold.[8]

What exactly does vitamin C do?

It enhances phagocyte activity. As a brief reminder, phagocytes are the trash trucks of the cell world; they move around the body, consuming and destroying foreign objects or perceived garbage.

Vitamin C enhances their migration potential, improves microbe engulfment, and it augments the ability to kill microbes by stimulating the contained production of reactive oxygen species. We also know that these cells actively accumulate vitamin C against a concentration gradient, with intracellular levels 50 to 100 times higher than in the plasma.

In addition, C enhances the differentiation and proliferation of B and T-cells.[3, 15] Because of this, there is enhanced antibody production in older folks[15] and younger, healthy people. A clinical trial with male university students, for example, demonstrated that supplementation significantly increased serum IgA and IgM concentrations.[15]

There is, in fact, a plethora of human evidence demonstrating immunologic benefits. Treatment of healthy subjects promoted and enhanced natural killer cell activities, lymphocyte proliferation, and chemotaxis.[15]

C also stands out as an anti-inflammatory. In cell culture, peripheral blood lymphocytes incubated with vitamin C decreased lipopolysaccharide-induced generation of pro-inflammatory cytokines TNF-$\alpha$ and IFN-$\gamma$ and increased anti-inflammatory IL-10 production.[14]

In a human study of community-acquired pneumonia, untreated patients had increased levels of free radicals, DNA damage, TNF-$\alpha$, and IL-6, with reduced superoxide dismutase levels. Vitamin C therapy reversed these issues.[4]

Lastly, there is evidence that C decreases circulating levels of histamine. In both guinea pigs and humans, vitamin C deficiency is associated with increased histamine levels, while increases of C decrease histamine. In humans, the effect was more apparent for those with allergic conditions.[3]

## 6) INDIVIDUAL CELL REQUIREMENTS: 2

Vitamin C has emerged as a critical regulator of stem cell identity and behavior, influencing plasticity, self-renewal, differentiation, and pluripotency, i.e., the ability to create different cell types. C enhances somatic cell reprogramming, nudging embryonic stem cells toward the naive state of pluripotency by modulating the cellular epigenetic profile.[5] In the differentiation category, it directs mesenchymal stem cells to create osteoblasts, chondrocytes, and tendons and triggers the proliferation of immunologically relevant T cells.[5]

How does this work? As a cofactor for γ-butyrobetaine dioxygenase (BBOX1) [yet another dioxygenase], C catalyzes the final step of L-carnitine biosynthesis. This enzyme, involved in the transport of fatty acids inside the mitochondria for β-oxidation, modulates osteogenesis and chondrogenesis in adipose- and bone marrow-derived stem cells.[5]

## 7) WASTE MANAGEMENT: 1

The evidence is a bit thin here, but one human study evaluated the effects of 1,250 mg ascorbic acid daily for eight weeks. In that time, researchers were able to demonstrate a significant reduction in AGEs. In addition, there was an improvement in HDL and LDL levels.[14]

## SYSTEMS KNOWN TO BENEFIT

### THE COMMON COLD

Despite all of this, the real question that most people want to know, of course, is does C actually help with the common cold.

The answer appears to be yes.

Examining 148 animal studies, researchers decided that vitamin C can alleviate or prevent infections caused by innumerable bacteria, viruses, and protozoa.

In humans, vitamin C administration doesn't decrease the average number of colds in the general population, but it does reduce the number of colds in physically active people by half. As well, regular administration of vitamin C shortens the duration of colds.

The amount one needs to take is sizable, however, with an intake of 6 to 8 g/day. Some studies using less reported no benefits.[7]

The other interesting twist is that infected cells deplete their C, so any potential reserves decline the longer a cold persists. As far back as 1973, it was reported that leukocyte vitamin C levels dropped 50% when subjects contracted a cold and rebounded one week after recovery.[7] Therefore, even if you are not deficient at the onset of a cold, you will be a few days into it. Thus, the intake of C during a cold can only be beneficial.

More consequential than the common cold, researchers examined the effects of vitamin C in twenty-one critically ill COVID-19 patients during the recent pandemic. Of note, most of these patients had low levels of vitamin C. In fact, older age and deficient levels appeared to be codependent risk factors for mortality.[1]

## BRAIN

In the central nervous system, vitamin C plays a complex role that is still being investigated. Cerebrospinal fluid ascorbic acid concentrations in the 200 to 400 mM range are higher than those in cerebral parenchyma or plasma (30 to 60 nM). It is transported into the brain by both the specific sodium-dependent vitamin C transporter 2 (SVCT2) as well as the non-specific GLUT1 transporter as dehydroascorbate (DHA), making vitamin C, along with glutathione, the most abundant antioxidant present in brain tissue. Of note, ascorbate is mainly present in nerve cells, while glutathione is in glial cells.[14]

The benefits of this accumulation are multiple. In the antioxidative category, vitamin C is especially beneficial after an ischemic event or cerebral reperfusion.[14] In cultured stem cells, ascorbic acid helps differentiate neuronal and astrocytes precursors, promoting synaptic maturation.[14] Ascorbic acid is also essential for the biosynthesis of catecholamines, peptide amination, myelin formation, synaptic function enhancement, and has neuroprotective activity against glutamate toxicity.[10, 14]

Turning to people, in the Rotterdam study established in the Netherlands in 1990, there was a consistent association between higher ascorbic acid intake and reduced relative risk for Alzheimer's disease. This was especially true in people that were the most deficient.[14]

More is not always better, however. Beneficial effects on cognition were noted at doses lower than 500 mg/day, while higher values (1 gram/day) were associated with poorer cognitive performance. This is important as the recommended doses for avoiding the common cold are significantly higher.[14]

We also know that vitamin C deficiency leads to depression and fatigue; and supplementing the vitamin is reported to improve the mood, at least in hospitalized patients.[7]

## SKIN

The skin is flush with vitamin C, actively pumped into the dermis and epidermis via the two sodium-dependent vitamin C transporters 1 and 2. Of note, the epidermis has five-fold more than in the dermis, except the outer-most layer of the epidermis, probably reflecting a depletion secondary to environmental exposure.[13]

The good news is that vitamin C has myriad effects in the skin.

It's an antioxidant and is anti-inflammatory.

It improves the epithelial barrier by assisting in the differentiation of keratinocytes, promoting the synthesis and organization of barrier lipids, and increasing the cornified envelope.

C alters gene expression within dermal fibroblasts, promoting fibroblast proliferation and migration, essential for tissue remodeling.

Lastly, but perhaps most importantly, vitamin C increases collagen gene expression in fibroblasts.

## COLLAGEN

Not just in the skin but all over the body, vitamin C is essential for collagen production. As a cofactor for prolyl 3-hydroxylase, prolyl 4-hydroxylase, and lysyl hydroxylase, C catalyzes the enzymatic post-translational modification of procollagen to produce collagen in fibroblasts and other collagen-producing cells.[5]

In the skin specifically, collagen formation is carried out by dermal fibroblasts, resulting in the generation of the basement membrane and the dermal collagen matrix. A deficiency of C results in both a decrease in the total collagen synthesis as well as a compromised tertiary collagen structure.[13]

Failure of collagen is the primary reason sailors with scurvy have no teeth.

## DEFICIENCY

Generally speaking, very few people get scurvy these days; there are people, however, that are clearly deficient. This is due to substandard dietary habits, increased requirements (e.g., smoking and alcohol or drug abuse), various diseases, and high pollutant exposure.[3]

Smoking is probably the worst culprit in terms of precipitating a deficiency; the mean serum concentrations in smokers are one-third lower than those of non-smokers. Sadly, vitamin C levels are also lower in children and adolescents exposed to tobacco smoke.[3]

It is estimated that even in industrial countries, there is a 5% prevalence of vitamin C deficiency and a 13% occurrence rate of suboptimal levels.[6]

## METABOLISM

Vitamin C is absorbed from the small intestine, achieving peak plasma concentrations roughly two hours after ingestion.

After absorption, as vitamin C is water-soluble, it is distributed from the bloodstream through the extracellular space. Tissues then accumulate C against a concentration gradient, which is dependent on dietary intake. The plasma concentration depends on the supply, absorption, and excretion, reaching the 50 μM range under normal conditions, but up to 150 μM with generous supplementation. Intracellular levels are even more elevated, with the highest in the adrenal and pituitary glands.[10]

## DOSE

Although the amount of vitamin C required to prevent scurvy is relatively low (about 10 mg/day), the recommended dietary intake is significantly higher. A standard diet supplying 100 to 200 mg/day is adequate for saturating plasma concentrations in healthy individuals and should reduce the risk of some chronic diseases.[11]

On the other hand, different health issues may require different doses. Recommendations for fighting a cold or illness are up to 5 or 6 grams daily.[14] However, daily treatment exceeding 500 mg may be associated with poor cognitive performance.

The other thing to note is because C is water-soluble, it does not get stored in the body. This means two things: it will not accumulate and cause long-term problems; however, a daily dose becomes mandatory to prevent hypovitaminosis.

If you prefer the natural approach, the highest levels of C are in the Kakadu plum, native to Australia, at 1000 to 5300 mg/100 gm, the camu camu plant from the Amazon at 2800 mg/ 100 gm, and Acerola from South and Central America at 1680 mg/100 gm.

The orange and lemon, by comparison, come in at a measly 53 mg/ 100 gm.

## TOPICAL

There is no doubt that oral vitamin C reaches the skin, but what about topical formulations?

Researchers believe that if plasma levels are saturated, very little topical C is actually absorbed. However, when this is not the case, vitamin C does get absorbed into the epidermal layer by topical application, although the efficacy depends on the particular cream or serum. The formulation must have a pH below 4 before C, in the form of ascorbic acid, gets absorbed.

On the other hand, very smart and determined researchers are busy at this very moment formulating mechanisms to increase topical absorption.[13]

# CHAPTER 31
# THE LAST CHAPTER

*"Science is magic that works."*

~ Kurt Vonnegut

Overwhelmed yet?
One can extract many concepts from this book, ranging from a more in-depth understanding of cellular functions to the appreciation of new, individual agents. But it's a lot of information, so please don't get discouraged!

To make it a bit easier...here are a few things that I think are important:

• There are innumerable ways to combat aging. Your longevity protocol should be based on as much information as possible and may be very different from anyone else's.

• In no way do I expect people to take every agent. [Of course, some of you will...and I already know who you are!]

• There is a lot of information in book one, and nothing here contradicts what has already been said; it only provides new incites and details.

• Consider these new agents as an addendum to the ones already presented. The goal is to have a continuously expanding encyclopedia of longevity options.

• There will be a continuing stream of scientific discoveries as time progresses, so keep in mind that whatever is written at this moment may be out of date by the time you get to read it.

• I said this in the opening chapter, but I will repeat it here. Daily oral agents are an absolute must for any longevity protocol, but they don't have to be the only therapy. There is absolutely no contraindication to combining this with other strategies.

Where to start?

In book one, I highly recommend the Panacea for those over forty without significant medical problems. This includes pterostilbene and/ or resveratrol, astaxanthin, nicotinamide in some form, curcumin, and carnosine. I believe this remains an excellent starting point as it covers the seven tenets of aging at a fundamental level. But there are now fantastic additions to our menu options, and more ways to determine what might work best for any individual.

Therefore, think about who you are and what you are trying to accomplish.

Some folks go with an assortment of the heavy hitters to get the most bang for their buck.

In this book, fisetin wins the day at 17 points. Spermidine and lactoferrin are close behind with 15 and 14 points apiece respectively, followed by salidroside, naringenin, and delphinidin at 13 points each. Keep in mind, however, that some of our ratings are rather conservative as key human studies have yet to be undertaken. Table one provides all of the ratings for an easy go-to guide.

If you are more interested in a particular skill set, I refer you to Table two, where the agents are cross-referenced by body part or desired effect. For example, many mountain-climber friends prefer a combination of salidroside, shilajit, and *Ganoderma Lucidum.*

Questions concerning glucose management are also something most people are curious about. In this arena, we need to shift our thinking from the individual cell to more of a systems approach. [The common sense answer is, of course, to consume less sugar, and many people do this to the extreme. But, where is the fun in that?]

We need to consider how we intake sugars, how our body processes them, and how they cause damage. Luckily, we can manipulate how our bodies deal with these sugars as there are many steps from ingestion to metabolism that we can exploit to our advantage. I have addressed these in the individual chapters, but to make it easier, it is organized below.

To start, carbohydrates get broken down at the intestinal wall in the GI tract to be absorbed as simple sugars.

It is possible to block at least two of these enzymes, i.e., alpha-glucosidase and alpha-amylase. Of the agents in this book, there are several that fall into this category.
>andrographolide
>berberine
>chlorogenic acid
>*Ecklonia cava*
>*Ganoderma lucidum*
>ellagic acid (amylase only)

Glucose absorption from the GI tract is also controlled by glucose-6-phosphate translocase (G-6-pase); inhibition of which reduces the quantity of absorbed simple sugars.
>chlorogenic acid

Once in the plasma, it is possible to transport more sugar into cells and out of the bloodstream, where it precipitates AGE formation. This can be accomplished by up-regulating the glucose transporter GLUT-4,
>andrographolide

or by increasing the production of insulin.
>andrographolide
>berberine
>black seed oil
>*Ecklonia cava*
>*Ganoderma lucidum*
>lactoferrin

In the liver, we can reduce gluconeogenesis, i.e., the production of even more glucose.
>berberine
>*Ganoderma lucidum*
>black seed oil
>ellagic acid
>chlorogenic acid
>*Ecklonia cava*

Another possibility is inhibiting the kidney's ability to retain glucose. By blocking the sodium-glucose co-transporter 2 (SGLT2) and the facilitated glucose transport 2 (GLUT2), the kidney excretes more glucose than it normally would.

*Kaempferia parviflora*

Quite another approach is the inhibition of AGEs; a general reduction of which has been documented with:
- *Aloe vera*
- chlorogenic acid
- *Ecklonia cava*
- ellagic acid
- *Ganoderma lucidum*
- fisetin

Lastly, there are agents with a high affinity to AGEs (lactoferrin) or glucose (arginine and lysine) that serve as decoys, binding the unwanted molecules and preventing them from adhering to more permanent structures.

Therefore, depending on how aggressive one wants to block glucose and its ill effects, any or all of these agents are respectable options.

Another commonly asked question revolves around fat or weight reduction. Much like controlling glucose, there are many ways to accomplish this.

One of my favorite methods is by increasing the relative ratio of brown to white fat. This occurs through the up-regulation of uncoupling protein 1 (UCP-1), which is great for weight loss, but not so great for energy conservation. [athletes should avoid this]

Many agents do this, including alpha-ketoglutarate, naringenin, ursolic acid, and leucine.

Some agents interfere with enzymes that are necessary to create fat in the first place. *Ganoderma lucidum*, for example, suppresses the expression of enzymes and proteins responsible for lipid synthesis, transport, and storage, i.e., fatty acid synthase, acyl-CoA synthetase-1, fatty acid binding protein-4, fatty acid transport protein-1, and perilipin.

We can also limit the breakdown of consumed fat within the GI tract and limit its absorption by inhibiting pancreatic lipase.
- ursolic acid
- *Aloe vera*
- chlorogenic acid

Much less specific, but agents recognized for overall fat reduction include the following:
- leucine
- *Kaempferia parviflora*
- *Ecklonia cava*
- chlorogenic acid
- ellagic acid

In the same general category of lipids where you don't want them, the following are known to improve the plasma lipid panel:
- *Kaempferia parviflora*
- *Ecklonia cava*
- berberine
- black seed oil
- ellagic acid

Thus, prior to embarking on a longevity protocol, think about who you are and precisely what you want to accomplish. Consider your family history, medical issues, and aging concerns specific to you.

The last thing to remember is that things take time. Give any protocol you decide on three to six months to have an effect. At that point, reevaluate your condition, and make appropriate alterations.

In closing, I wish you all the best and a very healthy, long lifespan!

*Dr. Sandy*

# CITATIONS

*"Sometimes the questions are complicated and the answers are simple."*

~ Dr. Seuss

**CHAPTER 2** - ORGAN SYSTEMS

BONE

Akbari, Solmaz, and Amir Alireza Rasouli-Ghahroudi. "Vitamin K and bone metabolism: a review of the latest evidence in preclinical studies." BioMed research international 2018 (2018).

Almeida, Maria. "Aging mechanisms in bone." BoneKEy reports 1 (2012).

Burr, David B. "Changes in bone matrix properties with aging." Bone 120 (2019): 85-93.

Cao, Jay J. "Effects of obesity on bone metabolism." Journal of orthopaedic surgery and research 6.1 (2011): 1-7.

Chapurlat, Roland D., and Cyrille B. Confavreux. "Novel biological markers of bone: from bone metabolism to bone physiology." Rheumatology 55.10 (2016): 1714-1725.

Doherty, Alison H., Cameron K. Ghalambor, and Seth W. Donahue. "Evolutionary physiology of bone: bone metabolism in changing environments." Physiology 30.1 (2015): 17-29.

Farr, Joshua N., and Sundeep Khosla. "Cellular senescence in bone." Bone 121 (2019): 121-133.

Grzibovskis, Maris, Mara Pilmane, and Ilga Urtane. "Today's understanding about bone aging." Stomatologija 12.4 (2010): 99-104.

Infante, Arantza, and Clara I. Rodríguez. "Osteogenesis and aging: lessons from mesenchymal stem cells." Stem cell research & therapy 9.1 (2018): 1-7.

Kini, Usha, and B. N. Nandeesh. "Physiology of bone formation, remodeling, and metabolism." Radionuclide and hybrid bone imaging. Springer, Berlin, Heidelberg, 2012. 29-57.

Klein-Nulend, Jenneke, Peter J. Nijweide, and Elisabeth H. Burger. "Osteocyte and bone structure." Current osteoporosis reports 1.1 (2003): 5-10.

Komaba, Hirotaka, et al. "Klotho expression in osteocytes regulates bone metabolism and controls bone formation." Kidney international 92.3 (2017): 599-611.

Shahi, Maryam, Amir Peymani, and Mehdi Sahmani. "Regulation of bone metabolism." Reports of biochemistry & molecular biology 5.2 (2017): 73.

## BRAIN

Bayliak, Maria M., et al. "Middle age as a turning point in mouse cerebral cortex energy and redox metabolism: Modulation by every-other-day fasting." Experimental Gerontology 145 (2021): 111182.

Castellano, Christian-Alexandre, et al. "Links between metabolic and structural changes in the brain of cognitively normal older adults: a 4-year longitudinal follow-up." Frontiers in aging neuroscience 11 (2019): 15.

Dienel, Gerald A. "Brain glucose metabolism: integration of energetics with function." Physiological reviews 99.1 (2019): 949-1045.

Erdő, Franciska, László Denes, and Elizabeth de Lange. "Age-associated physiological and pathological changes at the blood–brain barrier: a review." Journal of Cerebral Blood Flow & Metabolism 37.1 (2017): 4-24.

Garaschuk, Olga. "Age-related changes in microglial physiology: the role for healthy brain ageing and neurodegenerative disorders." Neuroforum 23.4 (2017): A182-A191.

Gomes-Osman, Joyce, et al. "Non-invasive brain stimulation: probing intracortical circuits and improving cognition in the aging brain." Frontiers in aging neuroscience 10 (2018): 177.

Goyal, Manu S., et al. "Loss of brain aerobic glycolysis in normal human aging." Cell metabolism 26.2 (2017): 353-360.

Goyal, Manu S., et al. "Persistent metabolic youth in the aging female brain." *Proceedings of the National Academy of Sciences* 116.8 (2019): 3251-3255.

Hadem, Ibanylla Kynjai Hynniewta, et al. "Beneficial effects of dietary restriction in aging brain." *Journal of chemical neuroanatomy* 95 (2019): 123-133.

Howieson, Diane B. "Cognitive skills and the aging brain: what to expect." *Cerebrum: the Dana forum on brain science.* Vol. 2015. Dana Foundation, 2015.

Jorge, Lilia, et al. "Is the retina a mirror of the aging brain? Aging of neural retina layers and primary visual cortex across the lifespan." *Frontiers in aging neuroscience* 11 (2020): 360.

Leung, Natalie TY, et al. "Neural plastic effects of cognitive training on aging brain." *Neural Plasticity* 2015 (2015).

Liu, Haijing, et al. "A voxel-based morphometric study of age-and sex-related changes in white matter volume in the normal aging brain." *Neuropsychiatric disease and treatment* 12 (2016): 453.

Mather, Mara. "The emotion paradox in the aging brain." (2012).

Mujica-Parodi, Lilianne R., et al. "Diet modulates brain network stability, a biomarker for brain aging, in young adults." *Proceedings of the National Academy of Sciences* 117.11 (2020): 6170-6177.

Palomar-Bonet, Miriam, et al. "Associations of Salivary Total Antioxidant Capacity With Cortical Amyloid-Beta Burden, Cortical Glucose Uptake, and Cognitive Function in Normal Aging." *The Journals of Gerontology: Series A* (2021).

Ritchie, Stuart J., et al. "Risk and protective factors for structural brain ageing in the eighth decade of life." *Brain Structure and Function* 222.8 (2017): 3477-3490.

Salminen, Antero, et al. "Astrocytes in the aging brain express characteristics of senescenceassociated secretory phenotype." *European Journal of Neuroscience* 34.1 (2011): 3-11.

Satoh, Akiko, Shin-ichiro Imai, and Leonard Guarente. "The brain, sirtuins, and ageing." *Nature Reviews Neuroscience* 18.6 (2017): 362.

Shankar, Susarla K. "Biology of aging brain." *Indian Journal of Pathology and Microbiology 53.4 (2010): 595.*

Stefanatos, Rhoda, and Alberto Sanz. "The role of mitochondrial ROS in the aging brain." *FEBS letters 592.5 (2018): 743-758.*

Sure, Venkata N., et al. "A novel high-throughput assay for respiration in isolated brain microvessels reveals impaired mitochondrial function in the aged mice." *Geroscience 40.4 (2018): 365-375.*

## SKIN

Gragnani, Alfredo, et al. "Review of major theories of skin aging." *Advances in Aging Research 2014 (2014).*

Krutmann, Jean, et al. "The skin aging exposome." *Journal of dermatological science 85.3 (2017): 152-161.*

Matsui, Takeshi, and Masayuki Amagai. "Dissecting the formation, structure and barrier function of the stratum corneum." *International immunology 27.6 (2015): 269-280.*

Pillai, S., C. Oresajo, and J. Hayward. "Ultraviolet radiation and skin aging: roles of reactive oxygen species, inflammation and protease activation, and strategies for prevention of inflammation-induced matrix degradation–a review." *International journal of cosmetic science 27.1 (2005): 17-34.*

Quan, Taihao, and Gary J. Fisher. "Role of age-associated alterations of the dermal extracellular matrix microenvironment in human skin aging: a mini-review." *Gerontology 61.5 (2015): 427-434.*

Rabionet, Mariona, Karin Gorgas, and Roger Sandhoff. "Ceramide synthesis in the epidermis." *Biochimica et Biophysica Acta (BBA)-Molecular and Cell Biology of Lipids 1841.3 (2014): 422-434.*

Tobin, Desmond J. "Introduction to skin aging." *Journal of tissue viability 26.1 (2017): 37-46.*

## GASTROINTESTINAL TRACT

Calvani, Riccardo, et al. "Of microbes and minds: a narrative review on the second brain aging." *Frontiers in medicine 5 (2018): 53.*

Kumar, Manish, et al. "Human gut microbiota and healthy aging: recent *developments and future prospective." Nutrition and Healthy aging 4.1 (2016): 3-16.*

Rémond, Didier, et al. *"Understanding the gastrointestinal tract of the elderly to develop dietary solutions that prevent malnutrition." Oncotarget 6.16 (2015): 13*

Siddappa, Pradeep K., and John W. Birk. *"Gastrointestinal Health and Healthy Aging." Healthy Aging. Springer, Cham, 2019. 67-79.*

## CHAPTER 3 - ALPHA KETOGLUTARATE

*1) Cai, Xingcai, et al. "Alpha-ketoglutarate promotes skeletal muscle hypertrophy and protein synthesis through Akt/mTOR signaling pathways." Scientific Reports 6.1 (2016): 1-11.*

*2) Chen, Shuai, et al. "Alpha-ketoglutarate (AKG) lowers body weight and affects intestinal innate immunity through influencing intestinal microbiota." Oncotarget 8.24 (2017): 38184.*

*3) He, Liuqin, et al. "Alpha-ketoglutarate suppresses the NF-κB-mediated inflammatory pathway and enhances the PXR-regulated detoxification pathway." Oncotarget 8.61 (2017): 102974.*

*4) Liu, Shaojuan, Liuqin He, and Kang Yao. "The antioxidative function of alpha-ketoglutarate and its applications." BioMed research international 2018 (2018).*

*5) Rhoads, Timothy W., and Rozalyn M. Anderson. "Alpha-Ketoglutarate, the Metabolite that Regulates Aging in Mice." Cell Metabolism 32.3 (2020): 323-325.*

*6) Shahmirzadi, Azar Asadi, et al. "Alpha-ketoglutarate, an endogenous metabolite, extends lifespan and compresses morbidity in aging mice." Cell Metabolism 32.3 (2020): 447-456.*

*7) Sharma, Rishi, and Arvind Ramanathan. "The Aging Metabolome—Biomarkers to Hub Metabolites." Proteomics 20.5-6 (2020): 1800407.*

*8) Son, Eui Dong, et al. "Alpha-ketoglutarate stimulates procollagen production in cultured human dermal fibroblasts, and decreases UVB-induced wrinkle formation following topical application on the dorsal skin of hairless mice." Biological and Pharmaceutical Bulletin 30.8 (2007): 1395-1399.*

9) Su, Yuan, et al. "Alpha-ketoglutarate extends Drosophila lifespan by inhibiting mTOR and activating AMPK." Aging (Albany NY) 11.12 (2019): 4183.

10) Tian, Qiyu, et al. "Dietary alpha-ketoglutarate promotes beige adipogenesis and prevents obesity in middle-aged mice." Aging cell 19.1 (2020): e13059.

11) Tian, Qiyu, Xiangdong Liu, and Min Du. "Alpha-ketoglutarate for adipose tissue rejuvenation." Aging (Albany NY) 12.14 (2020): 13845.

12) Wang, Lei, et al. "Dietary supplementation with α-ketoglutarate activates mTOR signaling and enhances energy status in skeletal muscle of lipopolysaccharide-challenged piglets." The Journal of nutrition 146.8 (2016): 1514-1520.

13) Wu, Nan, et al. "Alpha-ketoglutarate: physiological functions and applications." Biomolecules & therapeutics 24.1 (2016): 1.

14) Xiao, Dingfu, et al. "The glutamine-alpha-ketoglutarate (AKG) metabolism and its nutritional implications." Amino Acids 48.9 (2016): 2067-2080.

15) Yang, Qiyuan, et al. "AMPK/α-ketoglutarate axis dynamically mediates DNA demethylation in the Prdm16 promoter and brown adipogenesis." Cell metabolism 24.4 (2016): 542-554.

16) Zdzisińska, Barbara, Aleksandra Żurek, and Martyna Kandefer-Szerszeń. "Alphaketoglutarate as a molecule with pleiotropic activity: well-known and novel

17) Żurek, Aleksandra, et al. "Alpha ketoglutarate exerts a pro-osteogenic effect in osteoblast cell lines through activation of JNK and mTOR/S6K1/S6 signaling pathways." Toxicology and applied pharmacology 374 (2019): 53-64.

**CHAPTER 4** - ALOE VERA

1) Akhoondinasab, Mohammad Reza, Motahhare Akhoondinasab, and Mohsen Saberi. "Comparison of healing effect of aloe vera extract and silver sulfadiazine in burn injuries in experimental rat model." World journal of plastic surgery 3.1 (2014): 29.

2) Al-Madboly, Lamiaa A., et al. "Dietary cancer prevention with butyrate fermented by Aloe vera gel endophytic microbiota." Journal of Gastroenterology and Hepatology Research 6.2 (2017): 2312-2317.

3) Altincik, Ayca, et al. "Effects of Aloe vera leaf gel extract on rat peritonitis model." Indian journal of pharmacology 46.3 (2014): 322.

4) Boonyagul, Sani, et al. "Effect of acemannan, an extracted polysaccharide from Aloe vera, on BMSCs proliferation, differentiation, extracellular matrix synthesis, mineralization, and bone formation in a tooth extraction model." Odontology 102.2 (2014): 310-317.

5) Budai, Marietta M., et al. "Aloe vera downregulates LPS-induced inflammatory cytokine production and expression of NLRP3 inflammasome in human macrophages." Molecular immunology 56.4 (2013): 471-479.

6) Burusapat, Chairat, et al. "Topical Aloe vera gel for accelerated wound healing of splitthickness skin graft donor sites: a double-blind, randomized, controlled trial and systematic review." Plastic and reconstructive surgery 142.1 (2018): 217-226.

7) Cho, Soyun, et al. "Dietary Aloe vera supplementation improves facial wrinkles and elasticity and it increases the type I procollagen gene expression in human skin in vivo." Annals of dermatology 21.1 (2009): 6-11.

8) Cho, Soyun. "The Role of Functional Foods in Cutaneous Anti-aging." Journal of Lifestyle Medicine 4.1 (2014): 8.

9) Choi, Ho-Chun, et al. "Metabolic effects of aloe vera gel complex in obese prediabetes and early non-treated diabetic patients: Randomized controlled trial." Nutrition 29.9 (2013): 1110-1114.

10) Choi, Jin-Ho, et al. "Effect of Aloe on Learning and Memory Impairment Animal Model SAMP8 II. Feeding Effect of Aloe on Lipid Metabolism of SAMP8." Journal of Life Science 6.3 (1996): 178-184.

11) Devaraj, Sridevi, et al. "Effects of Aloe vera supplementation in subjects with prediabetes/ metabolic syndrome." Metabolic syndrome and related disorders 11.1 (2013): 35-40.

12) Dou, Fang, et al. "Aloe-emodin ameliorates renal fibrosis via inhibiting PI3K/Akt/mTOR signaling pathway in vivo and in vitro." Rejuvenation research 22.3 (2019): 218-229.

13) Guo, Xiaoqing, and Nan Mei. "Aloe vera: A review of toxicity and adverse clinical effects." Journal of Environmental Science and Health, Part C 34.2 (2016): 77-96.

14) Gupta, Akhilesh, and Swati Rawat. "Clinical importance of aloe vera." Research Journal of Topical and Cosmetic Sciences 8.1 (2017): 30-39.

15) Haniadka, Raghavendra, et al. "Review on the use of Aloe vera (Aloe) in dermatology." Bioactive Dietary Factors and Plant Extracts in Dermatology. Humana Press, Totowa, NJ, 2013. 125-133.

16) Haritha, Ketham, Bellamkonda Ramesh, and Desireddy Saralakumari. "Effect of Aloe vera gel on antioxidant enzymes in streptozotocin-induced cataractogenesis in male and female Wistar rats." Journal of Acute Medicine 4.1 (2014): 38-44.

17) Hassanpour, Hamid. "Effect of Aloe vera gel coating on antioxidant capacity, antioxidant enzyme activities and decay in raspberry fruit." LWT-Food Science and Technology 60.1 (2015): 495-501.

18) Hekmatpou, Davood, et al. "The effect of Aloe vera clinical trials on prevention and healing of skin wound: a systematic review." Iranian journal of medical sciences 44.1 (2019): 1.

19) Hęś, Marzanna, et al. "Aloe vera (L.) Webb.: Natural Sources of Antioxidants–A Review." Plant Foods for Human Nutrition (2019): 1-11.

20) Hosseini, Seyed Abdolhakim, et al. "Effect of Aloe vera on Albumin glycation reaction in vitro." American Journal of Drug Discovery and Development 3.4 (2013): 263-270.

21) Hu, Qiuhui, Yun Hu, and Juan Xu. "Free radical-scavenging activity of Aloe vera (Aloe barbadensis Miller) extracts by supercritical carbon dioxide extraction." Food Chemistry 91.1 (2005): 85-90.

22) Hu, Yun, Juan Xu, and Qiuhui Hu. "Evaluation of antioxidant potential of Aloe vera (Aloe barbadensis Miller) extracts."

23) Jittapiromsak, Nawaporn, et al. "Acemannan, an extracted product from Aloe vera, stimulates dental pulp cell proliferation, differentiation, mineralization, and dentin formation." Tissue Engineering Part A 16.6 (2010): 1997-2006.

24) Kaithwas, Gaurav, Prashant Singh, and Daksh Bhatia. "Evaluation of in vitro and in vivo antioxidant potential of polysaccharides from Aloe vera (Aloe barbadensis Miller) gel." Drug and chemical toxicology 37.2 (2014): 135-143.

25) Kaminaka, Chikako, et al. "Effects of low-dose Aloe sterol supplementation on skin moisture, collagen score and objective or subjective symptoms: 12-week, double-blind, randomized controlled trial." The Journal of dermatology 47.9 (2020): 998-1006.

26) Kang, Min-Cheol, et al. "In vitro and in vivo antioxidant activities of polysaccharide purified from aloe vera (Aloe barbadensis) gel." Carbohydrate polymers 99 (2014): 365-371.

27) Kang, Shimo, et al. "Main anthraquinone components in Aloe vera and their inhibitory effects on the formation of advanced glycation end-products." Journal of Food Processing and Preservation 41.5 (2017): e13160.

28) Khanam, Neelofar, and G. K. Sharma. "A Critical review on antioxidant and antimicrobial properties of aloe vera L." International Journal of Pharmaceutical Sciences and Research 4.9 (2013): 3304.

29) Kresnoadi, Utari, et al. "TLR2 signaling pathway in alveolar bone osteogenesis induced by aloe vera and xenograft (XCB)." Brazilian dental journal 28.3 (2017): 281-286.

30) Li, Chia-Yang, et al. "Aloe metabolites prevent LPS-induced sepsis and inflammatory response by inhibiting mitogen-activated protein kinase activation." The American journal of Chinese medicine 45.04 (2017): 847-861.

31) López, Zaira, et al. "Antioxidant and cytotoxicological effects of Aloe vera food supplements." Journal of Food Quality 2017 (2017).

32) Manvitha, Karkala, and Bhushan Bidya. "Aloe vera: a wonder plant its history, cultivation and medicinal uses." Journal of Pharmacognosy and Phytochemistry 2.5 (2014): 85-88.

33) Mohamed, Enas Ali Kamel. "Antidiabetic, antihypercholestermic and antioxidative effect of Aloe vera gel extract in alloxan induced diabetic rats." Aust J Basic Appl Sci 5.11 (2011): 1321-1327.

34) Molazem, Zahra, et al. "Aloe vera gel and cesarean wound healing; a randomized controlled clinical trial." Global journal of health science 7.1 (2015): 203.

35) Nejatzadeh-Barandozi, Fatemeh. "Antibacterial activities and antioxidant capacity of Aloe vera." Organic and medicinal chemistry letters 3.1 (2013): 5.

36) Park, Mi-Young, Hoon-Jeong Kwon, and Mi-Kyung Sung. "Dietary aloin,

*aloesin, or aloe-gel exerts anti-inflammatory activity in a rat colitis model." Life sciences 88.11-12 (2011): 486-492.*

37) Radha, Maharjan H., and Nampoothiri P. Laxmipriya. "Evaluation of biological properties and clinical effectiveness of Aloe vera: A systematic review." Journal of traditional and complementary medicine 5.1 (2015): 21-26.

38) Rahmani, Arshad H., et al. "Aloe vera: Potential candidate in health management via modulation of biological activities." Pharmacognosy reviews 9.18 (2015): 120.

39) Rahoui, Walid, et al. "Beneficial effects of Aloe vera gel on lipid profile, lipase activities and oxidant/antioxidant status in obese rats." Journal of Functional Foods 48 (2018): 525-532.

40) Raksha, Bawankar, Singh Pooja, and Subramanian Babu. "Bioactive compounds and medicinal properties of Aloe vera L.: An update." Journal of Plant Sciences 2.3 (2014): 102-107.

41) Rodriguez-Bigas, Miguel, Norma I. Cruz, and Albert Suarez. "Comparative evaluation of aloe vera in the management of burn wounds in guinea pigs." Plastic and reconstructive surgery 81.3 (1988): 386-389.

42) Sahu, Pankaj K., et al. "Therapeutic and medicinal uses of Aloe vera: a review." Pharmacology & Pharmacy 4.08 (2013): 599.

43) Saito, Marie, et al. "Oral administration of Aloe vera gel powder prevents UVB-induced decrease in skin elasticity via suppression of overexpression of MMPs in hairless mice." Bioscience, biotechnology, and biochemistry 80.7 (2016): 1416-1424.

44) Shahzad, Muhammad Naveed, and Naheed Ahmed. "Effectiveness of Aloe vera gel compared with 1% silver sulphadiazine cream as burn wound dressing in second degree burns." J Pak Med Assoc 63.2 (2013): 225-30.

45) Sogvar, Ommol Banin, Mahmoud Koushesh Saba, and Aryou Emamifar. "Aloe vera and ascorbic acid coatings maintain postharvest quality and reduce microbial load of strawberry fruit." Postharvest Biology and Technology 114 (2016): 29-35.

46) Suksomboon, N., N. Poolsup, and S. Punthanitisarn. "Effect of Aloe vera on glycaemic control in prediabetes and type 2 diabetes: A systematic review and meta-analysis." Journal of clinical pharmacy and therapeutics 41.2 (2016): 180-188.

47) Suyata, Astrinita Lestari, et al. "Aloe vera gel application for faster healing of split-thickness skin graft donor site on Wistar rats." IJBS 14.2 (2020): 104-107.

48) Tabandeh, Mohammad Reza, Ahmad Oryan, and Adel Mohammadalipour. "Polysaccharides of Aloe vera induce MMP-3 and TIMP-2 gene expression during the skin wound repair of rat." International journal of biological macromolecules 65 (2014): 424-430.

49) Tanaka, Miyuki, et al. "Effects of plant sterols derived from Aloe vera gel on human dermal fibroblasts in vitro and on skin condition in Japanese women." Clinical, cosmetic and investigational dermatology 8 (2015): 95.

50) Tanaka, Miyuki, et al. "Effects of Aloe sterol supplementation on skin elasticity, hydration, and collagen score: a 12-week double-blind, randomized, controlled trial." Skin pharmacology and physiology 29.6 (2016): 309-317.

51) Taukoorah, Urmeela, and M. Fawzi Mahomoodally. "Crude Aloe vera gel shows antioxidant propensities and inhibits pancreatic lipase and glucose movement in vitro." Advances in pharmacological sciences 2016 (2016).

52) Udupa, S. L., A. L. Udupa, and D. R. Kulkarni. "Anti-inflammatory and wound healing properties of Aloe vera." Fitoterapia 65.2 (1994): 141-145.

53) Vijayalakshmi, Damodharan, et al. "In vitro anti inflammatory activity of Aloe vera by down regulation of MMP-9 in peripheral blood mononuclear cells." Journal of ethnopharmacology 141.1 (2012): 542-546.

54) Williamson, Gary, et al. "Review of the efficacy of green tea, isoflavones and aloe vera supplements based on randomised controlled trials." Food & function 2.12 (2011): 753-759.

55) Woźniak, Anna, and Roman Paduch. "Aloe vera extract activity on human corneal cells." Pharmaceutical biology 50.2 (2012): 147-154.

56) Yagi, Akira, and P. Yu Byung. "Putative anti-cancer action of aloe vera via butyrate fermentation." Journal of Gastroenterology and Hepatology Research 6.5 (2017): 2419-2424.

57) Yagi, Akira, et al. "Dietary Aloe vera gel and microbiota interactions: Influence of butyrate and insulin sensitivity." Journal of Gastroenterology and Hepatology Research 6.4 (2017): 2376-2383.

58) Yagi, Akira, Amal Kabbash, and Lamiaa Al-Madboly. "Possible prophylaxes of Aloe vera gel ingestion to butyrate metabolism." Journal of Gastroenterology and Hepatology Research 5.5 (2016): 2158-2164.

59) Yao, Ruiqing, et al. "Daily Ingestion of Aloe Vera Gel Powder Containing Aloe Sterols Prevents Skin Photoaging in OVX Hairless Mice." Journal of food science 81.11 (2016).

60) Younus, Hina, and Shehwaz Anwar. "Antiglycating activity of Aloe vera gel extract and its active component Aloin. J." Proteins Proteom 9 (2018): 115-125.

61) Yuksel, Yasemin, et al. "Effects of aloe vera on spinal cord Ischemia–Reperfusion injury of rats." Journal of Investigative Surgery 29.6 (2016): 389-398.

## CHAPTER 5 - ANDROGRAPHOLIDE

1) Ahmed, Sahabuddin, et al. "Andrographolide suppresses NLRP3 inflammasome activation in microglia through induction of parkin-mediated mitophagy in in-vitro and in-vivo models of Parkinson disease." Brain, Behavior, and Immunity (2020).

2) Cisternas, Pedro Antonio, et al. "Presymptomatic treatment with andrographolide improves brain metabolic markers and cognitive behavior in a model of early-onset Alzheimer's disease." Frontiers in cellular neuroscience 13 (2019): 295.

3) Geng, Ji, et al. "Andrographolide sulfonate improves Alzheimer-associated phenotypes and mitochondrial dysfunction in APP/PS1 transgenic mice." Biomedicine & Pharmacotherapy 97 (2018): 1032-1039.

4) Geng, Ji, et al. "Andrographolide triggers autophagy-mediated inflammation inhibition and attenuates chronic unpredictable mild stress (CUMS)-induced depressive-like behavior in mice." Toxicology and applied pharmacology 379 (2019): 114688.

5) Gherardelli, Camila, et al. "Andrographolide restores glucose uptake in rat hippocampal neurons." Journal of Neurochemistry (2020).

6) Guan, S. P., et al. "Andrographolide protects against cigarette smoke-induced oxidative lung injury via augmentation of Nrf2 activity." British journal of pharmacology 168.7 (2013): 1707-1718.

7) Gupta, Swati, et al. "Andrographolide attenuates complete freund's adjuvant induced arthritis via suppression of inflammatory mediators and pro-inflammatory cytokines." Journal of ethnopharmacology 261 (2020): 113022.

8) Jayakumar, Thanasekaran, et al. "Experimental and clinical pharmacology of Andrographis paniculata and its major bioactive phytoconstituent andrographolide." Evidence-Based Complementary and Alternative Medicine 2013 (2013).

9) Kim, Nami, et al. "Andrographolide inhibits inflammatory responses in LPS-stimulated macrophages and murine acute colitis through activating AMPK." Biochemical pharmacology 170 (2019): 113646.

10) Li, Zun-zhong, et al. "Andrographolide benefits rheumatoid arthritis via inhibiting MAPK pathways." Inflammation 40.5 (2017): 1599-1605.

11) Li, Xiang, et al. "Andrographolide ameliorates intracerebral hemorrhage induced secondary brain injury by inhibiting neuroinflammation induction." Neuropharmacology 141 (2018): 305-315.

12) Lim, Jonathan Chee Woei, et al. "Andrographolide and its analogues: versatile bioactive molecules for combating inflammation and cancer." Clinical and Experimental Pharmacology and Physiology 39.3 (2012): 300-310.

13) Liu, Jianwei, et al. "Andrographolide prevents human nucleus pulposus cells against degeneration by inhibiting the NF-κB pathway." Journal of Cellular Physiology 234.6 (2019): 9631-9639.

14) Lu, Jiashu, et al. "A review for the neuroprotective effects of andrographolide in the central nervous system." Biomedicine & Pharmacotherapy 117 (2019): 109078.

15) Mishra, K. P. "Andrographolide: Regulating the Master Regulator NF-κB." Indian Journal of Clinical Biochemistry (2019): 1-3.

16) Mussard, Eugenie, et al. "Andrographolide, a natural antioxidant: an update." Antioxidants 8.12 (2019): 571.

17) Nugroho, Agung Endro, et al. "Antidiabetic and antihiperlipidemic effect of Andrographis paniculata (Burm. f.) Nees and andrographolide in high-fructose-fat-fed rats." Indian journal of pharmacology 44.3 (2012): 377.

18) Raghavan, Rahul, Sanith Cheriyamundath, and Joseph Madassery. "Exploring the mechanisms of cytotoxic and anti-inflammatory property of

*andrographolide and its derivatives." Pharmacognosy Reviews 12.23 (2018).*

19) Serrano, Felipe G., et al. "Andrographolide reduces cognitive impairment in young and mature AβPPswe/PS-1 mice." *Molecular neurodegeneration* 9.1 (2014): 61.

20) Su, Hongming, et al. "Andrographolide Exerts Antihyperglycemic Effect through Strengthening Intestinal Barrier Function and Increasing Microbial Composition of Akkermansia muciniphila." *Oxidative medicine and cellular longevity* 2020 (2020).

21) Tan, WS Daniel, et al. "Is there a future for andrographolide to be an anti-inflammatory drug? Deciphering its major mechanisms of action." *Biochemical pharmacology* 139 (2017): 71-81.

22) Tan, WS Daniel, et al. "Andrographolide simultaneously augments Nrf2 antioxidant defense and facilitates autophagic flux blockade in cigarette smoke-exposed human bronchial epithelial cells." *Toxicology and applied pharmacology* 360 (2018): 120-130.

23) Varela-Nallar, Lorena, et al. "Andrographolide stimulates neurogenesis in the adult hippocampus." *Neural plasticity* 2015 (2015).

24) Wang, Da-Peng, et al. "Andrographolide enhances hippocampal BDNF signaling and suppresses neuronal apoptosis, astroglial activation, neuroinflammation, and spatial memory deficits in a rat model of chronic cerebral hypoperfusion." *Naunyn-Schmiedeberg's Archives of Pharmacology* 392.10 (2019): 1277-1284.

25) Wong, Siew Ying, et al. "Andrographolide induces Nrf2 and heme oxygenase 1 in astrocytes by activating p38 MAPK and ERK." *Journal of neuroinflammation* 13.1 (2016): 251.

26) Wu, Ziqiang, et al. "Andrographolide promotes skeletal muscle regeneration after acute injury through epigenetic modulation." *European journal of pharmacology* 888 (2020): 173470.

27) Yang, Chih-Hao, et al. "Multi-targeting andrographolide, a novel NF-κB inhibitor, as a potential therapeutic agent for stroke." *International journal of molecular sciences* 18.8 (2017): 1638.

28) Zhang, Jing-Jing, et al. "Andrographolide exerts significant antidepressant-like effects involving the hippocampal BDNF system in mice." *International Journal of Neuropsychopharmacology* 22.9 (2019): 585-600.

29) Zhai, Z. J., et al. "Andrographolide suppresses RANKL-induced osteoclastogenesis in vitro and prevents inflammatory bone loss in vivo." British journal of pharmacology 171.3 (2014): 663-675.

30) Zhan, Janis Ya-Xian, et al. "Andrographolide sodium bisulfate prevents UV-induced skin photoaging through inhibiting oxidative stress and inflammation." Mediators of inflammation 2016 (2016).

31) Zhang, Shengmei, et al. "Andrographolide promotes pancreatic duct cells differentiation into insulin-producing cells by targeting PDX-1." Biochemical Pharmacology 174 (2020): 113785.

32) Zhu, Tao, et al. "Andrographolide protects against LPS-induced acute lung injury by inactivation of NF-κB." PLoS One 8.2 (2013): e56407.

## CHAPTER 6 - BERBERINE

1) Ahmed, Touqeer, et al. "Berberine and neurodegeneration: A review of literature." Pharmacological Reports 67.5 (2015): 970-979.

2) Ai, Fen, et al. "Berberine regulates proliferation, collagen synthesis and cytokine secretion of cardiac fibroblasts via AMPK-mTOR-p70S6K signaling pathway." International journal of clinical and experimental pathology 8.10 (2015): 12509.

3) Chen, Xing, et al. "Berberine Attenuates Cardiac Hypertrophy Through Inhibition of mTOR Signaling Pathway." Cardiovascular drugs and therapy 34 (2020): 463-473.

4) Chi, Liyi, et al. "The anti-atherogenic effects of berberine on foam cell formation are mediated through the upregulation of sirtuin 1." International journal of molecular medicine 34.4 (2014): 1087-1093.

5) Coelho, Ana R., et al. "Berberine-induced cardioprotection and Sirt3 modulation in doxorubicin-treated H9c2 cardiomyoblasts." Biochimica et Biophysica Acta (BBA)- Molecular Basis of Disease 1863.11 (2017): 2904-2923.

6) Fan, Jie, et al. "Pharmacological effects of berberine on mood disorders." Journal of cellular and molecular medicine 23.1 (2019): 21-28.

7) Fan, Xiaodi, et al. "Berberine alleviates ox-LDL induced inflammatory factors by upregulation of autophagy via AMPK/mTOR signaling pathway." Journal of translational medicine 13.1 (2015): 1-11.

8) Feng, Xiaojun, et al. "Berberine in cardiovascular and metabolic diseases: from mechanisms to therapeutics." Theranostics 9.7 (2019): 1923.

9) Gomes, Ana P., et al. "Berberine protects against high fat diet-induced dysfunction in muscle mitochondria by inducing SIRT1-dependent mitochondrial biogenesis." Biochimica et Biophysica Acta (BBA)-Molecular Basis of Disease 1822.2 (2012): 185-195.

10) Hang, Weijian, et al. "Berberine ameliorates high glucose-induced cardiomyocyte injury via AMPK signaling activation to stimulate mitochondrial biogenesis and restore autophagic flux." Frontiers in pharmacology 9 (2018): 1121.

11) Jang, Jaewoong, et al. "Berberine activates AMPK to suppress proteolytic processing, nuclear translocation and target DNA binding of SREBP-1c in 3T3-L1 adipocytes." Molecular medicine reports 15.6 (2017): 4139-4147.

12) Javadipour, Mansoureh, et al. "Metformin in contrast to berberine reversed arsenicinduced oxidative stress in mitochondria from rat pancreas probably via Sirt3-dependent pathway." Journal of biochemical and molecular toxicology 33.9 (2019): e22368.

13) Kong, Weijia, et al. "Berberine is a novel cholesterol-lowering drug working through a unique mechanism distinct from statins." Nature medicine 10.12 (2004): 1344-1351.

14) Lan, Jiarong, et al. "Meta-analysis of the effect and safety of berberine in the treatment of type 2 diabetes mellitus, hyperlipemia and hypertension." Journal of ethnopharmacology 161 (2015): 69-81.

15) Li, Wei, et al. "Berberine increases glucose uptake and intracellular ROS levels by promoting Sirtuin 3 ubiquitination." Biomedicine & Pharmacotherapy 121 (2020): 109563.

16) Ma, Xueling, et al. "The pathogenesis of diabetes mellitus by oxidative stress and inflammation: its inhibition by berberine." Frontiers in pharmacology 9 (2018): 782.

17) McCubrey, James A., et al. "Effects of resveratrol, curcumin, berberine and other nutraceuticals on aging, cancer development, cancer stem cells and microRNAs." Aging (Albany NY) 9.6 (2017): 1477.

18) Neag, Maria A., et al. "Berberine: Botanical occurrence, traditional uses, extraction methods, and relevance in cardiovascular, metabolic, hepatic, and renal disorders." Frontiers in pharmacology 9 (2018): 557.

19) Pang, Bing, et al. "Application of berberine on treating type 2 diabetes mellitus." International journal of endocrinology 2015 (2015).

20) Qin, Siru, et al. "AMPK and its Activator Berberine in the Treatment of Neurodegenerative Diseases." Current pharmaceutical design 26.39 (2020): 5054-5066.

21) Sadraie, Sepideh, et al. "Berberine ameliorates lipopolysaccharide-induced learning and memory deficit in the rat: insights into underlying molecular mechanisms." Metabolic brain disease 34.1 (2019): 245-255.

22) Teodoro, João Soeiro, et al. "Berberine reverts hepatic mitochondrial dysfunction in high-fat fed rats: a possible role for SirT3 activation." Mitochondrion 13.6 (2013): 637-646.

23) Wang, Haoran, et al. "Metformin and berberine, two versatile drugs in treatment of common metabolic diseases." Oncotarget 9.11 (2018): 10135.

24) Wang, Kun, et al. "The metabolism of berberine and its contribution to the pharmacological effects." Drug metabolism reviews 49.2 (2017): 139-157.

25) Wang, Zhixiang, et al. "Berberine acts as a putative epigenetic modulator by affecting the histone code." Toxicology in Vitro 36 (2016): 10-17.

26) Warowicka, Alicja, Robert Nawrot, and Anna Goździcka-Józefiak. "Antiviral activity of berberine." Archives of virology 165.9 (2020): 1935-1945.

27) Xu, Miao, et al. "Berberine promotes glucose consumption independently of AMPactivated protein kinase activation." PLoS One 9.7 (2014): e103702.

28) Xu, Zhifang, et al. "Rhizoma coptidis and berberine as a natural drug to combat aging and aging-related diseases via anti-oxidation and AMPK activation." Aging and disease 8.6 (2017): 760.

29) Yan, Xiao-Jin, et al. "Mitochondria play an important role in the cell proliferation suppressing activity of berberine." Scientific reports 7.1 (2017): 1-13.

30) Yerra, Veera Ganesh, et al. "Adenosine monophosphate-activated protein kinase modulation by berberine attenuates mitochondrial deficits and redox imbalance in experimental diabetic neuropathy." Neuropharmacology 131 (2018): 256-270.

31) Zhang, Xu, et al. "Modulation of gut microbiota by berberine and metformin during the treatment of high-fat diet-induced obesity in rats." Scientific reports

5.1 (2015): 1-10.

32) Zhang, Yu-pei, et al. "Berberine ameliorates high-fat diet-induced non-alcoholic fatty liver disease in rats via activation of SIRT3/AMPK/ACC pathway." Current medical science 39.1 (2019): 37-43.

33) Zhao, Chunhui, et al. "Berberine Alleviates Amyloid β-Induced Mitochondrial Dysfunction and Synaptic Loss." Oxidative medicine and cellular longevity 2019 (2019).

34) Zhu, Xiaofei, et al. "Hepatoprotection of berberine against hydrogen peroxideinduced apoptosis by upregulation of Sirtuin 1." Phytotherapy Research 27.3 (2013): 417-421.

35) Zhu, Xiaofei, et al. "Combination of berberine with resveratrol improves the lipidlowering efficacy." International journal of molecular sciences 19.12 (2018): 3903.

36) Zou, Kun, et al. "Advances in the study of berberine and its derivatives: a focus on anti-inflammatory and anti-tumor effects in the digestive system." Acta Pharmacologica Sinica 38.2 (2017): 157-167.

**CHAPTER 7** - BLACK SEED OIL

1) Ahmed Jawad, H., Y. Ibraheem Azhar, and I. Al-Hamdi Khalil. "Evaluation of efficacy, safety and antioxidant effect of Nigella sativa in patients with psoriasis: A randomized clinical trial." J Clin Exp Invest www. jceionline. org Vol 5.2 (2014).

2) Akhtar, Muhammad Tahir, et al. "Antidiabetic potential of Nigella sativa L seed oil in alloxaninduced diabetic rabbits." Tropical Journal of Pharmaceutical Research 19.2 (2020): 000-000.

3) Akbar, Shahid. "'Nigella sativa'(black seeds): Panacea or hyperbole?: A critical review of experimental and clinical observations." Australian Journal of Herbal and Naturopathic Medicine 30.4 (2018): 157.

4) Ali, Ahmad, and Dinesh Kumar. "Antiglycation and anti aggregation potential of thymoquinine." Natural Volatiles and Essential Oils 6.1: 25-33.

5) Amin, Bahareh, and Hossein Hosseinzadeh. "Black cumin (Nigella sativa) and its active constituent, thymoquinone: an overview on the analgesic and anti-inflammatory effects." Planta medica 82.01/02 (2016): 8-16.

6) Badary, Osama A., et al. "Thymoquinone is a potent superoxide anion scavenger." Drug and chemical toxicology 26.2 (2003): 87-98.

7) Beheshti, Farimah, Majid Khazaei, and Mahmoud Hosseini. "Neuropharmacological effects of Nigella sativa." Avicenna journal of phytomedicine 6.1 (2016): 104.

8) Bule, Mohammed, et al. "The antidiabetic effect of thymoquinone: A systematic review and meta-analysis of animal studies." Food Research International 127 (2020): 108736.

9) Darand, Mina, et al. "Nigella sativa and inflammatory biomarkers in patients with nonalcoholic fatty liver disease: Results from a randomized, double-blind, placebo-controlled, clinical trial." Complementary therapies in medicine 44 (2019): 204-209.

10) Eid, Ahmad M., et al. "A Review on the Cosmeceutical and External Applications of Nigella sativa." Journal of tropical medicine 2017 (2017).

11) El-Saleh, S. C. "Protection by Nigella Sativa [Black Seed] against Hyperhomocysteinemia in Rats." Vascular Disease Prevention 3.1 (2006): 73-78.

12) El-Tahir, Kamal El-Din Hussein, and Dana M. Bakeet. "The black seed Nigella sativa Linnaeus-A mine for multi cures: a plea for urgent clinical evaluation of its volatile oil." Journal of Taibah University Medical Sciences 1.1 (2006): 1-19.

13) Eltony, Sohair A., and S. A. Elgayar. "Histological study on effect of Nigella sativa on aged olfactory system of female albino rat." Rom J Morphol Embryol 55.2 (2014): 325-34.

14) Farkhondeh, Tahereh, Saeed Samarghandian, and Abasalt Borji. "An overview on cardioprotective and anti-diabetic effects of thymoquinone." Asian Pacific journal of tropical medicine 10.9 (2017): 849-854.

15) Heshmati, Javad, and Nazli Namazi. "Effects of black seed (Nigella sativa) on metabolic parameters in diabetes mellitus: A systematic review." Complementary therapies in medicine 23.2 (2015): 275-282.

16) Hosseini, M. S., et al. "Effects of Nigella sativa L. seed oil in type II diabetic Patients: a randomized, double-blind, placebo-controlled clinical trial." (2013): 93-99.

17) Khan, Md Asaduzzaman, Mousumi Tania, and Junjiang Fu. "Epigenetic role of thymoquinone: impact on cellular mechanism and cancer therapeutics." Drug discovery today (2019).

18) Kolahdooz, M., et al. "Effects of Nigella sativa L. seed oil on abnormal semen quality in infertile men: a randomized, double-blind, placebo-controlled clinical trial." Phytomedicine 21.6 (2014): 901-905.

19) Lee, Shu Ping, et al. "Thymoquinone activates imidazoline receptor to enhance glucagonlike peptide-1 secretion in diabetic rats." Archives of Medical Science 15.1 (2019).

20) Losso, Jack N., Hiba A. Bawadi, and Madhavi Chintalapati. "Inhibition of the formation of advanced glycation end products by thymoquinone." Food chemistry 128.1 (2011): 55-61.

21) Mahmoudi, Amin, et al. "Effects of Nigella sativa extracts on the lipid profile and uncoupling protein-1 gene expression in brown adipose tissue of mice." Advanced biomedical research 7 (2018).

22) Mustafa, Saad, Jogendra Singh Pawar, and Ilora Ghosh. "Thymoquinone Induces Epigenetic Modulation in ROS Dependent Manner." Research & Reviews: A Journal of Toxicology 7.1 (2017): 32-39.

23) Padhye, Subhash, et al. "From here to eternity-the secret of Pharaohs: Therapeutic potential of black cumin seeds and beyond." Cancer therapy 6.b (2008): 495.

24) Pandey, Rashmi, Dinesh Kumar, and A. L. I. Ahmad. "Nigella sativa seed extracts prevent the glycation of protein and DNA." Current Perspectives on Medicinal and Aromatic Plants (CUPMAP) 1.1 (2018): 1-7.

25) Pang, Jiuxia, et al. "Thymoquinone exerts potent growth-suppressive activity on leukemia through DNA hypermethylation reversal in leukemia cells." Oncotarget 8.21 (2017): 34453.

26) Perveen, Tahira, et al. "Increased 5-HT levels following repeated administration of Nigella sativa L.(black seed) oil produce antidepressant effects in rats." Scientia pharmaceutica 82.1 (2014): 161-170.

27) Qibi, Nazar M., Rami MA Al-Hayali, and Hazim A. Mohammad. "Effect of Nigella Sativa (Black seed) on the serum lipids of healthy individuals."

*Mustansiriya Medical Journal 6.2 (2006): 10-15.*

28) Rahmani, Arshad H., et al. "Therapeutic implications of black seed and its constituent thymoquinone in the prevention of cancer through inactivation and activation of molecular pathways." *Evidence-Based Complementary and Alternative Medicine 2014 (2014).*

29) Sahebkar, Amirhossein, et al. "A systematic review and meta-analysis of randomized controlled trials investigating the effects of supplementation with Nigella sativa (black seed) on blood pressure." *Journal of hypertension 34.11 (2016): 2127-2135.*

30) Tavakkoli, Alireza, et al. "Review on clinical trials of black seed (Nigella sativa) and its active constituent, thymoquinone." *Journal of pharmacopuncture 20.3 (2017): 179.*

31) Tesarova, Hana, et al. "Determination of oxygen radical absorbance capacity of black cumin (Nigella sativa) seed quinone compounds." *Natural product communications 6.2 (2011): 1934578X1100600214.*

32) Vaillancourt, France, et al. "Elucidation of molecular mechanisms underlying the protective effects of thymoquinone against rheumatoid arthritis." *Journal of cellular biochemistry 112.1 (2011): 107-117.*

33) Velagapudi, Ravikanth, et al. "AMPK and SIRT1 activation contribute to inhibition of neuroinflammation by thymoquinone in BV2 microglia." *Molecular and cellular biochemistry 435.1-2 (2017): 149-162.*

34) Yang, Yong, et al. "Upregulation of SIRT1-AMPK by thymoquinone in hepatic stellate cells ameliorates liver injury." *Toxicology letters 262 (2016): 80-91.*

35) Yarnell, Eric, N. D., and Kathy Abascal. "Nigella sativa." *Alternative and Complementary Therapies (2011).*

36) Yimer, Ebrahim M., et al. "Nigella sativa L.(Black Cumin): A Promising Natural Remedy for Wide Range of Illnesses." *Evidence-Based Complementary and Alternative Medicine 2019 (2019).*

37) Zafar, Hira, et al. "Glycation inhibition by Nigella sativa (Linn)–an in vitro model." *AsianJournal of Agricultural Biology 1 (2013): 187-189.*

**CHAPTER 8** - CENTELLA ASIATICA

1) Anand, T., et al. "Effect of asiaticosida rich extract from Centella asiatica (L.) Urb. on physical fatigue induced by weight loaded forced swim test." Asian Journal of Animal and Veterinary Advances 7.9 (2010): 832-841.

2) Anukunwithaya, Tosapol, et al. "Pharmacokinetics of a standardized extract of Centella asiatica ECa 233 in rats." Planta medica 83.08 (2017): 710-717.

3) Boondam, Yingrak, et al. "Inverted U-shaped response of a standardized extract of Centella asiatica (ECa 233) on memory enhancement." Scientific reports 9.1 (2019): 8404.

4) Bylka, Wiesława, et al. "Centella asiatica in cosmetology." Advances in Dermatology and Allergology/Postępy Dermatologii I Alergologii 30.1 (2013): 46.

5) Bylka, Wiesława, et al. "Centella asiatica in dermatology: an overview." Phytotherapy research 28.8 (2014): 1117-1124.

6) Chauhan, P. K., I. P. Pandey, and Vinod Kumar Dhatwalia. "Evaluation of the anti-diabetic effect of ethanolic and methanolic extracts of Centella asiatica leaves extract on alloxan induced diabetic rats." Adv Biol Res 4.1 (2010): 27-30.

7) Choi, Yeong Min, et al. "Titrated extract of Centella asiatica increases hair inductive property through inhibition of STAT signaling pathway in three-dimensional spheroid cultured human dermal papilla cells." Bioscience, biotechnology, and biochemistry 81.12 (2017): 2323-2329.

8) Gohil, Kashmira J., Jagruti A. Patel, and Anuradha K. Gajjar. "Pharmacological review on Centella asiatica: a potential herbal cure-all." Indian journal of pharmaceutical sciences 72.5 (2010): 546.

9) Gray, Nora E., et al. "Centella asiatica modulates antioxidant and mitochondrial pathways and improves cognitive function in mice." Journal of ethnopharmacology 180 (2016): 78-86.

10) Gray, Nora E., et al. "Centella asiatica increases hippocampal synaptic density and improves memory and executive function in aged mice." Brain and behavior 8.7 (2018): e01024.

11) Hao, Chunbo, et al. "Asiatic acid inhibits LPS-induced inflammatory

*response in human gingival fibroblasts." International immunopharmacology 50 (2017): 313-318.*

12) Hashim, Puziah, et al. "Triterpene composition and bioactivities of Centella asiatica." Molecules 16.2 (2011): 1310-1322.

13) Hashim, Puziah. "The effect of Centella asiatica, vitamins, glycolic acid and their mixtures preparations in stimulating collagen and fibronectin synthesis in cultured human skin fibroblast." Pakistan journal of pharmaceutical sciences 27.2 (2014).

14) He, Lilei, et al. "Asiaticoside, a component of Centella asiatica attenuates RANKL-induced osteoclastogenesis via NFATc1 and NF-κB signaling pathways." Journal of cellular physiology 234.4 (2019): 4267-4276.

15) Hsu, Yuan-Man, et al. "Anti-diabetic effects of madecassic acid and rotundic acid." Nutrients 7.12 (2015): 10065-10075.

16) Hu, Shu, et al. "Postpartum stretchmarks: repairing activity of an oral Centella asiatica supplementation (Centellicum®)." Minerva ginecologica 70.5 (2018): 629-634.

17) Huang, Shyh-Shyun, et al. "Antinociceptive activities and the mechanisms of antiinflammation of asiatic acid in mice." Evidence-Based Complementary and Alternative Medicine 2011 (2011).

18) James, Jacinda T., and Ian A. Dubery. "Pentacyclic triterpenoids from the medicinal herb, Centella asiatica (L.) Urban." Molecules 14.10 (2009): 3922-3941.

19) Jared, Silviya Rajakumari. "Enhancement of memory in rats with Centella asiatica." (2010).

20) Jenwitheesuk, Kamonwan, et al. "A prospective randomized, controlled, double-blind trial of the efficacy using centella cream for scar improvement." Evidence-Based Complementary and Alternative Medicine 2018 (2018).

21) Jiang, Hui, et al. "Identification of Centella asiatica's effective ingredients for inducing the neuronal differentiation." Evidence-Based Complementary and Alternative Medicine 2016 (2016).

22) Jieun Park, et al. "Effect of abdominal massage using Centella asiatica on body composition and changes in triglyceride concentration in blood." Korean Journal of Beauty Society 16.1 (2010): 169-175.

23) Kumar, MH Veerendra, and Y. K. Gupta. "Effect of different extracts of Centella asiatica on cognition and markers of oxidative stress in rats." Journal of ethnopharmacology 79.2 (2002): 253-260.

24) Legiawati, Lili, et al. "Oral and Topical Centella asiatica in Type 2 Diabetes Mellitus Patients with Dry Skin: A Three-Arm Prospective Randomized Double-Blind Controlled Trial." Evidence- Based Complementary and Alternative Medicine 2020 (2020).

25) Liu, Jun, et al. "Asiatic acid preserves beta cell mass and mitigates hyperglycemia in streptozocin-induced diabetic rats." Diabetes/metabolism research and reviews 26.6 (2010): 448-454.

26) Lokanathan, Yogeswaran, et al. "Recent updates in neuroprotective and neuroregenerative potential of Centella asiatica." The Malaysian journal of medical sciences: MJMS 23.1 (2016): 4.

27) Ma, Zhen-Guo, et al. "Asiatic acid protects against cardiac hypertrophy through activating AMPKα signalling pathway." International journal of biological sciences 12.7 (2016): 861.\

28) Mahmood, Amena, et al. "Triterpenoid saponin-rich fraction of Centella asiatica decreases IL-1β andNF-? B, and augments tissue regeneration and excision wound repair." Turkish Journal of Biology 40.2 (2016): 399-409.

29) Maquart, F. X., et al. "Triterpenes from Centella asiatica stimulate extracellular matrix accumulation in rat experimental wounds." European Journal of Dermatology 9.4 (1999): 289-96.

30) Maramaldi, Giada, et al. "Anti-inflammaging and antiglycation activity of a novel botanical ingredient from African biodiversity (Centevita™)." Clinical, cosmetic and investigational dermatology 7 (2014): 1.

31) Matthews, Donald G., et al. "exploring the effect of centella asiatica on mitochondrial biogenesis in the mouse brain." Alzheimer's & Dementia: The Journal of the Alzheimer's Association 13.7 (2017): P663.

32) Nema, Neelesh Kumar, et al. "Matrix metalloproteinase, hyaluronidase and elastase inhibitory potential of standardized extract of Centella asiatica." Pharmaceutical biology 51.9 (2013): 1182-1187.

33) Omar, Norazzila, et al. "The effects of Centella asiatica (L.) Urban on neural differentiation of human mesenchymal stem cells in vitro." BMC Complementary and Alternative Medicine 19.1 (2019): 167.

34) Orhan, Ilkay Erdogan. "Centella asiatica (L.) Urban: from traditional medicine to modern medicine with neuroprotective potential." Evidence-based complementary and alternative medicine 2012 (2012).

35) Oyenihi, Ayodeji B., et al. "Effects of Centella asiatica on skeletal muscle structure and key enzymes of glucose and glycogen metabolism in type 2 diabetic rats." Biomedicine & Pharmacotherapy 112 (2019): 108715.

36) Pakdeechote, Poungrat, et al. "Asiatic acid alleviates hemodynamic and metabolic alterations via restoring eNOS/iNOS expression, oxidative stress, and inflammation in dietinduced metabolic syndrome rats." Nutrients 6.1 (2014): 355-370.

37) Park, Ju, et al. "Anti-inflammatory effect of titrated extract of Centella asiatica in phthalic anhydride-induced allergic dermatitis animal model." International journal of molecular sciences 18.4 (2017): 738.

38) Paocharoen, Veeraya. "The efficacy and side effects of oral Centella asiatica extract for wound healing promotion in diabetic wound patients." J Med Assoc Thai 93.Suppl 7 (2010): S166-S170.

39) Puttarak, Panupong, et al. "Effects of Centella asiatica (L.) Urb. on cognitive function and mood related outcomes: A Systematic Review and Meta-analysis." Scientific reports 7.1 (2017): 1-12.

40) Raden, Abkar. "Pegagan (Centella asiatica) extract increases vagingal wall thickness in menopausal rats." Folia Medica Indonesiana 46.3 (2010).

41) Rahman, Sonia, et al. "Antidiabetic activity of centella asiatica (L.) urbana in alloxan induced type 1 diabetic model rats." Journal of Bio-Science 19 (2011): 23-27.

42) Ramachandran, Vinayagam, and Ramalingam Saravanan. "Asiatic acid prevents lipid peroxidation and improves antioxidant status in rats with streptozotocin-induced diabetes." Journal of functional foods 5.3 (2013): 1077-1087.

43) Ramachandran, Vinayagam, Ramalingam Saravanan, and Poomalai Senthilraja. "Antidiabetic and antihyperlipidemic activity of asiatic acid in diabetic rats, role of HMG CoA: in vivo and in silico approaches." Phytomedicine 21.3 (2014): 225-232.

44) Rao, Sulochana B., M. Chetana, and P. Uma Devi. "Centella asiatica treatment during postnatal period enhances learning and memory in mice."

*Physiology & behavior* 86.4 (2005): 449-457.

45) Ratz-Łyko, A., J. Arct, and K. Pytkowska. "Moisturizing and antiinflammatory properties of cosmetic formulations containing Centella asiatica extract." *Indian journal of pharmaceutical sciences* 78.1 (2016): 27.

46) Saansoomchai, Pahol, et al. "Enhanced VEGF Expression in Hair Follicle Dermal Papilla Cells by Centella asiatica Linn." *CMU J. Nat. Sci* 17.1 (2018): 25-37.

47) Seevaratnam, Vasantharuba, et al. "Functional properties of Centella asiatica (L.): a review." *Int J Pharm Pharm Sci* 4.5 (2012): 8-14.

48) Shen, Xueqing, et al. "Propionibacterium acnes related anti-inflammation and skin hydration activities of madecassoside, a pentacyclic triterpene saponin from Centella asiatica." *Bioscience, biotechnology, and biochemistry* 83.3 (2019): 561-568.

49) Shukla, A., et al. "In vitro and in vivo wound healing activity of asiaticoside isolated from Centella asiatica." *Journal of ethnopharmacology* 65.1 (1999): 1-11.

50) Soumyanath, Amala, et al. "Centella asiatica accelerates nerve regeneration upon oral administration and contains multiple active fractions increasing neurite elongation in-vitro." *Journal of Pharmacy and Pharmacology* 57.9 (2005): 1221-1229.

51) Somboonwong, Juraiporn, et al. "Wound healing activities of different extracts of Centella asiatica in incision and burn wound models: an experimental animal study." *BMC complementary and alternative medicine* 12.1 (2012): 103.

52) Sun, Boju, et al. "Therapeutic potential of Centella asiatica and its triterpenes: a review." *Frontiers in pharmacology* 11 (2020): 1373.

53) Togni, Stefano, et al. "Strengthening hair with Centella asiatica: a report of clinical and subjective efficacy of a local treatment with a 0.5% hair lotion."

54) Tsoukalas, Dimitris, et al. "Discovery of potent telomerase activators: Unfolding new therapeutic and anti-aging perspectives." *Molecular medicine reports* 20.4 (2019): 3701-3708.

55) Xu, Min-fang, et al. "Asiatic acid, a pentacyclic triterpene in Centella asiatica, attenuates glutamate-induced cognitive deficits in mice and apoptosis

*in SH-SY5Y cells." Acta Pharmacologica Sinica 33.5 (2012): 578-587.*

56) Yasurin, Patchanee, Malinee Sriariyanun, and Theerawut Phusantisampan. "The Bioavailability Activity of Centella asiatica." King Mongkut's University of Technology North Bangkok International Journal of Applied Science and Technology 9.1 (2016): 1-9.

57) Venesia, Nikko Fernando, Edy Fachrial, and I. Nyoman Ehrich Lister. "Effectiveness Test of Centella Asiatica Extract on Improvement of Collagen and Hydration in Female White Rat (Rattus Norwegicus Wistar)." American Scientific Research Journal for Engineering, Technology, and Sciences (ASRJETS) 65.1 (2020): 98-107.

58) Yunianto, I., Srijit Das, and M. Noor Mat. "Antispermatogenic and antifertility effect of Pegaga (Centella asiatica L) on the testis of male Sprague-Dawley rats." La Clinica Terapeutica 161.3 (2010): 235-239.

59) Wang, Ting, et al. "Intestinal interleukin-10 mobilization as a contributor to the anti-arthritis effect of orally administered madecassoside: a unique action mode of saponin compounds with poor bioavailability." Biochemical pharmacology 94.1 (2015): 30-38.

60) Wu, Fang, et al. "Identification of major active ingredients responsible for burn wound healing of Centella asiatica herbs." Evidence-Based Complementary and Alternative Medicine 2012 (2012).

61) Zahara, Kulsoom, Yamin Bibi, and Shaista Tabassum. "Clinical and therapeutic benefits of Centella asiatica." Pure and Applied Biology 3.4 (2014): 152.

62) Zainol, M. K., et al. "Antioxidative activity and total phenolic compounds of leaf, root and petiole of four accessions of Centella asiatica (L.) Urban." Food Chemistry 81.4 (2003): 575-581.

63) Zhao, Yun, et al. "Effect of centella asiatica on oxidative stress and lipid metabolism in hyperlipidemic animal models." Oxidative medicine and cellular longevity 2014 (2014).

**CHAPTER 9** - CHLOROGENIC ACID

1) Arçari, Demétrius Paiva, et al. "The in vitro and in vivo effects of yerba mate (Ilex paraguariensis) extract on adipogenesis." Food chemistry 141.2 (2013): 809-815.

*2) Bains, Yasmin, and Alejandro Gugliucci.* "Ilex paraguariensis and its main component chlorogenic acid inhibit fructose formation of advanced glycation endproducts with amino acids at conditions compatible with those in the digestive system." *Fitoterapia 117 (2017): 6-10.*

*3) Bains, Yasmin, Alejandro Gugliucci, and Russell Caccavello.* "Advanced glycation endproducts form during ovalbumin digestion in the presence of fructose: Inhibition by chlorogenic acid." *Fitoterapia 120 (2017): 1-5.*

*4) Bao, Liping, et al.* "Chlorogenic acid prevents diabetic nephropathy by inhibiting oxidative stress and inflammation through modulation of the Nrf2/ HO-1 and NF-κB pathways." *International immunopharmacology 54 (2018): 245-253.*

*5) Bhattacherjee, Abhishek, and Abhijit Datta.* "Mechanism of antiglycating properties of syringic and chlorogenic acids in in vitro glycation system." *Food Research International 77 (2015): 540-548.*

*6) Burgos-Morón, Estefanía, et al.* "The coffee constituent chlorogenic acid induces cellular DNA damage and formation of topoisomerase I–and II–DNA complexes in cells." *Journal of agricultural and food chemistry 60.30 (2012): 7384-7391.*

*7) Cai, Rui, Shuqing Chen, and Shenhua Jiang.* "Chlorogenic acid inhibits non-enzymatic glycation and oxidation of low density lipoprotein." *Zhejiang da xue xue bao. Yi xue ban= Journal of Zhejiang University. Medical sciences 47.1 (2018): 27-34.*

*8) Cha, Ji Won, et al.* "The polyphenol chlorogenic acid attenuates UVB-mediated oxidative stress in human HaCaT keratinocytes." *Biomolecules & therapeutics 22.2 (2014): 136.*

*9) Chen, Wei-Ping, et al.* "Anti-arthritic effects of chlorogenic acid in interleukin-1β-induced rabbit chondrocytes and a rabbit osteoarthritis model." *International immunopharmacology 11.1 (2011): 23-28.*

*10) Chen, Wei-Ping, and Li-Dong Wu.* "Chlorogenic acid suppresses interleukin-1β-induced inflammatory mediators in human chondrocytes." *International journal of clinical and experimental pathology 7.12 (2014): 8797.*

*11) Cho, Ae-Sim, et al.* "Chlorogenic acid exhibits anti-obesity property and improves lipid metabolism in high-fat diet-induced-obese mice." *Food and chemical toxicology 48.3 (2010): 937-943.*

12) Del Rio, Daniele, et al. "Bioavailability of coffee chlorogenic acids and green tea flavan-3- ols." Nutrients 2.8 (2010): 820-833.

13) Feng, Yan, et al. "Chlorogenic acid protects D-galactose-induced liver and kidney injury via antioxidation and anti-inflammation effects in mice." Pharmaceutical biology 54.6 (2016): 1027-1034.

14) Feng, Yingshu, et al. "Enhanced oral bioavailability and in vivo antioxidant activity of chlorogenic acid via liposomal formulation." International journal of pharmaceutics 501.1-2 (2016): 342-349.

15) Fernandez-Gomez, Beatriz, et al. "New knowledge on the antiglycoxidative mechanism of chlorogenic acid." Food & function 6.6 (2015): 2081-2090.

16) Fernandez-Gomez, Beatriz, et al. "Inhibitors of advanced glycation end products from coffee bean roasting by-product." European Food Research and Technology 244.6 (2018): 1101-1110.

17) Fuentes, Eduardo, et al. "Chlorogenic acid inhibits human platelet activation and thrombus formation." PloS one 9.3 (2014).

18) Gugliucci, A., et al. "Caffeic and chlorogenic acids in Ilex paraguariensis extracts are the main inhibitors of AGE generation by methylglyoxal in model proteins." Fitoterapia 80.6 (2009): 339-344.

19) Houben, Joyce MJ, et al. "Telomere shortening in COPD: potential protective effects of coffee consumption." Chronic oxidative stress and telomere shortening (2010): 99.

20) Jin, Shasha, et al. "Chlorogenic acid improves late diabetes through adiponectin receptor signaling pathways in db/db mice." PLoS One 10.4 (2015).

21) Kim, Hye-Rin, et al. "Chlorogenic acid suppresses pulmonary eosinophilia, IgE production, and Th2-type cytokine production in an ovalbumin-induced allergic asthma: activation of STAT-6 and JNK is inhibited by chlorogenic acid." International Immunopharmacology 10.10 (2010): 1242-1248.

22) Kim, Junghyun, et al. "Chlorogenic acid inhibits the formation of advanced glycation end products and associated protein cross-linking." Archives of pharmacal research 34.3 (2011): 495-500.

23) Kim, Junghyun, et al. "The extract of Aster Koraiensis prevents retinal Pericyte apoptosis in diabetic rats and its active compound, Chlorogenic acid inhibits AGE formation and AGE/RAGE interaction." Nutrients 8.9 (2016): 585.

24) Kong, Deqin, et al. "Chlorogenic acid prevents paraquat-induced apoptosis via Sirt1- mediated regulation of redox and mitochondrial function." Free radical research 53.6 (2019): 680-693.

25) Kremr, Daniel, et al. "Unremitting problems with chlorogenic acid nomenclature: a review." Química Nova 39.4 (2016): 530-533.

26) Lee, Kyungjin, et al. "Chlorogenic acid ameliorates brain damage and edema by inhibiting matrix metalloproteinase-2 and 9 in a rat model of focal cerebral ischemia." European journal of pharmacology 689.1-3 (2012): 89-95.

27) Lee, Won Jun, and Bao Ting Zhu. "Inhibition of DNA methylation by caffeic acid and chlorogenic acid, two common catechol-containing coffee polyphenols." Carcinogenesis 27.2 (2006): 269-277.

28) Liu, Jason J., et al. "Coffee consumption is positively associated with longer leukocyte telomere length in the nurses' health study." The Journal of nutrition 146.7 (2016): 1373-1378.

29) Meng, Shengxi, et al. "Roles of chlorogenic acid on regulating glucose and lipids metabolism: a review." Evidence-Based Complementary and Alternative Medicine 2013 (2013).

30) Meng, Zhao-Qing, et al. "Study on the anti-gout activity of chlorogenic acid: improvement on hyperuricemia and gouty inflammation." The American journal of Chinese medicine 42.06 (2014): 1471-1483.

31) Oboh, Ganiyu, et al. "Caffeic and chlorogenic acids inhibit key enzymes linked to type 2 diabetes (in vitro): a comparative study." Journal of Basic and Clinical Physiology and Pharmacology 26.2 (2015): 165-170.

32) Olthof, Margreet R., et al. "Consumption of high doses of chlorogenic acid, present in coffee, or of black tea increases plasma total homocysteine concentrations in humans." The American journal of clinical nutrition 73.3 (2001): 532-538.

33) Onakpoya, I. J., et al. "The effect of chlorogenic acid on blood pressure: a systematic review and meta-analysis of randomized clinical trials." Journal of human hypertension 29.2 (2015): 77.

34) Ong, Khang Wei, Annie Hsu, and Benny Kwong Huat Tan. "Anti-diabetic and anti-lipidemic effects of chlorogenic acid are mediated by ampk activation." Biochemical pharmacology 85.9 (2013): 1341-1351.

35) Ong, Khang Wei, Annie Hsu, and Benny Kwong Huat Tan. "Chlorogenic acid stimulates glucose transport in skeletal muscle via AMPK activation: a contributor to the beneficial effects of coffee on diabetes." PloS one 7.3 (2012).

36) Rani, MR Preetha, et al. "Chlorogenic acid attenuates glucotoxicity in H9c2 cells via inhibition of glycation and PKC α upregulation and safeguarding innate antioxidant status." Biomedicine & Pharmacotherapy 100 (2018): 467-477.

37) Ruan, Zheng, et al. "Metabolomic analysis of amino acid and energy metabolism in rats supplemented with chlorogenic acid." Amino acids 46.9 (2014): 2219-2229.

38) Ruifeng, Gao, et al. "Chlorogenic acid attenuates lipopolysaccharide-induced mice mastitis by suppressing TLR4-mediated NF-κB signaling pathway." European journal of pharmacology 729 (2014): 54-58.

39) Sadowska-Bartosz, Izabela, and Grzegorz Bartosz. "Prevention of protein glycation by natural compounds." Molecules 20.2 (2015): 3309-3334.

40) Santana-Gálvez, Jesús, Luis Cisneros-Zevallos, and Daniel A. Jacobo-Velázquez. "Chlorogenic acid: Recent advances on its dual role as a food additive and a nutraceutical against metabolic syndrome." Molecules 22.3 (2017): 358.

41) Shi, Haitao, et al. "Chlorogenic acid reduces liver inflammation and fibrosis through inhibition of toll-like receptor 4 signaling pathway." Toxicology 303 (2013): 107-114.

42) Tošović, Jelena, et al. "Antioxidative mechanisms in chlorogenic acid." Food chemistry 237 (2017): 390-398.

43) Tsai, Kun-Ling, et al. "Chlorogenic acid protects against oxLDL-induced oxidative damage and mitochondrial dysfunction by modulating SIRT1 in endothelial cells." Molecular nutrition & food research 62.11 (2018): 1700928.

44) Yang, Lele, et al. "Attenuation of Palmitic Acid–Induced Lipotoxicity by Chlorogenic Acid through Activation of SIRT1 in Hepatocytes." Molecular nutrition & food research 63.14 (2019): 1801432.

45) Yin, Huanshun, et al. "An electrochemical assay for DNA methylation, methyltransferase activity and inhibitor screening based on methyl binding domain protein." Biosensors and Bioelectronics 41 (2013): 492-497.

46) Zatorski, Hubert, et al. "Experimental colitis in mice is attenuated by

*topical administration of chlorogenic acid." Naunyn-Schmiedeberg's archives of pharmacology 388.6 (2015): 643-651.*

*47) Zheng, Zhiyong, et al. "The therapeutic detoxification of chlorogenic acid against acetaminophen-induced liver injury by ameliorating hepatic inflammation." Chemico-biological interactions 238 (2015): 93-101.*

*48) Zhou, Yan, et al. "Chlorogenic acid ameliorates intestinal mitochondrial injury by increasing antioxidant effects and activity of respiratory complexes." Bioscience, biotechnology, and biochemistry 80.5 (2016): 962-971.*

*49) Zhu, Bao Ting, et al. "Inhibition of human catechol-O-methyltransferase (COMT)-mediated O-methylation of catechol estrogens by major polyphenolic components present in coffee." The Journal of steroid biochemistry and molecular biology 113.1-2 (2009): 65-74.*

**CHAPTER 10** - CISTANCHE DESERTICOLA

*1) Al-Snafi, Ali Esmail. "Bioactive metabolites and pharmacology of Cistanche tubulosa-A review." IOSR Journal of Pharmacy 10.1 (2020): 37-46.*

*2) Cai, Run-Lan, et al. "Antifatigue activity of phenylethanoid-rich extract from Cistanche deserticola." Phytotherapy Research: An International Journal Devoted to Pharmacological and Toxicological Evaluation of Natural Product Derivatives 24.2 (2010): 313-315.*

*3) Chen, B. Q., Y. X. Liu, and W. Y. Kang. "Hepatoprotective effect and antioxidant activity of cultivate Cistanche deserticola YC Ma." Fine Chem 27 (2010): 342-345.*

*4) Chen, Chang, et al. "Echinacoside protects against MPTP/MPP+-induced neurotoxicity via regulating autophagy pathway mediated by Sirt1." Metabolic brain disease 34.1 (2019): 203-212.*

*5) Choi, Jin Gyu, et al. "Cistanches Herba enhances learning and memory by inducing nerve growth factor." Behavioural brain research 216.2 (2011): 652-658.*

*6) Gao, Yuan, et al. "Evaluation of the intestinal transport of a phenylethanoid glycoside-rich extract from Cistanche deserticola across the Caco-2 cell monolayer model." PloS one 10.2 (2015).*

*7) Gao, Yuqiu, et al. "Safety assessment of powdered Cistanche deserticola*

YC Ma by a 90-day feeding test in Sprague-Dawley rats." *Drug and chemical toxicology* 40.4 (2017): 383-389.

8) Gu, Caimei, Xianying Yang, and Linfang Huang. "Cistanches Herba: a neuropharmacology review." *Frontiers in pharmacology* 7 (2016): 289.

9) Gu, Li, et al. "Effects of Cistanche deserticola extract on penis erectile response in castrated rats." *Pakistan journal of pharmaceutical sciences* 29.2 (2016).

10) Jia, Yamin, et al. "Reduction of Inflammatory Hyperplasia in the Intestine in Colon Cancerprone Mice by Water-extract of Cistanche deserticola." *Phytotherapy Research* 26.6 (2012): 812-819

11) Leung, Hoi Yan, and Kam Ming Ko. "Herba cistanche extract enhances mitochondrial ATP generation in rat hearts and H9c2 cells." *Pharmaceutical Biology* 46.6 (2008): 418-424.

12) Leung, Hoi Yan. *Effect of Herba Cistanche on mitochondrial ATP generation: A pharmacological basis of 'Yang-invigoration'*. Diss. 2006.

13) Li, Te-Mao, et al. "Cistanche deserticola extract increases bone formation in osteoblasts." *Journal of Pharmacy and Pharmacology* 64.6 (2012): 897-907.

14) Li, Zhiming, et al. "Herba Cistanche (Rou Cong-Rong): One of the best pharmaceutical gifts of traditional Chinese medicine." *Frontiers in pharmacology* 7 (2016): 41.

15) Liu, M. H., G. J. Zhao, and Zhe Chen. "A study on the effect of phenylethanoid glycosides of the Cistanche deserticola on Scopolamine–induced impairment of learning memory in mice." *J. Baotou Med. Coll* 27 (2011): 9-10.

16) Lucius, Khara. "Botanical Medicine and Phytochemicals in Healthy Aging and Longevity— Part 1." *Alternative and Complementary Therapies* 26.1 (2020): 31-37.

17) Luo, Q. J., Y. S. Wang, and K. Q. Huang. "Protective effect of Cistanche deserticola on skeletal muscle oxidative injury in high intensity training rats." *J. Zhanjiang Norm. Coll* 33 (2012): 132-135.

18) Ma, Zhiguo, et al. "Determination of bioactive components of Cistanche deserticola (Roucongrong) by high-performance liquid chromatography with diode array and mass spectrometry detectors." *Analytical Letters* 47.17 (2014): 2783-2794.

19) Maruyama, Shinjiro, et al. "Cistanche salsa extract enhanced antibody production in human lymph node lymphocytes." Pharmacology 2 (2008): 341-348.

20) Nan, Ze-Dong, et al. "Chemical constituents from stems of Cistanche deserticola cultured in Tarim desert." Zhongguo Zhong yao za zhi= Zhongguo zhongyao zazhi= China journal of Chinese materia medica 38.16 (2013): 2665-2670.

21) Peng, Sheng, et al. "Cistanches alleviates sevoflurane-induced cognitive dysfunction by regulating PPAR-$\gamma$-dependent antioxidant and anti-inflammatory in rats." Journal of cellular and molecular medicine 24.2 (2020): 1345-1359.

22) Shi, Yimin, et al. "Fast repair of oxidative DNA damage by phenylpropanoid glycosides and their analogues." Mutagenesis 23.1 (2008): 19-26.

23) Siu, Ada Hoi-Ling, and Kam Ming Ko. "Herba Cistanche extract enhances mitochondrial glutathione status and respiration in rat hearts, with possible induction of uncoupling proteins." Pharmaceutical biology 48.5 (2010): 512-517.

24) Song, Dezhi, et al. "Cistanche deserticola polysaccharide attenuates osteoclastogenesis and bone resorption via inhibiting RANKL signaling and reactive oxygen species production." Journal of cellular physiology 233.12 (2018): 9674-9684.

25) Wang, Fujiang, et al. "Total glycosides of Cistanche deserticola promote neurological function recovery by inducing neurovascular regeneration via Nrf-2/Keap-1 pathway in MCAO/R rats." Frontiers in Pharmacology 11 (2020).

26) Wang, Ningqun, et al. "Herba Cistanches: anti-aging." Aging and disease 8.6 (2017): 740.

27) Wang, Tian, Xiaoying Zhang, and Wenyan Xie. "Cistanche deserticola YC Ma," Desert ginseng": a review." The American journal of Chinese medicine 40.06 (2012): 1123-1141.

28) Wang, Xiang-yan, et al. "The effect of Cistanche deserticola polysaccharides (CDPS) on marcrophages activation." Chin Pharmacol Bull 25.6 (2009): 787-790.

29) Wong, Hoi Shan, and Kam Ming Ko. "Herba Cistanches stimulates cellular glutathione redox cycling by reactive oxygen species generated from mitochondrial respiration in H9c2 cardiomyocytes." Pharmaceutical Biology 51.1 (2013): 64-73.

30) Xiong, Quanbo, et al. "Antioxidative effects of phenylethanoids from Cistanche deserticola." Biological and Pharmaceutical Bulletin 19.12 (1996): 1580-1585.

31) Xu, Xiaoxue, et al. "Therapeutic effect of cistanoside A on bone metabolism of ovariectomized mice." Molecules 22.2 (2017): 197.

32) Yonei, Yoshikazu, et al. "Effects of health food containing Cistanche deserticola extract on QOL and safety in elderly: An open pilot study of 12-week oral treatment." Anti-Aging Medicine 8.2 (2011): 7-14.

33) Zeng, J. C., et al. "Experimental study of directional differentiation of bone mesenchymal stem cells (BMSCs) to osteoblasts guided by serum containing cistanche deserticola." Zhongguo gu shang= China journal of orthopaedics and traumatology 23.8 (2010): 606-608.

34) Zhang, Ailian, et al. "Immunostimulatory activity of water-extractable polysaccharides from Cistanche deserticola as a plant adjuvant in vitro and in vivo." PloS one 13.1 (2018).

35) Zhang, H. Q., Yuan Li, and Y. Y. Song. "Effect of polysaccharides of Cistanche deserticola on immune cells and telomerase activity in aging mice." Chin. Pharm. J 46.14 (2011): 1081-1083.

36) Zhang, Ke, et al. "Extracts of Cistanche deserticola can antagonize immunosenescence and extend life span in senescence-accelerated mouse prone 8 (SAM-P8) mice." Evidence-Based Complementary and Alternative Medicine 2014 (2014).

37) Zhang, Tian, et al. "Study on comparing the Runchang purge and anti-fatigue effects of Cistanche deserticola and the tissue culture of Cistanche deserticola." Science and Technology of Food Industry 30 (2009): 155-156.

38) Zhao, Bing, et al. "Immunopotentiating effects of polysaccharides extracted from cultivated and wild Cistanche desertico-la in Xinjiang." Chinese Journal of Microbiology and Immunology 38.1 (2018): 7-13.

39) Zhou, Hai-Tao, Jian-min CAO, and Qiang LIN. "Effect of Cistanches Herba on the Swimming Ability and Oxidation Resistance of Mitochondrial in Rats." Chinese Journal of Experimental Traditional Medical Formulae 18.6 (2012): 229-235.

## CHAPTER 11 - COLLAGEN

1) Abreu-Velez, Ana Maria, and Michael S. Howard. "Collagen IV in normal skin and in pathological processes." North American journal of medical sciences 4.1 (2012): 1.

2) Argyropoulos, Angela J., et al. "Alterations of dermal connective tissue collagen in diabetes: molecular basis of aged-appearing skin." PloS one 11.4 (2016): e0153806.

3) Arseni, Lavinia, Anita Lombardi, and Donata Orioli. "From structure to phenotype: impact of collagen alterations on human health." International journal of molecular sciences 19.5 (2018): 1407.

4) Asserin, Jérome, et al. "The effect of oral collagen peptide supplementation on skin moisture and the dermal collagen network: evidence from an ex vivo model and randomized, placebocontrolled clinical trials." Journal of cosmetic dermatology 14.4 (2015): 291-301.

5) Avila Rodríguez, María Isabela, Laura G. Rodriguez Barroso, and Mirna Lorena Sánchez. "Collagen: A review on its sources and potential cosmetic applications." Journal of Cosmetic Dermatology 17.1 (2018): 20-26.

6) Bagi, C. M., et al. "Oral administration of undenatured native chicken type II collagen (UC-II) diminished deterioration of articular cartilage in a rat model of osteoarthritis (OA)." Osteoarthritis and cartilage 25.12 (2017): 2080-2090.

7) Baroni, Eloina do Rocio Valenga, et al. "Influence of aging on the quality of the skin of white women: the role of collagen." Acta cirurgica brasileira 27.10 (2012): 736-740.

8) Bolke, Liane, et al. "A collagen supplement improves skin hydration, elasticity, roughness, and density: Results of a randomized, placebo-controlled, blind study." Nutrients 11.10 (2019): 2494.

9) Cheng, Wang, et al. "The content and ratio of type I and III collagen in skin differ with age and injury." African Journal of Biotechnology 10.13 (2011): 2524-2529.

10) Choi, Franchesca D., et al. "Oral collagen supplementation: a systematic review of dermatological applications." Journal of drugs in dermatology: JDD 18.1 (2019): 9-16.

11) Choi, Sun Young, et al. "Effects of collagen tripeptide supplement on skin properties: A prospective, randomized, controlled study." Journal of Cosmetic and Laser Therapy 16.3 (2014): 132-137.

12) Coppola, Daniela, et al. "Marine collagen from alternative and sustainable sources: Extraction, processing and applications." Marine drugs 18.4 (2020): 214.

13) Deshmukh, Shrutal Narendra, et al. "Enigmatic insight into collagen." Journal of oral and maxillofacial pathology: JOMFP 20.2 (2016): 276.

14) Garnero, Patrick. "The role of collagen organization on the properties of bone." Calcified tissue international 97.3 (2015): 229-240.

15) Guillerminet, Fanny, et al. "Hydrolyzed collagen improves bone metabolism and biomechanical parameters in ovariectomized mice: an in vitro and in vivo study." Bone 46.3 (2010): 827-834.

16) Haratake, Akinori, et al. "Effects of oral administration of collagen peptides on skin collagen content and its underlying mechanism using a newly developed low collagen skin mice model." Journal of Functional Foods 16 (2015): 174-182.

17) Henriksen, K., and M. A. Karsdal. "Type I collagen." Biochemistry of collagens, laminins and elastin. Academic Press, 2016. 1-11.

18) Iba, Yoshinori, et al. "Oral administration of collagen hydrolysates improves glucose tolerance in normal mice through GLP-1-dependent and GLP-1-independent mechanisms." Journal of medicinal food 19.9 (2016): 836-843.

19) Kang, Min Cheol, Silvia Yumnam, and Sun Yeou Kim. "Oral intake of collagen peptide attenuates ultraviolet B irradiation-induced skin dehydration in vivo by regulating hyaluronic acid synthesis." International journal of molecular sciences 19.11 (2018): 3551.

20) Kim, Do-Un, et al. "Oral intake of low-molecular-weight collagen peptide improves hydration, elasticity, and wrinkling in human skin: a randomized, double-blind, placebo-controlled study." Nutrients 10.7 (2018): 826.

21) Kumar, Suresh, et al. "A double-blind, placebo-controlled, randomised, clinical study on the effectiveness of collagen peptide on osteoarthritis." Journal of the Science of Food and Agriculture 95.4 (2015): 702-707.

22) Leitinger, Birgit. "Transmembrane collagen receptors." Annual review of cell and developmental biology 27 (2011): 265-290.

23) Li, Yong, et al. "Age-associated increase in skin fibroblast–derived prostaglandin E2 contributes to reduced collagen levels in elderly human skin." Journal of Investigative Dermatology 135.9 (2015): 2181-2188.

24) Lin, Ping, et al. "Oral Collagen Drink for Antiaging: Antioxidation, Facilitation of the Increase of Collagen Synthesis, and Improvement of Protein Folding and DNA Repair in Human Skin Fibroblasts." Oxidative Medicine and Cellular Longevity 2020 (2020).

25) Liu, Zehua, et al. "Collagen peptides promote photoaging skin cell repair by activating the TGF-β/Smad pathway and depressing collagen degradation." Food & function 10.9 (2019): 6121-6134.

26) Proksch, E., et al. "Oral supplementation of specific collagen peptides has beneficial effects on human skin physiology: a double-blind, placebo-controlled study." Skin pharmacology and physiology 27.1 (2014): 47-55.

27) Proksch, Ehrhardt, et al. "Oral intake of specific bioactive collagen peptides reduces skin wrinkles and increases dermal matrix synthesis." Skin pharmacology and physiology 27.3 (2014): 113-119.

28) Pyun, Hee-Bong, et al. "Effects of collagen tripeptide supplement on photoaging and epidermal skin barrier in UVB-exposed hairless mice." Preventive Nutrition and Food Science 17.4 (2012): 245.

29) Quan, Taihao, et al. "Elevated matrix metalloproteinases and collagen fragmentation in photodamaged human skin: impact of altered extracellular matrix microenvironment on dermal fibroblast function." The Journal of investigative dermatology 133.5 (2013): 1362.

30) Raksha, Nataliia G., et al. "Prevention of diet-induced obesity in rats by oral application of collagen fragments." Archives of Biological Sciences 70.1 (2018): 77-86.

31) Sangsuwan, Wich, and Pravit Asawanonda. "Four-weeks daily intake of oral collagen hydrolysate results in improved skin elasticity, especially in sun-exposed areas: a randomized, double-blind, placebo-controlled trial." Journal of Dermatological Treatment (2020): 1-6.

32) Song, Hongdong, and Bo Li. "Beneficial effects of collagen hydrolysate: A review on recent developments." Biomed J Sci & Tech Res 1.2 (2017): 1-4.

33) Wang, Bin, et al. "Isolation and characterization of collagen and antioxidant collagen peptides from scales of croceine croaker (Pseudosciaena crocea)." Marine Drugs 11.11 (2013): 4641-4661.

34) Wang, Lin, et al. "Bioavailability and bioavailable forms of collagen after oral administration to rats." Journal of agricultural and food chemistry 63.14 (2015): 3752-3756.

35) Wang, Lin, et al. "Determination of bioavailability and identification of collagen peptide in blood after oral ingestion of gelatin." Journal of the Science of Food and Agriculture 95.13 (2015): 2712-2717.

36) Woo, T., et al. "Efficacy of oral collagen in joint pain-osteoarthritis and rheumatoid arthritis." Journal of Arthritis 6.2 (2017): 1-4.

37) Wu, Hui, et al. "Study on the scavenging activity of free radicals of collagen peptide from ox bone by enzymatic hydrolysis." Sci Technol Food Ind 4 (2010): 156-158.

38) Yazaki, Misato, et al. "Oral ingestion of collagen hydrolysate leads to the transportation of highly concentrated Gly-Pro-Hyp and its hydrolyzed form of Pro-Hyp into the bloodstream and skin." Journal of Agricultural and Food Chemistry 65.11 (2017): 2315-2322.\

39) Zague, Vivian, et al. "Collagen hydrolysate intake increases skin collagen expression and suppresses matrix metalloproteinase 2 activity." Journal of medicinal food 14.6 (2011): 618-624.

**CHAPTER 12** - COENZYME Q10

1) Abdollahzad, Hadi, et al. "Effects of coenzyme Q10 supplementation on inflammatory cytokines (TNF-α, IL-6) and oxidative stress in rheumatoid arthritis patients: a randomized controlled trial." Archives of medical research 46.7 (2015): 527-533.

2) Al Saadi, Tareq, et al. "Coenzyme Q10 for heart failure." Cochrane Database of Systematic Reviews 1 (2021).

3) Baburina, Yulia, et al. "Astaxanthin inhibits mitochondrial permeability transition pore opening in rat heart mitochondria." Antioxidants 8.12 (2019): 576.

4) Belviranli, Muaz, and Nilsel Okudan. "Well-known antioxidants and

*newcomers in sport nutrition: coenzyme Q10, quercetin, resveratrol, pterostilbene, pycnogenol and astaxanthin." (2015).*

5) Briston, Thomas, et al. "Mitochondrial permeability transition pore: sensitivity to opening and mechanistic dependence on substrate availability." Scientific reports 7.1 (2017): 1-13.

6) Dahri, Monireh, et al. "Oral coenzyme Q10 supplementation in patients with migraine: Effects on clinical features and inflammatory markers." Nutritional neuroscience 22.9 (2019): 607-615.

7) Del Pozo-Cruz, Jesús, et al. "Physical activity affects plasma coenzyme Q 10 levels differently in young and old humans." Biogerontology 15.2 (2014): 199-211.

8) Díaz-Castro, Javier, et al. "Coenzyme Q 10 supplementation ameliorates inflammatory signaling and oxidative stress associated with strenuous exercise." European journal of nutrition 51.7 (2012): 791-799.

9) DiNicolantonio, James J., et al. "Coenzyme Q10 for the treatment of heart failure: a review of the literature." Open Heart 2.1 (2015).

10) Dhingra, Rimpy, et al. "Impaired NF-κB signalling underlies cyclophilin D-mediated mitochondrial permeability transition pore opening in doxorubicin cardiomyopathy." Cardiovascular research 116.6 (2020): 1161-1174.

11) Doimo, Mara, et al. "Genetics of coenzyme q10 deficiency." Molecular syndromology 5.3-4 (2014): 156-162.

12) Fan, Li, et al. "Effects of coenzyme Q10 supplementation on inflammatory markers: a systematic review and meta-analysis of randomized controlled trials." Pharmacological research 119 (2017): 128-136.

13) Fang, Yi, et al. "Melatonin improves cryopreservation of ram sperm by inhibiting mitochondrial permeability transition pore opening." Reproduction in Domestic Animals 55.9 (2020): 1240-1249.

14) Greco, Tiffany, Jonathan Shafer, and Gary Fiskum. "Sulforaphane inhibits mitochondrial permeability transition and oxidative stress." Free Radical Biology and Medicine 51.12 (2011): 2164-2171.

15) Hernández-Camacho, Juan D., et al. "Coenzyme Q10 supplementation in aging and disease." Frontiers in physiology 9 (2018): 44.

16) Ilmarinen, Pinja, et al. "The polyamine spermine inhibits mitochondrial permeability transition (mPT) and apoptosis in human eosinophils." European Respiratory Journal 44.Suppl 58 (2014).

17) Kent, Andrew C., Khairat Bahgat Youssef El Baradie, and Mark W. Hamrick. "Targeting the Mitochondrial Permeability Transition Pore to Prevent Age-Associated Cell Damage and Neurodegeneration." Oxidative medicine and cellular longevity 2021 (2021).

18) Lee, Bor-Jen, et al. "Coenzyme Q10 supplementation reduces oxidative stress and increases antioxidant enzyme activity in patients with coronary artery disease." Nutrition 28.3 (2012): 250-255.

19) Lee, Bor-Jen, et al. "Effects of coenzyme Q10 supplementation (300 mg/day) on antioxidation and anti-inflammation in coronary artery disease patients during statins therapy: a randomized, placebo-controlled trial." Nutrition journal 12.1 (2013): 1-9.

20) Li, Lin, et al. "Pre-treatment with a combination of Shenmai and Danshen injection protects cardiomyocytes against hypoxia/reoxygenation-and H2O2-induced injury by inhibiting mitochondrial permeability transition pore opening." Experimental and therapeutic medicine 17.6 (2019): 4643-4652.

21) Martelli, Alma, et al. "Coenzyme Q10: Clinical applications in cardiovascular diseases." Antioxidants 9.4 (2020): 341.

22) Mishra, Jyotsna, et al. "Cyclosporin A increases mitochondrial buffering of calcium: an additional mechanism in delaying mitochondrial permeability transition pore opening." Cells 8.9 (2019): 1052.

23) Mortensen, Svend A., et al. "The effect of coenzyme Q10 on morbidity and mortality in chronic heart failure: results from Q-SYMBIO: a randomized double-blind trial." JACC: Heart Failure 2.6 (2014): 641-649.

24) Mortensen, Anne Louise, Franklin Rosenfeldt, and Krzysztof J. Filipiak. "Effect of coenzyme Q10 in Europeans with chronic heart failure: A sub-group analysis of the Q-SYMBIO randomized double-blind trial." Cardiology journal 26.2 (2019): 147-156.

25) Panel, Mathieu, Bijan Ghaleh, and Didier Morin. "Mitochondria and aging: A role for the mitochondrial transition pore?." Aging cell 17.4 (2018): e12793.

26) Raizner, Albert E. "Coenzyme Q10." Methodist DeBakey cardiovascular

*journal 15.3 (2019): 185.*

27) Rodick, Taylor C., et al. "Potential role of coenzyme Q 10 in health and disease conditions." *Nutrition and Dietary Supplements* 10 (2018): 1-11.

28) Rottenberg, Hagai, and Jan B. Hoek. "The path from mitochondrial ROS to aging runs through the mitochondrial permeability transition pore." *Aging cell* 16.5 (2017): 943-955.

29) Rottenberg, Hagai, and Jan B. Hoek. "The Mitochondrial Permeability Transition: Nexus of Aging, Disease and Longevity." *Cells* 10.1 (2021): 79.

30) Saini, Rajiv. "Coenzyme Q10: The essential nutrient." *J Pharm Bioallied Sci* 3.3 (2011): 466-467.

31) Sanoobar, Meisam, et al. "Coenzyme Q10 supplementation ameliorates inflammatory markers in patients with multiple sclerosis: a double blind, placebo, controlled randomized clinical trial." *Nutritional neuroscience* 18.4 (2015): 169-176.

32) Shen, Qiuhua, and Janet D. Pierce. "Supplementation of coenzyme Q10 among patients with type 2 diabetes mellitus." *Healthcare.* Vol. 3. No. 2. Multidisciplinary Digital Publishing Institute, 2015.

33) Tomasetti, Marco, Renata Alleva, and Andrew R. Collins. "In vivo supplementation with coenzyme Q10 enhances the recovery of human lymphocytes from oxidative DNA damage." *The FASEB Journal* 15.8 (2001): 1425-1427.

34) Varela-López, Alfonso, et al. "Coenzyme Q and its role in the dietary therapy against aging." *Molecules* 21.3 (2016): 373.

35) Wacquier, Benjamin, Laurent Combettes, and Geneviève Dupont. "Dual dynamics of mitochondrial permeability transition pore opening." *Scientific reports* 10.1 (2020): 1-10.

36) Wang, Hong, et al. "Ochratoxin A-induced apoptosis of IPEC-J2 cells through ROSmediated mitochondrial permeability transition pore opening pathway." *Journal of agricultural and food chemistry* 65.48 (2017): 10630-10637.

37) Zhai, Junya, et al. "Effects of coenzyme Q10 on markers of inflammation: a systematic review and meta-analysis." *PloS one* 12.1 (2017): e0170172.

## CHAPTER 13 - DELPHINIDIN

1) Afaq, Farrukh, et al. "Delphinidin, an anthocyanidin in pigmented fruits and vegetables, protects human HaCaT keratinocytes and mouse skin against UVB-mediated oxidative stress and apoptosis." Journal of Investigative Dermatology 127.1 (2007): 222-232.

2) Alvarado, Jorge L., et al. "Delphinidin-rich maqui berry extract (Delphinol®) lowers fasting and postprandial glycemia and insulinemia in prediabetic

3) Chamcheu, Jean Christopher, et al. "Dual inhibition of PI3K/Akt and mTOR by the dietary antioxidant, delphinidin, ameliorates psoriatic features in vitro and in an imiquimod-induced psoriasis-like disease in mice." Antioxidants & redox signaling 26.2 (2017): 49-69.

4) Chen, Chun-ye, et al. "Delphinidin attenuates stress injury induced by oxidized low-density lipoprotein in human umbilical vein endothelial cells." Chemico-biological interactions 183.1 (2010): 105-112.

5) Chen, Youming, et al. "Delphinidin attenuates pathological cardiac hypertrophy via the AMPK/ NOX/MAPK signaling pathway." Aging (Albany NY) 12.6 (2020): 5362.

6) Daveri, Elena, et al. "Cyanidin and delphinidin modulate inflammation and altered redox signaling improving insulin resistance in high fat-fed mice." Redox biology 18 (2018): 16-24.

7) Goszcz, Katarzyna, et al. "Bioavailable concentrations of delphinidin and its metabolite, gallic acid, induce antioxidant protection associated with increased intracellular glutathione in cultured endothelial cells." Oxidative medicine and cellular longevity 2017 (2017).

8) Haseeb, Abdul, Dongxing Chen, and Tariq M. Haqqi. "Delphinidin inhibits IL-1β-induced activation of NF-κB by modulating the phosphorylation of IRAK-1Ser376 in human articular chondrocytes." Rheumatology 52.6 (2013): 998-1008.

9) Hwang, Mun Kyung, et al. "Fyn kinase is a direct molecular target of delphinidin for the inhibition of cyclooxygenase-2 expression induced by tumor necrosis factor-α." Biochemical pharmacology 77.7 (2009): 1213-1222.

10) Jin, Xin, et al. "Delphinidin-3-glucoside protects human umbilical vein endothelial cells against oxidized low-density lipoprotein-induced injury by

*autophagy upregulation via the AMPK/ SIRT1 signaling pathway." Molecular nutrition & food research 58.10 (2014): 1941-1951.*

11) Kanakis, C. D., et al. "DNA interaction with naturally occurring antioxidant flavonoids quercetin, kaempferol, and delphinidin." Journal of Biomolecular Structure and Dynamics 22.6 (2005): 719-724.

12) Kanakis, Charalabos D., et al. "An overview of DNA and RNA bindings to antioxidant flavonoids." Cell Biochemistry and Biophysics 49.1 (2007): 29-36.

13) Kuo, Hsiao-Chen Dina, et al. "Anthocyanin Delphinidin Prevents Neoplastic Transformation of Mouse Skin JB6 P+ Cells: Epigenetic Re-activation of Nrf2-ARE Pathway." The AAPS journal 21.5 (2019): 83.

14) Kwon, Jung Yeon, et al. "Delphinidin suppresses ultraviolet B-induced cyclooxygenases-2 expression through inhibition of MAPKK4 and PI-3 kinase." Carcinogenesis 30.11 (2009): 1932-1940.

15) Lee, Dong-Yeong, et al. "Cytoprotective Effects of Delphinidin for Human Chondrocytes against Oxidative Stress through Activation of Autophagy." Antioxidants 9.1 (2020): 83.

16) Lim, Tae-Gyu, et al. "NADPH oxidase is a novel target of delphinidin for the inhibition of UVB-induced MMP-1 expression in human dermal fibroblasts." Experimental dermatology 22.6 (2013): 428-430.

17) Mas, Thierry, et al. "DNA triplex stabilization property of natural anthocyanins." Phytochemistry 53.6 (2000): 679-687.

18) Moriwaki, Sawako, et al. "Delphinidin, one of the major anthocyanidins, prevents bone loss through the inhibition of excessive osteoclastogenesis in osteoporosis model mice." PLoS One 9.5 (2014): e97177.

19) Murata, Motoki, et al. "Delphinidin prevents disuse muscle atrophy and reduces stressrelated gene expression." Bioscience, Biotechnology, and Biochemistry 80.8 (2016): 1636-1640.

20 ) Murata, Motoki, et al. "Delphinidin prevents muscle atrophy and upregulates miR-23a expression." Journal of agricultural and food chemistry 65.1 (2017): 45-50.

21) Ni, Timin, Wanju Yang, and Yiqiao Xing. "Protective effects of delphinidin against H2O2– induced oxidative injuries in human retinal pigment epithelial cells." Bioscience reports 39.8 (2019).

22) Pal, H. C., et al. "Topical application of delphinidin reduces psoriasiform lesions in the flaky skin mouse model by inducing epidermal differentiation and inhibiting inflammation." British Journal of Dermatology 172.2 (2015): 354-364.

23) Saulite, Liga, et al. "Effects of malvidin, cyanidin and delphinidin on human adipose mesenchymal stem cell differentiation into adipocytes, chondrocytes and osteocytes." Phytomedicine 53 (2019): 86-95.

24) Seong, Ah-Reum, et al. "Delphinidin, a specific inhibitor of histone acetyltransferase, suppresses inflammatory signaling via prevention of NF-κB acetylation in fibroblast-like synoviocyte MH7A cells." Biochemical and biophysical research communications 410.3 (2011): 581-586.

25) Skemiene, Kristina, Julius Liobikas, and Vilmante Borutaite. "Anthocyanins as substrates for mitochondrial complex I–protective effect against heart ischemic injury." The FEBS journal 282.5 (2015): 963-971.

26) Sobiepanek, Anna, et al. "The effect of delphinidin on the mechanical properties of keratinocytes exposed to UVB radiation." Journal of Photochemistry and Photobiology B: Biology 164 (2016): 264-270.

27) Tanaka, Junji, et al. "Maqui berry (Aristotelia chilensis) and the constituent delphinidin glycoside inhibit photoreceptor cell death induced by visible light." Food chemistry 139.1-4 (2013): 129-137.

28) Tsuyuki, Satoshi, et al. "Delphinidin Induces Autolysosome as well as Autophagosome Formation and Delphinidin-Induced Autophagy Exerts a Cell Protective Role." Journal of biochemical and molecular toxicology 26.11 (2012): 445-453.

29) Vergara, Daniela, et al. "The intake of maqui (Aristotelia chilensis) berry extract normalizes $H_2O_2$ and IL-6 concentrations in exhaled breath condensate from healthy smokers-an explorative study." Nutrition journal 14.1 (2015): 27.

30) Watson, Ronald R., and F. Schönlau. "Nutraceutical and antioxidant effects of a delphinidinrich maqui berry extract Delphinol®: a review." Minerva Cardioangiol 63.2 Suppl (2015): 1-12.

31) Xin, J. I. N., et al. "Delphinidin-3-glucoside inhibits Ox-LDL-induced injury in vascular endothelial cells: roles of membrane potential and MAPK phosphorylation [J]." Acta Academiae Medicinae Militaris Tertiae 19 (2009).

32) Yang, Yan, et al. "Plant food delphinidin-3-glucoside significantly inhibits platelet activation and thrombosis: novel protective roles against cardiovascular diseases." PloS one 7.5 (2012): e37323.

## CHAPTER 14 - ECKLONIA CAVA

1) Ahn, Byul-Nim, et al. "Dioxinodehydroeckol enhances the differentiation of osteoblasts by regulating the expression of Phospho-Smad1/5/8." Marine drugs 14.9 (2016): 168.

2) Ahn, Gin-Nae, et al. "Antioxidant activities of phlorotannins purified from Ecklonia cava on free radical scavenging using ESR and $H_2O_2$-mediated DNA damage." European Food Research and Technology 226.1-2 (2007): 71-79.

3) Athukorala, Yasantha, and You-Jin Jeon. "Screening for angiotensin 1-converting enzyme inhibitory activity of Ecklonia cava." Preventive nutrition and food science 10.2 (2005): 134-139.

4) Athukorala, Yasantha, Kil-Nam Kim, and You-Jin Jeon. "Antiproliferative and antioxidant properties of an enzymatic hydrolysate from brown alga, Ecklonia cava." Food and chemical toxicology 44.7 (2006): 1065-1074.

5) Bak, S. S., et al. "Ecklonia cava promotes hair growth." Clinical and experimental dermatology 38.8 (2013): 904-910.

6) Chang, Mun Young, et al. "Protective effect of a purified polyphenolic extract from Ecklonia cava against noise-induced hearing loss: Prevention of temporary threshold shift." International journal of pediatric otorhinolaryngology 87 (2016): 178-184.

7) Cho, Suengmok, et al. "Phlorotannins of the edible brown seaweed Ecklonia cava Kjellman induce sleep via positive allosteric modulation of gamma-aminobutyric acid type A– benzodiazepine receptor: A novel neurological activity of seaweed polyphenols." Food chemistry 132.3 (2012): 1133-1142.

8) Cho, Suengmok, et al. "Depressive effects on the central nervous system and underlying mechanism of the enzymatic extract and its phlorotannin-rich fraction from Ecklonia cava edible brown seaweed." Bioscience, biotechnology, and biochemistry 76.1 (2012): 163-168.

9) Cho, Su-Hyeon, et al. "Eckol from Ecklonia cava ameliorates TNF-α/IFN-γ-induced inflammatory responses via regulating MAPKs and NF-κB signaling

*pathway in HaCaT cells." International Immunopharmacology 82 (2020): 106146.*

10) Choi, Eun-Kyung, et al. "Clinical trial of the hypolipidemic effects of a brown alga Ecklonia cava extract in patients with hypercholesterolemia." *International Journal of Pharmacology 11.7 (2015): 798-805.*

11) Choi, Hye-Jin, et al. "Suppression of NF-κB by Dieckol extracted from Ecklonia cava negatively regulates LPS induction of inducible nitric oxide synthase gene." *Applied biochemistry and biotechnology 173.4 (2014): 957-967.*

12) Choi, Hyeon-Son, et al. "Dieckol, a major phlorotannin in Ecklonia cava, suppresses lipid accumulation in the adipocytes of high-fat diet-fed zebrafish and mice: Inhibition of early adipogenesis via cell-cycle arrest and AMPKα activation." *Molecular nutrition & food research 59.8 (2015): 1458-1471.*

13) Eo, Hyeyoon, et al. "Brown Alga Ecklonia cava polyphenol extract ameliorates hepatic lipogenesis, oxidative stress, and inflammation by activation of AMPK and SIRT1 in high-fat diet-induced obese mice." *Journal of agricultural and food chemistry 63.1 (2015): 349-359.*

14) Han, Eui Jeong, et al. "Eckol from Ecklonia cava Suppresses Immunoglobulin E-mediated Mast Cell Activation and Passive Cutaneous Anaphylaxis in Mice (Running Title: Anti-Allergic Activity of Ecklonia cava)." *Nutrients 12.5 (2020): 1361.*

15) Heo, Soo-Jin, et al. "Effect of phlorotannins isolated from Ecklonia cava on melanogenesis and their protective effect against photo-oxidative stress induced by UV-B radiation." *Toxicology in vitro 23.6 (2009): 1123-1130.*

16) Heo, Soo-Jin, et al. "Antioxidant activity of enzymatic extracts from a brown seaweed Ecklonia cava by electron spin resonance spectrometry and comet assay." *European Food Research and Technology 221.1-2 (2005): 41-47.*

17) Hwang, Hye Jeong. "Skin elasticity and sea polyphenols." *Seanol Sci. Centre Rev 1.110 (2010): 17.*

18) Jeon, Hui-Jeon, et al. "Seapolynol extracted from Ecklonia cava inhibits adipocyte differentiation in vitro and decreases fat accumulation in vivo." *Molecules 20.12 (2015): 21715-21731.*

19) Jeon, H. J., S. H. Kim, and B. Y. Lee. "Oral Glucose and Maltose Tolerance Test and Inhibition Effect of alpha-Glucosidase of Ecklonia cava Extract (Seapolynol) and Dieckol." *Journal of The Korean Society of Food Science*

*and Nutrition (2018).*

20) Jesumani, Valentina, et al. "Potential use of seaweed bioactive compounds in skincare—A review." *Marine drugs* 17.12 (2019): 688.

21) Kang, Changkeun, et al. "Brown alga Ecklonia cava attenuates type 1 diabetes by activating AMPK and Akt signaling pathways." *Food and chemical toxicology* 48.2 (2010): 509-516.

22) Kang, Jung-Il, et al. "Effect of Dieckol, a component of Ecklonia cava, on the promotion of hair growth." *International journal of molecular sciences* 13.5 (2012): 6407-6423.

23) Kang, Kyoung Ah, et al. "Eckol isolated from Ecklonia cava attenuates oxidative stress induced cell damage in lung fibroblast cells." *FEBS letters* 579.28 (2005): 6295-6304.

24) Kang, Sung-Myung, et al. "Molecular docking studies of a phlorotannin, dieckol isolated from Ecklonia cava with tyrosinase inhibitory activity." *Bioorganic & medicinal chemistry* 20.1 (2012): 311-316.

25) Kang, Sung-Myung, et al. "Isolation and identification of new compound, 2, 7 ''- phloroglucinol-6, 6'-bieckol from brown algae, Ecklonia cava and its antioxidant effect." *Journal of functional foods* 4.1 (2012): 158-166.

26) Kang, Sung-Myung, et al. "Neuroprotective effects of phlorotannins isolated from a brown alga, Ecklonia cava, against H2O2-induced oxidative stress in murine hippocampal HT22 cells." *Environmental Toxicology and Pharmacology* 34.1 (2012): 96-105.

27) Kim, Haejin, et al. "Evaluation of inhibitory effect of phlorotannins from Ecklonia cava on triglyceride accumulation in adipocyte." *Journal of agricultural and food chemistry* 61.36 (2013): 8541-8547.

28) Kim, In-Hye, et al. "Anti-obesity effects of pectinase and cellulase enzyme-treated Ecklonia cava extract in high-fat diet-fed C57BL/6N mice." *International journal of molecular medicine* 41.2 (2018): 924-934.

29) Kim, Hayeon, et al. "The Potential Application of Ecklonia cava Extract in Scalp Protection." *Cosmetics* 7.1 (2020): 9.

30) Kim, Hye Kyung. "Ecklonia cava inhibits glucose absorption and stimulates insulin secretion in streptozotocin-induced diabetic mice." *Evidence-Based Complementary and Alternative Medicine* 2012 (2012).

31) Kim, Hye Kyung. "Role of ERK/MAPK signalling pathway in anti-inflammatory effects of Ecklonia cava in activated human mast cell line-1 cells." Asian Pacific journal of tropical medicine 7.9 (2014): 703-708.

32) Kim, Ki Cheon, et al. "Up-regulation of Nrf2-mediated heme oxygenase-1 expression by eckol, a phlorotannin compound, through activation of Erk and PI3K/Akt." The international journal of biochemistry & cell biology 42.2 (2010): 297-305.

33) Kim, Ki Cheon, et al. "Triphlorethol-A from Ecklonia cava up-regulates the oxidant sensitive 8-oxoguanine DNA glycosylase 1." Marine drugs 12.11 (2014): 5357-5371.

34) Kim, Koth Bong Woo Ri, et al. "Lipase inhibitory activity of ethyl acetate fraction from Ecklonia cava extracts." Biotechnology and bioprocess engineering 17.4 (2012): 739-745.

35) Kim, Mi-Ja, and Hye Kyung Kim. "Insulinotrophic and hypolipidemic effects of Ecklonia cava in streptozotocin–induced diabetic mice." Asian Pacific journal of tropical medicine 5.5 (2012): 374-379.

36) Kim, Moon-Moo, et al. "Phlorotannins in Ecklonia cava extract inhibit matrix metalloproteinase activity." Life Sciences 79.15 (2006): 1436-1443.

37) Kim, Se-Kwon, and Chang-Suk Kong. "Anti-adipogenic effect of dioxinodehydroeckol via AMPK activation in 3T3-L1 adipocytes." Chemico-biological interactions 186.1 (2010): 24-29.

38) Kim, Seul-Young, et al. "The effects of anti-obesity on enzyme-treated Ecklonia cava extracts." Korean Journal of Fisheries and Aquatic Sciences 47.4 (2014): 363-369.

39) Kim, Seonyoung, et al. "Ecklonia cava extract containing dieckol suppresses RANKLinduced osteoclastogenesis via MAP kinase/NF-κB pathway inhibition and heme oxygenase-1 induction." J. Microbiol. Biotechnol 29.1 (2019): 11-20.

40) Kim, Tae Hoon, and Jong-Sup Bae. "Ecklonia cava extracts inhibit lipopolysaccharide induced inflammatory responses in human endothelial cells." Food and Chemical Toxicology 48.6 (2010): 1682-1687.

41) Kim, Yunsoo, et al. "Comparison of postprandial hypoglycemic effect of Ecklonia cava extract in healthy subjects using three different challenge models." The FASEB Journal 31.1_supplement (2017): lb216-lb216.

42) Ko, Seok-Chun, et al. "Dieckol, a phlorotannin isolated from a brown seaweed, Ecklonia cava, inhibits adipogenesis through AMP-activated protein kinase (AMPK) activation in 3T3-L1 preadipocytes." Environmental toxicology and pharmacology 36.3 (2013): 1253-1260.

43) Kong, Chang-Suk, et al. "1-(3', 5'-dihydroxyphenoxy)-7-(2 ", 4 ", 6-trihydroxyphenoxy)-2, 4, 9-trihydroxydibenzo-1, 4-dioxin Inhibits Adipocyte Differentiation of 3T3-L1 Fibroblasts." Marine biotechnology 12.3 (2010): 299-307.

44) Kong, Chang-Suk, Haejin Kim, and Youngwan Seo. "Edible Brown Alga E cklonia cava Derived Phlorotannin-Induced Anti-Adipogenic Activity in Vitro." Journal of Food Biochemistry 39.1 (2015): 1-10.

45) Le, Quang-To, et al. "Inhibitory effects of polyphenols isolated from marine alga Ecklonia cava on histamine release." Process Biochemistry 44.2 (2009): 168-176.

46) Lee, Dong Hyeon, et al. "Effects of Ecklonia cava polyphenol in individuals with hypercholesterolemia: a pilot study." Journal of medicinal food 15.11 (2012): 1038-1044.

47) Lee, Hyo Jin, Oran Kwon, and Ji Yeon Kim. "Supplementation of a polyphenol extract from Ecklonia cava reduces body fat, oxidative and inflammatory stress in overweight healthy subjects with abdominal obesity: A randomized, placebo-controlled, double-blind trial." Journal of Functional Foods 46 (2018): 356-364.

48) Lee, Hyun-Ah, Ji-Hyeok Lee, and Ji-Sook Han. "A phlorotannin constituent of Ecklonia cava alleviates postprandial hyperglycemia in diabetic mice." Pharmaceutical biology 55.1 (2017): 1149-1154.

49) Lee, Sang-Hoon, et al. "α-Glucosidase and α-amylase inhibitory activities of phloroglucinal derivatives from edible marine brown alga, Ecklonia cava." Journal of the Science of Food and Agriculture 89.9 (2009): 1552-1558.

50) Lee, Seung-Hong, et al. "Bioactive compounds extracted from Gamtae (Ecklonia cava) by using enzymatic hydrolysis, a potent α-glucosidase and α-amylase inhibitor, alleviates postprandial hyperglycemia in diabetic mice." Food Science and Biotechnology 21.4 (2012): 1149-1155.

51) Lee, Seung-Hong, et al. "Molecular characteristics and anti-inflammatory activity of the fucoidan extracted from Ecklonia cava." Carbohydrate polymers 89.2 (2012): 599-606.

52) Lee, Seung-Hong, et al. "Effects of brown alga, Ecklonia cava on glucose and lipid metabolism in C57BL/KsJ-db/db mice, a model of type 2 diabetes mellitus." Food and chemical toxicology 50.3-4 (2012): 575-582.

53) Lee, Seung-Hong, et al. "Cellular activities and docking studies of eckol isolated from Ecklonia cava (Laminariales, Phaeophyceae) as potential tyrosinase inhibitor." Algae 30.2 (2015): 163.

54) Lee, Seung-Hong, and You-Jin Jeon. "Efficacy and safety of a dieckol-rich extract (AGdieckol) of brown algae, Ecklonia cava, in pre-diabetic individuals: A double-blind, randomized, placebo-controlled clinical trial." Food & function 6.3 (2015): 853-858.

55) Li, Yong, et al. "Chemical components and its antioxidant properties in vitro: an edible marine brown alga, Ecklonia cava." Bioorganic & Medicinal Chemistry 17.5 (2009): 1963-1973.

56) Moon, Hee Jung, et al. "Fucoidan inhibits UVB-induced MMP-1 promoter expression and down regulation of type I procollagen synthesis in human skin fibroblasts." European Journal of Dermatology 19.2 (2009): 129-134.

57) Oh, Jung Hwan, et al. "Phlorofucofuroeckol A from Edible Brown Alga Ecklonia Cava Enhances Osteoblastogenesis in Bone Marrow-Derived Human Mesenchymal Stem Cells." Marine drugs 17.10 (2019): 543.

58) Park, Sae Rom, et al. "Inhibitory activity of minor phlorotannins from Ecklonia cava on α- glucosidase." Food chemistry 257 (2018): 128-134.

59) Shin, Hyeon-Cheol, et al. "Effects of 12-week oral supplementation of Ecklonia cava polyphenols on anthropometric and blood lipid parameters in overweight Korean individuals: a double-blind randomized clinical trial." Phytotherapy research 26.3 (2012): 363-368.

60) Shin, Hyoseung, et al. "Enhancement of human hair growth using Ecklonia cava polyphenols." Annals of dermatology 28.1 (2016): 15-21.

61) Song, Eu-Jin, et al. "Effect of Ecklonia cava water extracts on inhibition of IgE in food allergy mouse model." Journal of the Korean Society of Food Science and Nutrition 39.12 (2010): 1776-1782.

62) Sugiura, Shingo, et al. "Evaluation of Anti-glycation Activities of Phlorotannins in Human and Bovine Serum Albumin-glyceraldehyde Models." Natural Product Communications 13.8 (2018): 1934578X1801300820.

63) Vo, Thanh Sang, et al. "The suppressive activity of Fucofuroeckol-A derived from brown algal ecklonia stolonifera okamura on UVB-induced mast cell degranulation." Marine drugs 16.1 (2018): 1.

64) Wang, Lei, et al. "Dieckol, an Algae-Derived Phenolic Compound, Suppresses UVB-Induced Skin Damage in Human Dermal Fibroblasts and Its Underlying Mechanisms." Antioxidants 10.3 (2021): 352.

65) Wijesekara, Isuru, Na Young Yoon, and Se-Kwon Kim. "Phlorotannins from Ecklonia cava (Phaeophyceae): Biological activities and potential health benefits." Biofactors 36.6 (2010): 408-414.

66) Yang, Yeong-In, et al. "6, 6'-Bieckol, isolated from marine alga Ecklonia cava, suppressed LPS-induced nitric oxide and PGE2 production and inflammatory cytokine expression in macrophages: The inhibition of NFκB." International immunopharmacology 12.3 (2012): 510-517.

67) Yeo, A-Reum, et al. "Anti-hyperlipidemic effect of polyphenol extract (Seapolynol™) and dieckol isolated from Ecklonia cava in in vivo and in vitro models." Preventive nutrition and food science 17.1 (2012): 1.

68) Yoon, Ji-Young, Hojung Choi, and Hee-Sook Jun. "The effect of phloroglucinol, a component of Ecklonia cava extract, on hepatic glucose production." Marine drugs 15.4 (2017): 106.

69) You, Han-Nui, et al. "Phlorofucofuroeckol A isolated from Ecklonia cava alleviates postprandial hyperglycemia in diabetic mice." European journal of pharmacology 752 (2015): 92-96.

70) Yun, Jun-Won, et al. "Enzymatic extract from Ecklonia cava: Acute and subchronic oral toxicity and genotoxicity studies." Regulatory Toxicology and Pharmacology 92 (2018): 46-54.

71) Zhang, Chen, and Se-Kwon Kim. "Matrix metalloproteinase inhibitors (MMPIs) from marine natural products: the current situation and future prospects." Marine drugs 7.2 (2009): 71-84.

**CHAPTER 15** - ELLAGIC ACID

1) Aiyer, Harini S., et al. "Dietary berries and ellagic acid prevent oxidative DNA damage and modulate expression of DNA repair genes." International journal of molecular sciences 9.3 (2008): 327-341.

2) Amin, Mohamed M., and Mahmoud S. Arbid. "Estimation of Ellagic acid and/or Repaglinide Effects on Insulin Signaling, Oxidative Stress and Inflammatory Mediators of Liver, Pancreas, Adipose Tissue and Brain in Insulin Resistant/Type 2 Diabetic Rats." Applied Physiology, Nutrition, and Metabolism ja (2016).

3) Anwar, Shehwaz, and Hina Younus. "Antiglycating potential of ellagic acid against glucose and methylglyoxal-induced glycation of superoxide dismutase." Journal of Proteins & Proteomics 8.1 (2017).

4) Bae, Ji-Young, et al. "Dietary compound ellagic acid alleviates skin wrinkle and inflammation induced by UV-B irradiation." Experimental dermatology 19.8 (2010).

5) Baek, Sung Hee, and Keun Il Kim. "Epigenetic control of autophagy: nuclear events gain
more attention." Molecular cell 65.5 (2017): 781-785.

6) Chao, Che-yi, et al. "Anti-glycative and anti-inflammatory effects of caffeic acid and ellagic acid in kidney of diabetic mice." Molecular nutrition & food research 54.3 (2010): 388-395.

7) Chao, Pei-chun, Cheng-chin Hsu, and Mei-chin Yin. "Anti-inflammatory and anti-coagulatory activities of caffeic acid and ellagic acid in cardiac tissue of diabetic mice." Nutrition & metabolism 6.1 (2009): 33.

8) Choi, Yun Ho, and Guang Hai Yan. "Ellagic acid attenuates immunoglobulin E-mediated allergic response in mast cells." Biological and Pharmaceutical Bulletin 32.6 (2009): 1118-1121.

9) Číž, Milan, et al. "The Role of Dietary Phenolic Compounds in Epigenetic Modulation Involved in Inflammatory Processes." Antioxidants 9.8 (2020): 691.

10) Dahl, Amanda, et al. "Tolerance and efficacy of a product containing ellagic and salicylic acids in reducing hyperpigmentation and dark spots in comparison with 4% hydroquinone." Journal of drugs in dermatology: JDD 12.1 (2013): 52-58.

11) Espín, Juan Carlos, et al. "Biological significance of urolithins, the gut microbial ellagic acidderived metabolites: the evidence so far." Evidence-Based Complementary and Alternative
Medicine 2013 (2013).

12) Galano, Annia, Misaela Francisco Marquez, and Adriana Pérez-González.

*"Ellagic acid: an unusually versatile protector against oxidative stress."* Chemical research in toxicology 27.5 (2014): 904-918.

13) Goswami, Sumanta Kumar, et al. *"Efficacy of ellagic acid and sildenafil in diabetes-induced sexual dysfunction."* Pharmacognosy magazine 10.Suppl 3 (2014): S581.

14) Han, Dong Hoon, Min Jeon Lee, and Jeong Hee Kim. *"Antioxidant and apoptosis-inducing activities of ellagic acid."* Anticancer research 26.5A (2006): 3601-3606.

15) Kang, Inhae, et al. *"Improvements in Metabolic Health with Consumption of Ellagic Acid and Subsequent Conversion into Urolithins: Evidence and Mechanisms."* Advances in Nutrition: An International Review Journal 7.5 (2016): 961-972.

16) Kasai, Kouichi, et al. *"Effects of oral administration of ellagic acid-rich pomegranate extract on ultraviolet-induced pigmentation in the human skin."* Journal of nutritional science and vitaminology 52.5 (2006): 383-388.

17) Khodaei, Forouzan, et al. *"Ellagic acid improves muscle dysfunction in cuprizone-induced demyelinated mice via mitochondrial Sirt3 regulation."* Life sciences 237 (2019): 116954.

18) Larrosa, Mar, et al. *"Ellagitannins, ellagic acid and vascular health."* Molecular aspects of medicine 31.6 (2010): 513-539.

19) Lee, Ki Hoon, et al. *"Unripe Rubus coreanus Miquel Extract Containing Ellagic Acid Regulates AMPK, SREBP-2, HMGCR, and INSIG-1 Signaling and Cholesterol Metabolism In Vitro and In Vivo."* Nutrients 12.3 (2020): 610.

20) Lipińska, Lidia, Elżbieta Klewicka, and Michał Sójka. *"The structure, occurrence and biological activity of ellagitannins: a general review."* Acta Scientiarum Polonorum Technologia Alimentaria 13.3 (2014): 289-299.

21) Liu, Qing-Shan, et al. *"Ellagic acid improves endogenous neural stem cells proliferation and neurorestoration through Wnt/β-catenin signaling in vivo and in vitro."* Molecular Nutrition & Food Research 61.3 (2017): 1600587.

22) Liu, Ying, et al. *"Chronic administration of ellagic acid improved the cognition in middle-aged overweight men."* Applied Physiology, Nutrition, and Metabolism 43.3 (2018): 266-273.

23) Mo, Jiao, et al. *"Topical Anti-inflammatory Potential of Standardized*

*Pomegranate Rind Extract and Ellagic Acid in Contact Dermatitis."* Phytotherapy research 28.4 (2014): 629-632.

24) Mo, Jiao, et al. "Wound healing activities of standardized pomegranate rind extract and its major antioxidant ellagic acid in rat dermal wounds." Journal of natural medicines 68.2 (2014): 377-386.

25) Muthenna, Puppala, Chandrasekhar Akileshwari, and G. Bhanuprakash Reddy. "Ellagic acid, a new antiglycating agent: its inhibition of Nϵ-(carboxymethyl) lysine." Biochemical Journal 442.1 (2012): 221-230.

26) Raghu, G., et al. "Ellagic acid inhibits non-enzymatic glycation and prevents proteinuria in diabetic rats." Food & function 7.3 (2016): 1574-1583.

27) Rahnasto-Rilla, Minna, et al. "Effects of galloflavin and ellagic acid on sirtuin 6 and its antitumorigenic activities." Biomedicine & Pharmacotherapy 131 (2020): 110701.

28) Ríos, José-Luis, et al. "A pharmacological update of ellagic acid." Planta Med 84.15 (2018): 1068-1093.

29) Shimogaki, H., et al. "In vitro and in vivo evaluation of ellagic acid on melanogenesis inhibition." International journal of cosmetic science 22.4 (2000): 291-304.

30) Toney, Ashley Mulcahy, et al. "Urolithin A, a gut metabolite, improves insulin sensitivity through augmentation of mitochondrial function and biogenesis." Obesity 27.4 (2019): 612-620.

31) Velagapudi, Ravikanth, et al. "Induction of Autophagy and Activation of SIRT-1 Deacetylation Mechanisms Mediate Neuroprotection by the Pomegranate Metabolite Urolithin A in BV2 Microglia and Differentiated 3D Human Neural Progenitor Cells." Molecular nutrition & food research 63.10 (2019): 1801237.

32) Xu, Ziheng, and Daniel J. Klionsky. "The AMPK-SKP2-CARM1 axis links nutrient sensing to transcriptional and epigenetic regulation of autophagy." Annals of translational medicine 4.Suppl 1 (2016).

33) Yoshimura, Mineka, et al. "Inhibitory effect of an ellagic acid-rich pomegranate extract on tyrosinase activity and ultraviolet-induced pigmentation." Bioscience, biotechnology, and biochemistry 69.12 (2005): 2368-2373.

34) Zarfeshany, Aida, Sedigheh Asgary, and Shaghayegh Haghjoo Javanmard. "Potent health effects of pomegranate." Advanced biomedical research 3 (2014).

## CHAPTER 16 - FISETIN

1) Adhami, Vaqar M., Rahul K. Lall, and Hasan Mukhtar. "Fisetin, a dietary flavonoid and novel mTOR inhibitor for treatment and prevention of prostate cancer." (2012): 612-612.

2) Antika, Lucia Dwi, and Rita Marleta Dewi. "Pharmacological aspects of fisetin." Asian Pacific Journal of Tropical Biomedicine 11.1 (2021): 1.

3) Bhattacharjee, Snehasish, et al. "Exploring the interactions of the dietary plant flavonoids fisetin and naringenin with G-quadruplex and duplex DNA, showing contrasting binding behavior: spectroscopic and molecular modeling approaches." The Journal of Physical Chemistry B 120.34 (2016): 8942-8952.

4) Choi, Sik-Won, et al. "Fisetin inhibits osteoclast differentiation via downregulation of p38 and c-Fos-NFATc1 signaling pathways." Evidence-Based Complementary and Alternative Medicine 2012 (2012).

5) Colman, Ricki, et al. "Effect of Combined Dasatinib and Fisetin Treatment on Senescent Cell Clearance in Monkeys." Innovation in Aging 4.Suppl 1 (2020): 131.

6) Currais, Antonio, et al. "Fisetin reduces the impact of aging on behavior and physiology in the rapidly aging SAMP8 mouse." The Journals of Gerontology: Series A 73.3 (2017): 299-307.

7) Demchuk, Oleg, and Grzegorz Grynkiewicz Grynkiewicz. "New perspectives for fisetin." Frontiers in Chemistry 7 (2019): 697.

8) Farsad-Naeimi, Alireza, et al. "Effect of fisetin supplementation on inflammatory factors and matrix metalloproteinase enzymes in colorectal cancer patients." Food & function 9.4 (2018): 2025-2031.

9) Goh, Fera Y., et al. "Fisetin, a bioactive flavonol, attenuates allergic airway inflammation through negative regulation of NF-κB." European journal of pharmacology 679.1-3 (2012): 109-116.

10) Hambright, Sealy, et al. "The senolytic drug fisetin mitigates age-related bone density loss in the progeroid mouse model Zmpste24−/−." The FASEB

*Journal 34.S1 (2020): 1-1.*

11) Higa, Shinji, et al. "Fisetin, a flavonol, inhibits TH2-type cytokine production by activated human basophils." *Journal of Allergy and Clinical Immunology* 111.6 (2003): 1299-1306.

12) Jin, Taewon, et al. "Fisetin up-regulates the expression of adiponectin in 3T3-L1 adipocytes via the activation of silent mating type information regulation 2 homologue 1 (SIRT1)- deacetylase and peroxisome proliferator-activated receptors (PPARs)." *Journal of agricultural and food chemistry* 62.43 (2014): 10468-10474.

13) Jo, Woo-Ri, and Hye-Jin Park. "Antiallergic effect of fisetin on IgE-mediated mast cell activation in vitro and on passive cutaneous anaphylaxis (PCA)." *The Journal of nutritional biochemistry* 48 (2017): 103-111.

14) Jung, Chang Hwa, et al. "Fisetin regulates obesity by targeting mTORC1 signaling." *The Journal of nutritional biochemistry* 24.8 (2013): 1547-1554.

15) Kang, Kyoung Ah, et al. "Fisetin attenuates hydrogen peroxide-induced cell damage by scavenging reactive oxygen species and activating protective functions of cellular glutathione system." *In Vitro Cellular & Developmental Biology-Animal* 50.1 (2014): 66-74.

16) Kim, Hye Joo, Seong Hwan Kim, and Jung-Mi Yun. "Fisetin inhibits hyperglycemia-induced proinflammatory cytokine production by epigenetic mechanisms." *Evidence-Based Complementary and Alternative Medicine* 2012 (2012).

17) Kim, Sang Chon, et al. "Fisetin induces Sirt1 expression while inhibiting early adipogenesis in 3T3-L1 cells." *Biochemical and biophysical research communications* 467.4 (2015): 638-644.

18) Krasieva, Tatiana B., et al. "Cell and brain tissue imaging of the flavonoid fisetin using labelfree two-photon microscopy." *Neurochemistry international* 89 (2015): 243-248.

19) Léotoing, Laurent, et al. "The flavonoid fisetin promotes osteoblasts differentiation through Runx2 transcriptional activity." *Molecular nutrition & food research* 58.6 (2014): 1239-1248.

20) Léotoing, Laurent, et al. "The polyphenol fisetin protects bone by repressing NF-κB and MKP-1-dependent signaling pathways in osteoclasts." *PLoS One* 8.7 (2013): e68388.

21) Liou, Chian-Jiun, et al. "Fisetin Protects Against Hepatic Steatosis Through Regulation of the Sirt1/AMPK and Fatty Acid β-Oxidation Signaling Pathway in High-Fat Diet-Induced Obese Mice." Cellular Physiology and Biochemistry 49.5 (2018): 1870-1884.

22) Maher, Pamela, Tatsuhiro Akaishi, and Kazuho Abe. "Flavonoid fisetin promotes ERKdependent long-term potentiation and enhances memory." Proceedings of the National Academy of Sciences 103.44 (2006): 16568-16573.

23) Maher, Pamela. "A comparison of the neurotrophic activities of the flavonoid fisetin and some of its derivatives." Free Radical Research 40.10 (2006): 1105-1111.

24) Maher, Pamela. "Modulation of multiple pathways involved in the maintenance of neuronal function during aging by fisetin." Genes & nutrition 4.4 (2009): 297.

25) Maher, Pamela, et al. "Fisetin lowers methylglyoxal dependent protein glycation and limits the complications of diabetes." PLoS One 6.6 (2011): e21226.

26) Maher, Pamela. "Fisetin acts on multiple pathways to reduce the impact of age and disease on CNS function." Frontiers in bioscience (Scholar edition) 7 (2015): 58.

27) Maher, Pamela. "Preventing and treating neurological disorders with the flavonol fisetin." Brain Plasticity Preprint (2020): 1-12.

28) Naeimi, Alireza Farsad, and Mohammad Alizadeh. "Antioxidant properties of the flavonoid fisetin: An updated review of in vivo and in vitro studies." Trends in food science & technology 70 (2017): 34-44.

29) Pal, Harish Chandra, et al. "Fisetin inhibits UVB-induced cutaneous inflammation and activation of PI3K/AKT/NFκB signaling pathways in SKH-1 hairless mice." Photochemistry and photobiology 91.1 (2015): 225-234.

30) Park, Hyo-Hyun, et al. "Anti-inflammatory activity of fisetin in human mast cells (HMC-1)." Pharmacological research 55.1 (2007): 31-37.

31) Park, Jun Hyoung, et al. "Fisetin inhibits matrix metalloproteinases and reduces tumor cell invasiveness and endothelial cell tube formation." Nutrition and cancer 65.8 (2013): 1192-1199.

32) Prasath, Gopalan Sriram, Chinnakrishnan Shanmuga Sundaram, and

Sorimuthu Pillai Subramanian. "Fisetin averts oxidative stress in pancreatic tissues of streptozotocin-induced diabetic rats." Endocrine 44.2 (2013): 359-368.

33) Prasath, Gopalan Sriram, and Sorimuthu Pillai Subramanian. "Antihyperlipidemic Effect of Fisetin, a Bioflavonoid of Strawberries, Studied in Streptozotocin-Induced Diabetic Rats." Journal of biochemical and molecular toxicology 28.10 (2014): 442-449.

34) Sahu, Bidya Dhar, Jerald Mahesh Kumar, and Ramakrishna Sistla. "Fisetin, a dietary flavonoid, ameliorates experimental colitis in mice: Relevance of NF-κB signaling." The Journal of nutritional biochemistry 28 (2016): 171-182.

35) Sandireddy, Reddemma, et al. "Fisetin imparts neuroprotection in experimental diabetic neuropathy by modulating Nrf2 and NF-κB pathways." Cellular and molecular neurobiology 36.6 (2016): 883-892.

36) Sakai, Eiko, et al. "Fisetin inhibits osteoclastogenesis through prevention of RANKL-induced ROS production by Nrf2-mediated up-regulation of phase II antioxidant enzymes." Journal of pharmacological sciences 121.4 (2013): 288-298.

37) Sengupta, Bidisha, et al. "Prospect of bioflavonoid fisetin as a quadruplex DNA ligand: a biophysical approach." PLoS One 8.6 (2013): e65383.

38) Shon, Myung-Soo, et al. "Potential of fisetin as a nutri-cosmetics material through evaluating anti-oxidant and anti-adipogenic activities." Asian Journal of Beauty and Cosmetology 14.1 (2016): 6-17.

39) Singh, Sandeep, et al. "Fisetin as a caloric restriction mimetic protects rat brain against aging induced oxidative stress, apoptosis and neurodegeneration." Life sciences 193 (2018): 171-179.

40) Sung, Bokyung, Manoj K. Pandey, and Bharat B. Aggarwal. "Fisetin, an inhibitor of cyclindependent kinase 6, down-regulates nuclear factor-κB-regulated cell proliferation, antiapoptotic and metastatic gene products through the suppression of TAK-1 and receptor-interacting protein-regulated IκBα kinase activation." Molecular pharmacology 71.6 (2007): 1703-1714.

41) Wang, Limin, et al. "Fisetin Prolongs Therapy Window of Brain Ischemic Stroke Using Tissue Plasminogen Activator: A Double-Blind Randomized Placebo-Controlled Clinical Trial." Clinical and Applied Thrombosis/Hemostasis 25 (2019): 1076029619871359.

42) Wu, Mei-Yao, Shih-Kai Hung, and Shu-Ling Fu. "Immunosuppressive effects of fisetin in ovalbumin-induced asthma through inhibition of NF-κB activity." Journal of agricultural and food chemistry 59.19 (2011): 10496-10504.

43) Yousefzadeh, Matthew J., et al. "Fisetin is a senotherapeutic that extends health and lifespan." EBioMedicine 36 (2018): 18-28.

44) Zhang, Jiaqi, et al. "Fisetin Prevents Acetaminophen-Induced Liver Injury by Promoting Autophagy." Frontiers in Pharmacology 11 (2020): 162.

45) Zhen, Linlin, et al. "The antidepressant-like effect of fisetin involves the serotonergic and noradrenergic system." Behavioural brain research 228.2 (2012): 359-366.

46) Zheng, Wenhao, et al. "Fisetin inhibits IL-1β-induced inflammatory response in human osteoarthritis chondrocytes through activating SIRT1 and attenuates the progression of osteoarthritis in mice." International immunopharmacology 45 (2017): 135-147.

47) Zhu, Yi, et al. "New agents that target senescent cells: the flavone, fisetin, and the BCL-XL inhibitors, A1331852 and A1155463." Aging (Albany NY) 9.3 (2017): 955.

## CHAPTER 17 - GANODERMA LUCIDUM

1) Ahmad, Faruque, et al. "Ganoderma lucidum: A potential source to surmount viral infections through β-glucans immunomodulatory and triterpenoids antiviral properties." International Journal of Biological Macromolecules (2021).

2) Ajith, T. A., et al. "Effect of Ganoderma lucidum on the activities of mitochondrial dehydrogenases and complex I and II of electron transport chain in the brain of aged rats." Experimental gerontology 44.3 (2009): 219-223.

3) Bhardwaj, Anuja, et al. "Lingzhi or Reishi Medicinal Mushroom, Ganoderma lucidum (Agaricomycetes), Mycelium Aqueous Extract Modulates High-Altitude–Induced Stress." International Journal of Medicinal Mushrooms 21.5 (2019).

4) Bhardwaj, Anuja, and Kshipra Misra. "Ganoderma sp.: The Royal Mushroom for High-Altitude Ailments." Management of High Altitude Pathophysiology.

*Academic Press, 2018. 115-152.*

5) Bhardwaj, Neha, Priya Katyal, and Anil K Sharma. "Suppression of inflammatory and allergic responses by pharmacologically potent fungus Ganoderma lucidum." *Recent patents on inflammation & allergy drug discovery* 8.2 (2014): 104-117.

6) Boh, Bojana, et al. "Ganoderma lucidum and its pharmaceutically active compounds." *Biotechnology annual review* 13 (2007): 265-301.

7) Chang, Chih-Jung, et al. "Ganoderma lucidum reduces obesity in mice by modulating the composition of the gut microbiota." *Nature communications* 6.1 (2015): 1-19.

8) Chen, Yang, et al. "Effects of Ganoderma lucidum polysaccharides on advanced glycation end products and receptor of aorta pectoralis in T2DM rats." *Zhongguo Zhong yao za zhi= Zhongguo zhongyao zazhi= China journal of Chinese materia medica* 36.5 (2011): 624-627.

9) Cheng, Poh-Guat, et al. "Polysaccharides-rich extract of Ganoderma lucidum (MA Curtis: Fr.) P. Karst accelerates wound healing in streptozotocin-induced diabetic rats." *Evidence-Based Complementary and Alternative Medicine* 2013 (2013).

10) Diling, Chen, et al. "Metabolic regulation of Ganoderma lucidum extracts in high sugar and fat diet-induced obese mice by regulating the gut-brain axis." *Journal of Functional Foods* 65 (2020): 103639.

11) Dudhgaonkar, Shailesh, Anita Thyagarajan, and Daniel Sliva. "Suppression of the inflammatory response by triterpenes isolated from the mushroom Ganoderma lucidum." *International immunopharmacology* 9.11 (2009): 1272-1280.

12) Gao, Yihuai, et al. "A phase I/II study of Ling Zhi mushroom Ganoderma lucidum (W. Curt.: Fr.) Lloyd (Aphyllophoromycetideae) extract in patients with type II diabetes mellitus." *International Journal of Medicinal Mushrooms* 6.1 (2004).

13) Gupta, Asheesh, et al. "Wound healing activity of an aqueous extract of the Lingzhi or Reishi medicinal mushroom Ganoderma lucidum (Higher Basidiomycetes)." *International journal of medicinal mushrooms* 16.4 (2014).

14) Henao, Sandra Lorena Duque, et al. "Randomized clinical trial for the evaluation of immune modulation by yogurt enriched with β-glucans from

*lingzhi or reishi medicinal mushroom, Ganoderma lucidum (Agaricomycetes), in children from Medellin, Colombia." International journal of medicinal mushrooms 20.8 (2018).*

15) Huang, Qing, et al. "GPP (Composition of Ganoderma lucidum polysaccharides and Polyporus umbellatus poly-saccharides) enhances innate immune function in mice." Nutrients 11.7 (2019): 1480.

16) Jia, Jie, et al. "Evaluation of in vivo antioxidant activities of Ganoderma lucidum polysaccharides in STZ-diabetic rats." Food Chemistry 115.1 (2009): 32-36.

17) Jung, Kyung Hee, et al. "Ganoderma lucidum extract stimulates glucose uptake in L6 rat skeletal muscle cells." Acta Biochimica Polonica 53.3 (2006): 597-601.

18) Koganti, Praveen, et al. "Role of Hydroalcoholic Extract of Lingzhi or Reishi Medicinal Mushroom, Ganoderma lucidum (Agaricomycetes), in Facilitating Cellular Acclimatization in a Low-Oxygen Microenvironment." International journal of medicinal mushrooms 20.5 (2018).

19) Kohda, Hiroshi, et al. "The biologically active constituents of Ganoderma lucidum (Fr.) Karst. Histamine release-inhibitory triterpenes." Chemical and Pharmaceutical Bulletin 33.4 (1985): 1367-1374.

20) Kubota, Atsuhito, et al. "Reishi mushroom Ganoderma lucidum Modulates IgA production and alpha-defensin expression in the rat small intestine." Journal of ethnopharmacology 214 (2018): 240-243.

21) Lai, Guoxiao, et al. "Alcohol extracts from Ganoderma lucidum delay the progress of Alzheimer's disease by regulating DNA methylation in rodents." Frontiers in Pharmacology 10 (2019): 272.

22) Lee, Jong-Min, et al. "Inhibition of lipid peroxidation and oxidative DNA damage by Ganoderma lucidum." Phytotherapy Research: An International Journal Devoted to Pharmacological and Toxicological Evaluation of Natural Product Derivatives 15.3 (2001): 245-249.

23) Li, Koukou, et al. "Effects of Ganoderma lucidum polysaccharides on chronic pancreatitis and intestinal microbiota in mice." International journal of biological macromolecules 93 (2016): 904-912.

24) Lin, Zhi-Bin. "Cellular and molecular mechanisms of immuno-modulation by Ganoderma lucidum." Journal of Pharmacological Sciences 99.2 (2005):

144-153.

25) Liu, Tao, et al. "Effects of sporoderm-broken spores of Ganoderma lucidum on growth performance, antioxidant function and immune response of broilers." Animal Nutrition 6.1 (2020): 39-46.

26) Ma, Haou-Tzong, Jung-Feng Hsieh, and Shui-Tein Chen. "Anti-diabetic effects of Ganoderma lucidum." Phytochemistry 114 (2015): 109-113.

27) Meng, Guoliang, et al. "Attenuating effects of Ganoderma lucidum polysaccharides on myocardial collagen cross-linking relates to advanced glycation end product and antioxidant enzymes in high-fat-diet and streptozotocin-induced diabetic rats." Carbohydrate Polymers 84.1 (2011): 180-185.

28) Pillai, Thulasi G., C. K. K. Nair, and Devi Uma. "Post radiation protection and enhancement of DNA repair of beta glucan isolated from Ganoderma lucidum." Proceedings of the thirty eighth annual conference of Environmental Mutagen Society of India and national conference on current perspectives on environmental mutagenesis and human health: abstracts. 2013.

29) Pillai, Thulasi G., C. K. K. Nair, and K. K. Janardhanan. "Enhancement of repair of radiation induced DNA strand breaks in human cells by Ganoderma mushroom polysaccharides." Food Chemistry 119.3 (2010): 1040-1043.

30) Rahman, Mohammad Azizur, et al. "Validation of Ganoderma lucidum against hypercholesterolemia and Alzheimer's disease." European Journal of Biological Research 10.4 (2020): 314-325.

31) Sarker, Md Moklesur Rahman. "Antihyperglycemic, insulin-sensitivity and antihyperlipidemic potential of Ganoderma lucidum, a dietary mushroom, on alloxan-and glucocorticoid-induced diabetic Long-Evans rats." Functional Foods in Health and Disease 5.12 (2015): 450-466.

32) Sharma, Purva, Rajkumar Tulsawani, and Usha Agrawal. "Pharmacological effects of Ganoderma lucidum extract against high-altitude stressors and its subchronic toxicity assessment." Journal of Food Biochemistry 43.12 (2019): e13081.

33) Siwulski, Marek, et al. "Ganoderma lucidum (Curt.: Fr.) Karst.–health-promoting properties. A review." Herba Polonica 61.3 (2015): 105-118.

34) Smina, T. P., et al. "Antioxidant activity and toxicity profile of total triterpenes isolated from Ganoderma lucidum (Fr.) P. Karst occurring in South

*India." Environmental toxicology and pharmacology 32.3 (2011): 438-446.*

35) Sudheesh, N. P., T. A. Ajith, and K. K. Janardhanan. "Ganoderma lucidum (Fr.) P. Karst enhances activities of heart mitochondrial enzymes and respiratory chain complexes in the aged rat." *Biogerontology 10.5 (2009): 627-636.*

36) Sun, Xin-zhi, et al. "Neuroprotective effects of ganoderma lucidum polysaccharides against oxidative stress-induced neuronal apoptosis." *Neural regeneration research 12.6 (2017): 953.*

37) Tasaka, K., et al. "Anti-allergic constituents in the culture medium of Ganoderma lucidum.(II) The inhibitory effect of cyclooctasulfur on histamine release." *Agents and actions 23.3-4 (1988): 157-160.*

38) Thyagarajan-Sahu, Anita, Brandon Lane, and Daniel Sliva. "ReishiMax, mushroom based dietary supplement, inhibits adipocyte differentiation, stimulates glucose uptake and activates AMPK." *BMC complementary and alternative medicine 11.1 (2011): 74.*

39) Vitak, Taras Y., et al. "The effect of the medicinal mushrooms Agaricus brasiliensis and Ganoderma lucidum (higher Basidiomycetes) on the erythron system in normal and streptozotocin-induced diabetic rats." *International journal of medicinal mushrooms 17.3 (2015).*

40) Vu Thi, et al. "The anti-oxidation and anti-aging effects of Ganoderma lucidum in Caenorhabditis elegans." *Experimental gerontology 117 (2019): 99-105.*

41) Wachtel-Galor, Sissi, et al. "Ganoderma lucidum (Lingzhi or Reishi)." *Herbal Medicine: Biomolecular and Clinical Aspects. 2nd edition. CRC Press/ Taylor & Francis, 2011.*

42) Wachtel-Galor, Sissi, et al. "Antioxidant Power and DNA Repair Effects of Lingzhi or Reishi Medicinal Mushroom, Ganoderma lucidum (W. Curt.: Fr.) P. Karst.(Aphyllophoromycetideae), in Human Acute Post-ingestion Study." *International Journal of Medicinal Mushrooms 12.4 (2010).*

43) Wang, Jue, et al. "Emerging roles of Ganoderma Lucidum in anti-aging." *Aging and disease 8.6 (2017): 691.*

44) Wang, Xin, and Zhibin Lin. "Immunomodulating effect of Ganoderma (lingzhi) and possible mechanism." *Ganoderma and Health (2019): 1-37.*

45) Xiao, Chun, et al. "Hypoglycemic effects of Ganoderma lucidum polysaccharides in type 2 diabetic mice." Archives of pharmacal research 35.10 (2012): 1793-1801.

46) Xiao, Chun, et al. "Antidiabetic activity of Ganoderma lucidum polysaccharides F31 downregulated hepatic glucose regulatory enzymes in diabetic mice." Journal of ethnopharmacology 196 (2017): 47-57.

47) Yang, Zhou, et al. "Modulation of energy metabolism and mitochondrial biogenesis by a novel proteoglycan from Ganoderma lucidum." RSC advances 9.5 (2019): 2591-2598.

48) Zhang, Hui-Na, and Zhi-Bin Lin. "Hypoglycemic effect of Ganoderma lucidum polysaccharides." Acta Pharmacologica Sinica 25.2 (2004): 191-195.

49) Zhang, Wang-xin, et al. "Effect of ganoderma on the SOD and MDA in the brain and serum after cerebral ischemic reperfusion in gerbils [J]." Journal of Taishan Medical College 1 (2010): 003.

50) Zhu, Min, et al. "Triterpene antioxidants from Ganoderma lucidum." Phytotherapy Research: An International Journal Devoted to Pharmacological and Toxicological Evaluation of Natural Product Derivatives 13.6 (1999): 529-531.

**CHAPTER 18** - HYALURONIC ACID

1) Abatangelo, G., et al. "Hyaluronic acid: redefining its role." Cells 9.7 (2020): 1743.

2) Allegra, Luigi, Sabrina Della Patrona, and Giuseppe Petrigni. "Hyaluronic Acid." Heparin-A Century of Progress. Springer, Berlin, Heidelberg, 2012. 385-401.

3) Chistyakov, Dmitry V., et al. "High and low molecular weight hyaluronic acid differentially influences Oxylipins synthesis in course of Neuroinflammation." International journal of molecular sciences 20.16 (2019): 3894.

4) Dahiya, Parveen, and Reet Kamal. "Hyaluronic acid: a boon in periodontal therapy." North American journal of medical sciences 5.5 (2013): 309.

5) Fallacara, Arianna, et al. "Hyaluronic acid in the third millennium." Polymers 10.7 (2018): 701.

6) Göllner, Imke, et al. "Ingestion of an oral hyaluronan solution improves skin hydration, wrinkle reduction, elasticity, and skin roughness: Results of a clinical study." *Journal of evidence-based complementary & alternative medicine* 22.4 (2017): 816-823.

7) Gupta, Ramesh Chandra, et al. "Hyaluronic acid: molecular mechanisms and therapeutic trajectory." *Frontiers in Veterinary Science* 6 (2019): 192.

8) Ke, Chunlin, et al. "Antioxidant acitivity of low molecular weight hyaluronic acid." *Food and chemical toxicology* 49.10 (2011): 2670-2675.

9) Kim, Yeseul, et al. "Oral Hyaluronic Acid Supplementation for the Treatment of Dry Eye Disease: A Pilot Study." *Journal of ophthalmology* 2019 (2019).

10) Kimura, Mamoru, et al. "Absorption of orally administered hyaluronan." *Journal of medicinal food* 19.12 (2016): 1172-1179.

11) Kochlamazashvili, Gaga, et al. "The extracellular matrix molecule hyaluronic acid regulates hippocampal synaptic plasticity by modulating postsynaptic L-type Ca2+ channels." *Neuron* 67.1 (2010): 116-128.

12) Lotfi, Saied, et al. "Hyaluronic acid improves frozen-thawed sperm quality and fertility potential in rooster." *Animal reproduction science* 184 (2017): 204-210.

13) Nelson, F. R., et al. "The effects of an oral preparation containing hyaluronic acid (Oralvisc®) on obese knee osteoarthritis patients determined by pain, function, bradykinin, leptin, inflammatory cytokines, and heavy water analyses." *Rheumatology international* 35.1 (2015): 43-52.

14) Oe, Mariko, et al. "Dietary hyaluronic acid migrates into the skin of rats." *The Scientific World Journal* 2014 (2014).

15) Papakonstantinou, Eleni, Michael Roth, and George Karakiulakis. "Hyaluronic acid: A key molecule in skin aging." *Dermato-endocrinology* 4.3 (2012): 253-258.

16) Qian, Li, Sijiu Yu, and Yan Zhou. "Protective effect of hyaluronic acid on cryopreserved boar sperm." *International journal of biological macromolecules* 87 (2016): 287-289.

17) Rayahin, Jamie E., et al. "High and low molecular weight hyaluronic acid differentially influence macrophage activation." *ACS biomaterials science &*

*engineering 1.7 (2015): 481-493.*

*18) Salwowska, Natalia M., et al. "Physiochemical properties and application of hyaluronic acid: a systematic review." Journal of cosmetic dermatology 15.4 (2016): 520-526.*

*19) Tashiro, Toshiyuki, et al. "Oral administration of polymer hyaluronic acid alleviates symptoms of knee osteoarthritis: a double-blind, placebo-controlled study over a 12-month period." The Scientific World Journal 2012 (2012).*

**CHAPTER 19** - KAEMPFERIA PARVIFLORA

*1) Butkhup, L., and S. Samappito. "In vitro free radical scavenging and antimicrobial activity of some selected Thai medicinal plants." Research Journal of Medicinal Plant 5.3 (2011): 254-265.*

*2) Chaiyasut, Chaiyavat, and Sunee Chansakaow. "Inhibitory effects of some thai plant extracts on AAPH-induced protein oxidation and protein glycation." Naresuan University Journal: Science and Technology (NUJST) 15.1 (2013): 35-41.*

*3) Chen, Dalin, et al. "Kaempferia parviflora and Its Methoxyflavones: Chemistry and Biological Activities." Evidence-Based Complementary and Alternative Medicine 2018 (2018).*

*4) Chivapat, Songpol, et al. "Chronic toxicity study of Kaempferia parviflora Wall ex. extract." The Thai Journal of Veterinary Medicine 40.4 (2010): 377-383.*

*5) Hirsh, Steven, et al. "An open label study to evaluate the effect of Kaempferia parviflora in support of erectile function and male sexual health among overall healthy males 50–70." The FASEB Journal 31.1_supplement (2017): 636-1.*

*6) Horigome, Satoru, et al. "Identification and evaluation of anti-inflammatory compounds from Kaempferia parviflora." Bioscience, biotechnology, and biochemistry 78.5 (2014): 851-860.*

*7) Jin, Solee, and Mi-Young Lee. "Kaempferia parviflora extract as a potential anti-acne agent with anti-inflammatory, sebostatic and anti-propionibacterium acnes activity." International journal of molecular sciences 19.11 (2018): 3457.*

*8) Kim, Jae Kyung, et al. "5, 7-Dimethoxyflavone, an activator of PPARα/γ, inhibits UVB-induced MMP expression in human skin fibroblast cells."*

*Experimental dermatology 21.3 (2012): 211-216.*

9) Kim, Mi-Bo, et al. "Standardized Kaempferia parviflora extract enhances exercise performance through activation of mitochondrial biogenesis." *Journal of medicinal food 21.1 (2018): 30-38.*

10) Kobayashi, Hiroko, et al. "Effects of ethyl acetate extract of Kaempferia parviflora on brown adipose tissue." *Journal of natural medicines 70.1 (2016): 54-61.*

11) Kobayashi, Hiroko, et al. "Effect of Kaempferia parviflora extract on knee osteoarthritis." *Journal of natural medicines 72.1 (2018): 136-144.*

12) Lee, Myung-hee, et al. "Antiskin Inflammatory Activity of Black Ginger (Kaempferia parviflora) through Antioxidative Activity." *Oxidative medicine and cellular longevity 2018 (2018).*

13) Lee, Sunkyu, et al. "Standardized Kaempferia parviflora Wall. ex Baker (Zingiberaceae) extract inhibits fat accumulation and muscle atrophy in ob/ob mice." *Evidence-Based Complementary and Alternative Medicine 2018 (2018).*

14) Lert-Amornpat, T., C. Maketon, and W. Fungfuang. "Effect of Kaempferia parviflora on sexual performance in streptozotocin-induced diabetic male rats." *Andrologia 49.10 (2017): e12770.*

15) Matsushita, Mami, et al. "Kaempferia parviflora extract increases whole-body energy expenditure in humans: roles of brown adipose tissue." *Journal of nutritional science and vitaminology 61.1 (2015): 79-83.*

16) Murata, Kazuya, et al. "Suppression of benign prostate hyperplasia by Kaempferia parviflora rhizome." *Pharmacognosy research 5.4 (2013): 309.*

17) Ochiai, Masaru, et al. "Kaempferia parviflora Ethanol Extract, a Peroxisome Proliferator- Activated Receptor γ Ligand-binding Agonist, Improves Glucose Tolerance and Suppresses Fat Accumulation in Diabetic NSY Mice." *Journal of food science 84.2 (2019): 339-348.*

18) Ochiai, Masaru, Yoshiyuki Azuma, and Tatsuhiro Matsuo. "Dietary effect of Kaempferia parviflora, a PPARγ agonist, on glucose tolerance in non-obese type diabetic mice." *The FASEB Journal 31.1_supplement (2017): 974-20.*

19) Okabe, Yui, et al. "Suppression of adipocyte hypertrophy by polymethoxyflavonoids isolated from Kaempferia parviflora." *Phytomedicine 21.6 (2014): 800-806.*

20) Ono, Shintaro, et al. "5-Hydroxy-7-methoxyflavone derivatives from Kaempferia parviflora induce skeletal muscle hypertrophy." Food science & nutrition 7.1 (2019): 312-321.

21) Park, Ji-Eun, et al. "The protective effect of K aempferia parviflora extract on UVB-induced skin photoaging in hairless mice." Photodermatology, photoimmunology & photomedicine 30.5 (2014): 237-245.

22) Park, Ji-Eun, et al. "Standardized Kaempferia parviflora extract inhibits intrinsic aging process in human dermal fibroblasts and hairless mice by inhibiting cellular senescence and mitochondrial dysfunction." Evidence-Based Complementary and Alternative Medicine 2017 (2017).

23) Mekjaruskul, Catheleeya, Michael Jay, and Bungorn Sripanidkulchai. "Pharmacokinetics, bioavailability, tissue distribution, excretion, and metabolite identification of methoxyflavones in Kaempferia parviflora extract in rats." Drug Metabolism and Disposition 40.12 (2012): 2342-2353.

24) Miyata, Yoshiki, et al. "Potential Therapeutic Agents, Polymethoxylated Flavones Isolated from Kaempferia parviflora for Cataract Prevention through Inhibition of Matrix Metalloproteinase-9 in Lens Epithelial Cells." Biological and Pharmaceutical Bulletin 42.10 (2019): 1658-1664.

25) Nakata, Asami, et al. "Potent SIRT1 enzyme-stimulating and anti-glycation activities of polymethoxyflavonoids from Kaempferia parviflora." Natural product communications 9.9 (2014): 1934578X1400900918.

26) Saokaew, Surasak, et al. "Clinical effects of Krachaidum (Kaempferia parviflora): a systematic review." Journal of evidence-based complementary & alternative medicine 22.3 (2017): 413-428.

27) Stein, Richard A., et al. "Kaempferia parviflora ethanol extract improves self-assessed sexual health in men: A pilot study." Journal of integrative medicine 16.4 (2018): 249-254.

28) Sutthanut, K., et al. "Simultaneous identification and quantitation of 11 flavonoid constituents in Kaempferia parviflora by gas chromatography." Journal of Chromatography A 1143.1-2 (2007): 227-233.

29) Tewtrakul, Supinya, et al. "Anti-inflammatory effects of compounds from Kaempferia parviflora and Boesenbergia pandurata." Food Chemistry 115.2 (2009): 534-538.

30) Tewtrakul, Supinya, Sanan Subhadhirasakul, and Sopa Kummee. "Anti-

*allergic activity of compounds from Kaempferia parviflora." Journal of ethnopharmacology 116.1 (2008): 191-193.*

31) Thao, Nguyen Phuong, et al. "Anti-osteoporotic and antioxidant activities by rhizomes of Kaempferia parviflora Wall. ex baker." *Natural Product Sciences* 22.1 (2016): 13-19.

32) Thipboonchoo, Natechanok, and Sunhapas Soodvilai. "Effects of Kaempferia parviflora Extract on Glucose Transporters in Human Renal Proximal Tubular Cells." *J Physiol* 30.2 (2017): 57-61.

33) Toda, Kazuya, et al. "Black ginger extract increases physical fitness performance and muscular endurance by improving inflammation and energy metabolism." *Heliyon* 2.5 (2016): e00115.

34) Wasuntarawat, Chanchira, et al. "No effect of acute ingestion of Thai ginseng (Kaempferia parviflora) on sprint and endurance exercise performance in humans." *Journal of sports sciences* 28.11 (2010): 1243-1250.

35) Wattanathorn, Jintanaporn, et al. "Aphrodisiac activity of Kaempferia parviflora." *American Journal of Agricultural and Biological Science* (2011).

36) Wattanathorn, Jintanaporn, et al. "Positive modulation effect of 8-week consumption of Kaempferia parviflora on health-related physical fitness and oxidative status in healthy elderly volunteers." *Evidence-Based Complementary and Alternative Medicine* 2012 (2012).

37) Wattanathorn, Jintanaporn, et al. "Evaluation of the anxiolytic and antidepressant effects of alcoholic extract of Kaempferia parviflora in aged rats." *American Journal of Agricultural and Biological Science* (2007).

38) Wen, Kuo-Ching, et al. "Ixora parviflora protects against UVB-induced photoaging by inhibiting the expression of MMPs, MAP kinases, and COX-2 and by promoting type I procollagen synthesis." *Evidence-Based Complementary and Alternative Medicine* 2012 (2012).

39) Yagi, Masayuki, et al. "Antiglycative effect of Kaempferia parviflora Wall. Ex. Baker (Zingiberaceae): Prevention of advanced glycation end product formation." *Glycative Stress Research* 5.4 (2018): 163.

40) Yoshino, Susumu, et al. "Daily intake of Kaempferia parviflora extract decreases abdominal fat in overweight and preobese subjects: A randomized, double-blind, placebo-controlled clinical study." *Diabetes, metabolic syndrome and obesity: targets and therapy* 11 (2018): 447.

41) Yoshino, Susumu, et al. "Evaluation of the Safety of Daily Consumption of Kaempferia parviflora Extract (KPFORCE): A Randomized Double-Blind Placebo-Controlled Trial." Journal of medicinal food 22.11 (2019): 1168-1174.

42) Yorsin, Somruedee, et al. "Effects of Kaempferia parviflora rhizomes dichloromethane extract on vascular functions in middle-aged male rat." Journal of ethnopharmacology 156 (2014): 162-174.

**CHAPTER 20** - LACTOFERRIN

1) A Amini, A., and L. S Nair. "Lactoferrin: a biologically active molecule for bone regeneration." Current medicinal chemistry 18.8 (2011): 1220-1229.

2) Chen, Jie-Hua, et al. "Role of advanced glycation end products in mobility and considerations in possible dietary and nutritional intervention strategies." Nutrition & metabolism 15.1 (2018): 1-18.

3) Cornish, Jillian, et al. "Lactoferrin is a potent regulator of bone cell activity and increases bone formation in vivo." Endocrinology 145.9 (2004): 4366-4374.

4) Cornish, J., et al. "Lactoferrin and bone; structure–activity relationships." Biochemistry and Cell Biology 84.3 (2006): 297-302.

5) Fan, Linlin, et al. "Lactoferrin could alleviate liver injury caused by Maillard reaction products with furan ring through regulating necroptosis pathway." Food Science & Nutrition (2021).

6) Fleet, James C. "A new role for lactoferrin: DNA binding and transcription activation." Nutrition reviews 53.8 (1995): 226-227.

7) H Albar, Abdulgader, et al. "Structural heterogeneity and multifunctionality of lactoferrin." Current Protein and Peptide Science 15.8 (2014): 778-797.

8) He, Jianglin, and Philip Furmanski. "Sequence specificity and transcriptional activation in the binding of lactoferrin to DNA." Nature 373.6516 (1995): 721-724.

9) Hsu, Yung-Ho, et al. "Lactoferrin contributes a renoprotective effect in acute kidney injury and early renal fibrosis." Pharmaceutics 12.5 (2020): 434.

10) Huang, Hsiu-Chin, Hsuan Lin, and Min-Chuan Huang. "Lactoferrin promotes hair growth in mice and increases dermal papilla cell proliferation through Erk/Akt and Wnt signaling

*pathways." Archives of dermatological research 311.5 (2019): 411-420.*

*11) Hwang, Shen-An, et al. "A novel recombinant human lactoferrin augments the BCG vaccine and protects alveolar integrity upon infection with Mycobacterium tuberculosis in mice." Vaccine 27.23 (2009): 3026-3034.*

*12) Iijima, Hironobu, et al. "Genetic and epigenetic inactivation of LTF gene at 3p21. 3 in lung cancers." International journal of cancer 118.4 (2006): 797-801.*

*13) Kanwar, Jagat R., et al. "Multifunctional iron bound lactoferrin and nanomedicinal approaches to enhance its bioactive functions." Molecules 20.6 (2015): 9703-9731.*

*14) Kawakami, H., et al. "Effects of enteric-coated lactoferrin supplementation on the immune function of elderly individuals: a randomised, double-blind, placebo-controlled trial." International Dairy Journal 47 (2015): 79-85.*

*15) Kawashima, Motoko, et al. "Calorie restriction (CR) and CR mimetics for the prevention and treatment of age-related eye disorders." Experimental Gerontology 48.10 (2013): 1096-1100.*

*16) Kim, Chan Woo, et al. "Human lactoferrin suppresses TNF-α-induced intercellular adhesion molecule-1 expression via competition with NF-κB in endothelial cells." FEBS letters 586.3 (2012): 229-234.*

*17) Kruzel, Marian L., Michal Zimecki, and Jeffrey K. Actor. "Lactoferrin in a context of inflammation-induced pathology." Frontiers in immunology 8 (2017): 1438.*

*18) Li, Wenyang, Songsong Zhu, and Jing Hu. "Bone regeneration is promoted by orally administered bovine lactoferrin in a rabbit tibial distraction osteogenesis model." Clinical Orthopaedics and Related Research® 473.7 (2015): 2383-2393.*

*19) Li, Yong Ming, Annie X. Tan, and Helen Vlassara. "Antibacterial activity of lysozyme and lactoferrin is inhibited by binding of advanced glycation–modified proteins to a conserved motif." Nature medicine 1.10 (1995): 1057-1061.*

*20) Li, Yong Ming. "Glycation ligand binding motif in lactoferrin." Advances in Lactoferrin Research (1998): 57-63.*

*21) Lönnerdal, Bo, Xiaogu Du, and Rulan Jiang. "Biological activities of commercial bovine lactoferrin sources." Biochemistry and Cell Biology 99.1 (2021): 35-46.*

22) Mohamed, Waleed A., and Mona F. Schaalan. "Antidiabetic efficacy of lactoferrin in type 2 diabetic pediatrics; controlling impact on PPAR-γ, SIRT-1, and TLR4 downstream signaling pathway." Diabetology & metabolic syndrome 10.1 (2018): 1-12.

23) Mulder, Ann M., et al. "Bovine lactoferrin supplementation supports immune and antioxidant status in healthy human males." Nutrition Research 28.9 (2008): 583-589.

24) Naidu, Sreus AG, et al. "Lactoferrin for Mental Health: Neuro-Redox Regulation and Neuroprotective Effects across the Blood-Brain Barrier with Special Reference to Neuro-COVID-19." Journal of Dietary Supplements (2021): 1-35.

25) Naot, Dorit, Kate Palmano, and Jillian Cornish. Lactoferrin–A Potential Anabolic Intervention in Osteoporosis. IntechOpen, 2012.

26) Niaz, Bushra, et al. "Lactoferrin (LF): a natural antimicrobial protein." International Journal of Food Properties 22.1 (2019): 1626-1641.

27) Porter, Corey M., et al. "Lactoferrin CpG Island Hypermethylation and decoupling of mRNA and protein expression in the early stages of prostate carcinogenesis." The American journal of pathology 189.11 (2019): 2311-2322.

28) Qari, Sameer H., and Kamal Attia. "Gene expression of renal lactoferrin and glycemic homeostasis in diabetic rats with reference to the protective role of exogenous bovine lactoferrin." The Journal of Basic and Applied Zoology 81.1 (2020): 1-6.

29) Rasheed, Naila, Abdullah Alghasham, and Zafar Rasheed. "Lactoferrin from Camelus dromedarius inhibits nuclear transcription factor-kappa B activation, cyclooxygenase-2 expression and prostaglandin E2 production in stimulated human chondrocytes." Pharmacognosy research 8.2 (2016): 135.

30) Sharrouf, Kinda Ali, and Irina O. Suchkova. "The influence of lactoferrin on the epigenetic characteristics of mammalian cells of different types." Medical academic journal 21.1 (2021): 85-95.

31) Shi, Jin, et al. "Metabolic effects of lactoferrin during energy restriction and weight regain in diet-induced obese mice." Journal of Functional Foods 4.1 (2012): 66-78.

32) Shiga, Yuki, et al. "Recombinant human lactoferrin-Fc fusion with an improved plasma halflife." European Journal of Pharmaceutical Sciences 67 (2015): 136-143.

33) Silva, André MN, et al. "Human transferrin: An inorganic biochemistry perspective." Coordination Chemistry Reviews 449 (2021): 214186.

34) Superti, Fabiana. "Lactoferrin from bovine milk: a protective companion for life." Nutrients 12.9 (2020): 2562.

35) Takayama, Yoshiharu, and Toshiaki Takezawa. "Lactoferrin promotes collagen gel contractile activity of fibroblasts mediated by lipoprotein receptors." Biochemistry and cell biology 84.3 (2006): 268-274.

36) Vega-Bautista, Alan, et al. "The impact of lactoferrin on the growth of intestinal inhabitant bacteria." International journal of molecular sciences 20.19 (2019): 4707.

37) Zakharova, Elena T., et al. "Human apo-lactoferrin as a physiological mimetic of hypoxia stabilizes hypoxia-inducible factor-1 alpha." Biometals 25.6 (2012): 1247-1259.

38) Zemankova, Nada, et al. "Bovine lactoferrin free of lipopolysaccharide can induce a proinflammatory response of macrophages." BMC veterinary research 12.1 (2016): 1-9.

39) Zhang, T-N., and N. Liu. "Effect of bovine lactoferricin on DNA methyltransferase 1 levels in Jurkat T-leukemia cells." Journal of dairy science 93.9 (2010): 3925-3930.

40) Zhao, Xiurong, et al. "Optimized lactoferrin as a highly promising treatment for intracerebral hemorrhage: pre-clinical experience." Journal of Cerebral Blood Flow & Metabolism 41.1 (2021): 53-66.

41) Zheng, Jiaping, et al. "Lactoferrin improves cognitive function and attenuates brain senescence in aged mice." Journal of Functional Foods 65 (2020): 103736.

**CHAPTER 21** - LEUCINE

1) Borack, Michael S., and Elena Volpi. "Efficacy and safety of leucine supplementation in the elderly." The Journal of nutrition 146.12 (2016): 2625S-2629S.

2) Deng, Huiling, et al. "Activation of mammalian target of rapamycin signaling in skeletal muscle of neonatal chicks: Effects of dietary leucine and age." Poultry Science 93.1 (2014): 114-121.

3) Dodd, Kayleigh M., and Andrew R. Tee. "Leucine and mTORC1: a complex relationship." American Journal of Physiology-Endocrinology and Metabolism (2012).

4) Elango, Rajavel, Betina Rasmussen, and Kenneth Madden. "Safety and tolerability of leucine supplementation in elderly men." The Journal of nutrition 146.12 (2016): 2630S-2634S.

5) English, Kirk L., et al. "Leucine partially protects muscle mass and function during bed rest in middle-aged adults, 2." The American journal of clinical nutrition 103.2 (2016): 465-473.

6) Garlick, Peter J. "The role of leucine in the regulation of protein metabolism." The Journal of nutrition 135.6 (2005): 1553S-1556S.

7) Jackman, Sarah R., et al. "Branched-chain amino acid ingestion stimulates muscle myofibrillar protein synthesis following resistance exercise in humans." Frontiers in physiology 8 (2017).

8) Kato, Hiroyuki, et al. "Leucine-enriched essential amino acids attenuate inflammation in rat muscle and enhance muscle repair after eccentric contraction." Amino acids 48.9 (2016): 2145-2155.

9) Komar, B., L. Schwingshackl, and Georg Hoffmann. "Effects of leucine-rich protein supplements on anthropometric parameter and muscle strength in the elderly: a systematic review and meta-analysis." The journal of nutrition, health & aging 19.4 (2015): 437-446.

10) Li, Hongliang, et al. "Leucine supplementation increases SIRT1 expression and prevents mitochondrial dysfunction and metabolic disorders in high-fat diet-induced obese mice." American Journal of Physiology-Endocrinology and Metabolism 303.10 (2012): E1234-E1244.

11) Liang, Chunzi, et al. "Leucine modulates mitochondrial biogenesis and SIRT1-AMPK signaling in C2C12 myotubes." Journal of nutrition and metabolism 2014 (2014).

12) Pedroso, João AB, Thais T. Zampieri, and Jose Donato. "Reviewing the effects of L-leucine supplementation in the regulation of food intake, energy balance, and glucose homeostasis." Nutrients 7.5 (2015): 3914-3937.

13) Pereira, Marcelo G., et al. "Leucine supplementation accelerates connective tissue repair of injured tibialis anterior muscle." Nutrients 6.10 (2014): 3981-4001. Pereira, Marcelo G., et al. "Leucine supplementation improves

*regeneration of skeletal muscles from old rats." Experimental gerontology 72 (2015): 269-277.*

14) Perry Jr, Richard A., et al. *"Differential effects of leucine supplementation in young and aged mice at the onset of skeletal muscle regeneration." Mechanisms of ageing and development 157 (2016): 7-16.*

16) Rieu, Isabelle, et al. *"Increased availability of leucine with leucine-rich whey proteins improves postprandial muscle protein synthesis in aging rats." Nutrition 23.4 (2007): 323-331.*

17) Sato, Yoriko, et al. *"Acute oral administration of L-leucine upregulates slow-fiber–and mitochondria-related genes in skeletal muscle of rats." Nutrition Research 57 (2018): 36-44.*

18) Schnuck, Jamie K., et al. *"Leucine stimulates PPARβ/δ-dependent mitochondrial biogenesis and oxidative metabolism with enhanced GLUT4 content and glucose uptake in myotubes." Biochimie 128 (2016): 1-7.*

19) Stancliffe, Renee Ashley. *"Role of beta-hydroxy-beta-methylbutyrate (HMB) in leucine stimulation of mitochondrial biogenesis and fatty acid oxidation." (2012).*

20) Sun, Xiaocun, and Michael B. Zemel. *"Leucine modulation of mitochondrial mass and oxygen consumption in skeletal muscle cells and adipocytes." Nutrition & metabolism 6.1 (2009): 26.*

21) Theis, Nicola, et al. *"Leucine Supplementation Increases Muscle Strength and Volume, Reduces Inflammation, and Affects Wellbeing in Adults and Adolescents with Cerebral Palsy." The Journal of Nutrition (2020).*

22) Valerio, Alessandra, Giuseppe D'Antona, and Enzo Nisoli. *"Branched-chain amino acids, mitochondrial biogenesis, and healthspan: an evolutionary perspective." Aging (Albany NY) 3.5 (2011): 464.*

23) Van Loon, Luc JC. *"Leucine as a pharmaconutrient in health and disease." Current Opinion in Clinical Nutrition & Metabolic Care 15.1 (2012): 71-77.*

24) Xia, Zhi, et al. *"Targeting inflammation and downstream protein metabolism in sarcopenia: a brief up-dated description of concurrent exercise and leucine-based multimodal intervention." Frontiers in Physiology 8 (2017): 434.*

25) Xu, Zhe-rong, et al. *"The effectiveness of leucine on muscle protein synthesis, lean body mass and leg lean mass accretion in older people: a*

*systematic review and meta-analysis." British Journal of Nutrition 113.1 (2015): 25-34.*

26) Yao, Kang, et al. "Leucine in obesity: therapeutic prospects." *Trends in Pharmacological Sciences* 37.8 (2016): 714-727.

27) Yudkoff, Marc, et al. "Brain amino acid requirements and toxicity: the example of leucine." *The Journal of nutrition* 135.6 (2005): 1531S-1538S.

28) Zeanandin, Gilbert, et al. "Differential effect of long-term leucine supplementation on skeletal muscle and adipose tissue in old rats: an insulin signaling pathway approach." *Age* 34.2 (2012): 371-387.

29) Zemel, Michael B. "Modulation of Energy Sensing by Leucine Synergy with Natural Sirtuin Activators: Effects on Health Span." *Journal of Medicinal Food* (2020).

30) Zhang, Lingyu, et al. "Leucine Supplementation: A Novel Strategy for Modulating Lipid Metabolism and Energy Homeostasis." *Nutrients* 12.5 (2020): 1299.

## CHAPTER 22 - MAGNESIUM THREONATE

1) Abumaria, Nashat, et al. "Effects of elevation of brain magnesium on fear conditioning, fear extinction, and synaptic plasticity in the infralimbic prefrontal cortex and lateral amygdala." *Journal of Neuroscience* 31.42 (2011): 14871-14881.

2) Al-Ghazali, Kateba, et al. "Serum Magnesium and Cognitive Function Among Qatari Adults." *Frontiers in Aging Neuroscience* 12 (2020): 101.

3) Altura, Burton M., et al. "Magnesium deficiency results in oxidation and fragmentation of DNA, down regulation of telomerase activity, and ceramide release in cardiovascular tissues and cells: potential relationship to atherogenesis, cardiovascular diseases and aging." (2016).

4) Barbagallo, Mario, and Ligia J. Dominguez. "Magnesium Role in Health and Longevity." *Trace Elements and Minerals in Health and Longevity*. Springer, Cham, 2018. 235-264.

5) Billard, Jean-Marie. "Brain free magnesium homeostasis as a target for reducing cognitive aging." *Magnesium in the Central Nervous System*

*[Internet]. University of Adelaide Press, 2011.*

6) Đurić, Vedrana, et al. "A single dose of magnesium, as well as chronic administration, enhances long-term memory in novel object recognition test, in healthy and ACTH-treated rats." *Magnesium research 31.1 (2018): 24-32.*

7) Jia, Shanshan, et al. "Elevation of brain magnesium potentiates neural stem cell proliferation in the hippocampus of young and aged mice." *Journal of Cellular Physiology 231.9 (2016): 1903-1912.*

8) Liu, Guosong, et al. "Efficacy and safety of MMFS-01, a synapse density enhancer, for treating cognitive impairment in older adults: a randomized, double-blind, placebo-controlled trial." *Journal of Alzheimer's Disease 49.4 (2016): 971-990.*

9) Li, Wei, et al. "Elevation of brain magnesium prevents synaptic loss and reverses cognitive deficits in Alzheimer's disease mouse model." *Molecular brain 7.1 (2014): 65.*

10) Lou, Zhi-Yi, et al. "Dietary intake of magnesium-L-threonate alleviates memory deficits induced by developmental lead exposure in rats." *RSC advances 7.14 (2017): 8241-8249.*

11) Maguire, Donogh, et al. "Telomere homeostasis: interplay with magnesium." *International Journal of Molecular Sciences 19.1 (2018): 157.*

12) Maier, Jeanette A., et al. "Magnesium and inflammation: Advances and perspectives." *Seminars in Cell & Developmental Biology. Academic Press, 2020.*

13) Nielsen, Forrest H. "Magnesium deficiency and increased inflammation: current perspectives." *Journal of inflammation research 11 (2018): 25.*

14) Razzaque, Mohammed S. "Magnesium: are we consuming enough?." *Nutrients 10.12 (2018): 1863.*

15) Sadir, Sadia, et al. "Neurobehavioral and biochemical effects of magnesium chloride (MgCl 2), magnesium sulphate (MgSO 4) and magnesium-L-threonate (MgT) supplementation in rats: A dose dependent comparative study." *Pakistan journal of pharmaceutical sciences 32 (2019).*

16) Schwalfenberg, Gerry K., and Stephen J. Genuis. "The importance of magnesium in clinical healthcare." *Scientifica 2017 (2017).*

17) Shen, Yanling, et al. "Treatment of magnesium-L-threonate elevates the magnesium level in the cerebrospinal fluid and attenuates motor deficits and dopamine neuron loss in a mouse model of Parkinson's disease." Neuropsychiatric Disease and Treatment 15 (2019): 3143.

18) Slutsky, Inna, et al. "Enhancement of learning and memory by elevating brain magnesium." Neuron 65.2 (2010): 165-177.

19) Sun, Qifeng, et al. "Regulation of structural and functional synapse density by L-threonate through modulation of intraneuronal magnesium concentration." Neuropharmacology 108 (2016): 426-439.

20) Toffa, Dènahin Hinnoutondji, et al. "Can magnesium reduce central neurodegeneration in Alzheimer's disease? Basic evidences and research needs." Neurochemistry international 126 (2019): 195-202.

21) Villa-Bellosta, Ricardo. "Dietary magnesium supplementation improves lifespan in a mouse model of progeria." EMBO molecular medicine 12.10 (2020): e12423.

22) Vink, Robert. "Magnesium in the CNS: recent advances and developments." Magnesium Research 29.3 (2016): 95-101.

23) Wang, Jun, et al. "Magnesium L-threonate prevents and restores memory deficits associated with neuropathic pain by inhibition of TNF-alpha." Pain physician 16.5 (2013): E563-575.

24) Wang, Pu, et al. "Magnesium ion influx reduces neuroinflammation in Aβ precursor protein/ Presenilin 1 transgenic mice by suppressing the expression of interleukin-1β." Cellular & molecular immunology 14.5 (2017): 451-464.

25) Xiong, Wenxiang, et al. "Erythrocyte intracellular Mg 2+ concentration as an index of recognition and memory." Scientific reports 6 (2016): 26975.

26) Yamanaka, Ryu, Yutaka Shindo, and Kotaro Oka. "Magnesium is a key player in neuronal maturation and neuropathology." International journal of molecular sciences 20.14 (2019): 3439.

27) Yu, Xin, et al. "Magnesium ions inhibit the expression of tumor necrosis factor α and the activity of γ-secretase in a β-amyloid protein-dependent mechanism in APP/PS1 transgenic mice." Frontiers in molecular neuroscience 11 (2018): 172.

28) Zhou, Xin, et al. "Chronic Oral Administration of Magnesium-L-Threonate Prevents Oxaliplatin-Induced Memory and Emotional Deficits by

Normalization of TNF-α/NF-κB Signaling in Rats." *Neuroscience bulletin* (2020): 1-15.

## CHAPTER 23 - NARINGENIN

1) Alam, M. Ashraful, et al. "Effect of citrus flavonoids, naringin and naringenin, on metabolic syndrome and their mechanisms of action." *Advances in Nutrition: An International Review Journal* 5.4 (2014): 404-417.

2) Annadurai, Thangaraj, et al. "Antihyperglycemic and antioxidant effects of a flavanone, naringenin, in streptozotocin–nicotinamide-induced experimental diabetic rats." *Journal of physiology and biochemistry* 68.3 (2012): 307-318.

3) Bharti, Saurabh, et al. "Preclinical evidence for the pharmacological actions of naringin: a review." *Planta medica* 80.06 (2014): 437-451.

4) Cao, Xvhai, et al. "Naringin rescued the TNF-α-induced inhibition of osteogenesis of bone marrow-derived mesenchymal stem cells by depressing the activation of NF-κB signaling pathway." *Immunologic research* 62.3 (2015): 357-367.

5) Carswell, Savion. "Anti-inflammatory and osteogenic effects of naringenin." (2020).

6) Carswell, Savion, et al. "Osteogenic Effects of Naringenin Are Mediated Through the Activation of AMPK Pathway." *Current Developments in Nutrition* 4.Supplement_2 (2020): 12-12.

7) Cavia-Saiz, Monica, et al. "Antioxidant properties, radical scavenging activity and biomolecule protection capacity of flavonoid naringenin and its glycoside naringin: a comparative study." *Journal of the Science of Food and Agriculture* 90.7 (2010): 1238-1244.

8) Da Pozzo, Eleonora, et al. "The citrus flavanone naringenin protects myocardial cells against age-associated damage." *Oxidative Medicine and Cellular Longevity* 2017 (2017).

9) De Oliveira, Marcos Roberto, Flávia Bittencourt Brasil, and Cláudia Marlise Balbinotti Andrade. "Naringenin attenuates $H_2O_2$-induced mitochondrial dysfunction by an Nrf2- dependent mechanism in SH-SY5Y cells." *Neurochemical Research* 42.11 (2017): 3341-3350.

10) El-Mahdy, Mohamed A., et al. "Naringenin protects HaCaT human keratinocytes against UVB-induced apoptosis and enhances the removal of cyclobutane pyrimidine dimers from the genome." Photochemistry and photobiology 84.2 (2008): 307-316.

11) Fan, Jifeng, Jie Li, and Qinbo Fan. "Naringin promotes differentiation of bone marrow stem cells into osteoblasts by upregulating the expression levels of microRNA-20a and downregulating the expression levels of PPARγ." Molecular medicine reports 12.3 (2015): 4759-4765.

12) Gao, Kun, et al. "The citrus flavonoid naringenin stimulates DNA repair in prostate cancer cells." The Journal of nutritional biochemistry 17.2 (2006): 89-95.

13) Guo-Chung, D. O. N. G., et al. "A study of Drynaria fortunei in modulation of BMP–2 signalling by bone tissue engineering." Turkish journal of medical sciences 50.5 (2020): 1444.

14) Hernández-Aquino, Erika, and Pablo Muriel. "Beneficial effects of naringenin in liver diseases: Molecular mechanisms." World journal of gastroenterology 24.16 (2018): 1679.

15) Huang, J., et al. "Dietary supplementation of total flavonoids from Rhizoma Drynariae improves bone health in older caged laying hens." Poultry Science 99.10 (2020): 5047-5054.

16) Kampa, Rafal Pawel, et al. "Naringenin as an opener of mitochondrial potassium channels in dermal fibroblasts." Experimental dermatology 28.5 (2019): 543-550.

17) Li, Nianhu, et al. "Naringin promotes osteoblast differentiation and effectively reverses ovariectomy-associated osteoporosis." Journal of Orthopaedic Science 18.3 (2013): 478-485.

18) Lou, Haiyan, et al. "Naringenin protects against 6-OHDA-induced neurotoxicity via activation of the Nrf2/ARE signaling pathway." Neuropharmacology 79 (2014): 380-388.

19) Nyane, Annah Nsoaki, et al. "Metformin-like antidiabetic, cardio-protective and non-glycemic effects of naringenin: molecular and pharmacological insights." European journal of pharmacology (2017).

20) Raza, S. S., et al. "Neuroprotective effect of naringenin is mediated through suppression of NF-κB signaling pathway in experimental stroke."

*Neuroscience 230 (2013): 157-171.*

21) Rebello, Candida J., et al. "Naringenin promotes thermogenic gene expression in human white adipose tissue." *Obesity 27.1 (2019): 103-111.*

22) Rebello, Candida J., et al. "Safety and pharmacokinetics of naringenin: A randomized, controlled, single-ascending-dose clinical trial." *Diabetes, Obesity and Metabolism 22.1 (2020): 91-98.*

23) Rivoira, M., et al. "Naringin prevents bone loss in a rat model of type 1 Diabetes mellitus." *Archives of biochemistry and biophysics 637 (2018): 56-63.*

24) Salehi, Bahare, et al. "The therapeutic potential of naringenin: a review of clinical trials." *Pharmaceuticals 12.1 (2019): 11.*

25) Sippl, W., and F. Ntie-Kang. "Natural Products as Modulators of Sirtuins." *Molecules (Basel, Switzerland) 25.14 (2020).*

26) Song, Nan, et al. "Naringin promotes fracture healing through stimulation of angiogenesis by regulating the VEGF/VEGFR-2 signaling pathway in osteoporotic rats." *Chemico-biological interactions 261 (2017): 11-17.*

27) Song, Shuang-hong, et al. "Effects of total flavonoids from Drynariae Rhizoma prevent bone loss in vivo and in vitro." *Bone reports 5 (2016): 262-273.*

28) Testai, Lara, et al. "The Citrus Flavonoid Naringenin Protects the Myocardium from Ageing- Dependent Dysfunction: Potential Role of SIRT1." *Oxidative Medicine and Cellular Longevity 2020 (2020).*

29) Tsai, Shih-Jei, et al. "Anti-inflammatory and antifibrotic effects of naringenin in diabetic mice." *Journal of agricultural and food chemistry 60.1 (2012): 514-521.*

30) Venkateswara Rao, P., et al. "Flavonoid: a review on naringenin." *J. Pharmacogn. Phytochem 6 (2017): 2778-2783.*

31) Wang, Kaihua, et al. "Naringenin reduces oxidative stress and improves mitochondrial dysfunction via activation of the Nrf2/ARE signaling pathway in neurons." *International Journal of Molecular Medicine 40.5 (2017): 1582-1590.*

32) Wang, Wengang, et al. "Naringenin inhibits osteoclastogenesis through modulation of helper T cells-secreted IL-4." *Journal of cellular biochemistry 119.2 (2018): 2084-2093.*

33) Xu, Tong, et al. "The function of naringin in inducing secretion of osteoprotegerin and inhibiting formation of osteoclasts." Evidence-Based Complementary and Alternative Medicine 2016 (2016).

34) Yi, Li-Tao, et al. "Involvement of monoaminergic system in the antidepressant-like effect of the flavonoid naringenin in mice." Progress in Neuro-Psychopharmacology and Biological Psychiatry 34.7 (2010): 1223-1228.

35) Yin, Lihua, et al. "Effects of naringin on proliferation and osteogenic differentiation of human periodontal ligament stem cells in vitro and in vivo." Stem cells international 2015 (2015).

36) Yin, F. M., L. B. Xiao, and Y. Zhang. "Research progress on Drynaria fortunei naringin on inflammation and bone activity." Zhongguo gu shang= China journal of orthopaedics and traumatology 28.2 (2015): 182-186.

37) Yu, Li-Ming, et al. "Naringenin improves mitochondrial function and reduces cardiac damage following ischemia-reperfusion injury: the role of the AMPK-SIRT3 signaling pathway." Food & function 10.5 (2019): 2752-2765.

38) Zaidun, Nurul Hannim, Zar Chi Thent, and Azian Abd Latiff. "Combating oxidative stress disorders with citrus flavonoid: Naringenin." Life sciences 208 (2018): 111-122.

39) Zeng, Wenfeng, et al. "Naringenin as a potential immunomodulator in therapeutics." Pharmacological research 135 (2018): 122-126.

40) Zhai, Yuan-Kun, et al. "Effects of naringin on proliferation, differentiation and maturation of rat calvarial osteoblasts in vitro." Zhongguo Zhong yao za zhi= Zhongguo zhongyao zazhi= China journal of Chinese materia medica 38.1 (2013): 105-111.

41) Zhang, Yili, et al. "Total flavonoids from Rhizoma Drynariae (Gusuibu) for treating osteoporotic fractures: implication in clinical practice." Drug Design, Development and Therapy 11 (2017): 1881.

42) Zhao, Yunpeng, et al. "Naringin Protects Against Cartilage Destruction in Osteoarthritis Through Repression of NF-κB Signaling Pathway." Inflammation 39.1 (2016): 385-392.

43) Zygmunt, Katarzyna, et al. "Naringenin, a citrus flavonoid, increases muscle cell glucose uptake via AMPK." Biochemical and biophysical research communications 398.2 (2010): 178-183.

## CHAPTER 24 - POLYPODIUM LEUCOTOMOS

1) Berman, Brian, Charles Ellis, and Craig Elmets. "Polypodium leucotomos-an overview of basic investigative findings." Journal of drugs in dermatology: JDD 15.2 (2016): 224.

2) Del Rosso, James Q. "Polypodium Leucotomos Extract (PLE): New Study Gives Evidencebased Insight into "Ain't Nothing Like the Real Thing"." The Journal of clinical and aesthetic dermatology 12.8 (2019): 45.

3) Delgado-Wicke, Pablo, et al. "Fernblock® Upregulates NRF2 Antioxidant Pathway and Protects Keratinocytes from PM2. 5-Induced Xenotoxic Stress." Oxidative Medicine and Cellular Longevity 2020 (2020).

4) Emanuel, Patrick, and Noah Scheinfeld. "A review of DNA repair and possible DNA-repair adjuvants and selected natural anti-oxidants." Dermatol Online J 13.3 (2007): 10.

5) Fernandez-Novoa, L., et al. "Effects of anapsos on the activity of the enzyme Cu-Znsuperoxide dismutase in an animal model of neuronal degeneration." Methods and findings in experimental and clinical pharmacology 19.2 (1997): 99.

6) Gonzalez, S., et al. "An extract of the fern Polypodium leucotomos (Difur®) modulates Th1/ Th2 cytokines balance in vitro and appears to exhibit anti-angiogenic activities in vivo: pathogenic relationships and therapeutic implications." Anticancer research 20.3A (2000): 1567-1576.

7) Gonzalez, Salvador, Yolanda Gilaberte, and Neena Philips. "Mechanistic insights in the use of a Polypodium leucotomos extract as an oral and topical photoprotective agent." Photochemical & Photobiological Sciences 9.4 (2010): 559-563.

8) Jiménez, D., et al. "Anapsos, an antipsoriatic drug, in atopic dermatitis." Allergologia et Immunopathologia 15.4 (1987): 185.

9) López-Abán, Julio, et al. "Immunomodulation of the response to excretory/ secretory antigens of Fasciola hepatica by Anapsos® in Balb/C mice and rat alveolar macrophages." Journal of Parasitology 93.2 (2007): 428-432.

10) Nestor, Mark, et al. "Polypodium leucotomos as an adjunct treatment of pigmentary disorders." The Journal of clinical and aesthetic dermatology 7.3 (2014): 13.

11) Nestor, Mark S., Brian Berman, and Nicole Swenson. "Safety and efficacy of oral Polypodium leucotomos extract in healthy adult subjects." The Journal of clinical and aesthetic dermatology 8.2 (2015): 19.

12) Parrado, Concepcion, et al. "Fernblock (Polypodium leucotomos extract): molecular mechanisms and pleiotropic effects in light-related skin conditions, photoaging and skin cancers, a review." International journal of molecular sciences 17.7 (2016): 1026.

13) Philips, Neena, et al. "Predominant effects of Polypodium leucotomos on membrane integrity, lipid peroxidation, and expression of elastin and matrixmetalloproteinase-1 in ultraviolet radiation exposed fibroblasts, and keratinocytes." Journal of dermatological science 32.1 (2003): 1-9.

14) Philips, Neena, et al. "Beneficial regulation of matrixmetalloproteinases and their inhibitors, fibrillar collagens and transforming growth factor-β by Polypodium leucotomos, directly or in dermal fibroblasts, ultraviolet radiated fibroblasts, and melanoma cells." Archives of dermatological research 301.7 (2009): 487.

15) Philips, Neena, David Bynum, and Hyeondo Hwang. "Counteraction of skin inflammation and aging or cancer by polyphenols and flavonoids from Polypodium leucotomos and xanthohumol." Anti-Inflammatory & Anti-Allergy Agents in Medicinal Chemistry (Formerly Current Medicinal Chemistry-Anti-Inflammatory and Anti-Allergy Agents) 9.2 (2010): 142-149.

16) Sempere-Ortells, J. M., et al. "Anapsos (Polypodium leucotomos) modulates lymphoid cells and the expression of adhesion molecules." Pharmacological research 46.2 (2002): 185-190.

17) Sempere, J. M., et al. "Effect of Anapsos (Polypodium leucotomos extract) on in vitro production of cytokines." British journal of clinical pharmacology 43.1 (1997): 85-89.

18) Solivellas, Bartolomé Marí, and Teo Cabanes Martín. "Polypodium leucotomos Extract use to prevent and reduce the risk of infectious diseases in high performance athletes." Infection and drug resistance 5 (2012): 149.

19) Winkelmann, Richard R., James Del Rosso, and Darrell S. Rigel. "Polypodium leucotomos extract: a status report on clinical efficacy and safety." Journal of Drugs in Dermatology: JDD 14.3 (2015): 254-261.

20) Zamarrón, Alicia, et al. "Fernblock prevents dermal cell damage induced by visible and infrared a radiation." International journal of molecular sciences 19.8 (2018): 2250.

21) Zattra, Edoardo, et al. "Polypodium leucotomos extract decreases UV-induced Cox-2 expression and inflammation, enhances DNA repair, and decreases mutagenesis in hairless mice." The American journal of pathology 175.5 (2009): 1952-1961.

## CHAPTER 25 - PYRROLOQUINOLINE QUINONE

1) Akagawa, Mitsugu, Masahiko Nakano, and Kazuto Ikemoto. "Recent progress in studies on the health benefits of pyrroloquinoline quinone." Bioscience, biotechnology, and biochemistry 80.1 (2016): 13-22.

2) Cheng, Qiong, et al. "Pyrroloquinoline quinone promotes mitochondrial biogenesis in rotenone-induced Parkinson's disease model via AMPK activation." Acta Pharmacologica Sinica (2020): 1-14.

3) Geng, Qinghe, et al. "Pyrroloquinoline quinone prevents estrogen deficiency-induced osteoporosis by inhibiting oxidative stress and osteocyte senescence." International Journal of Biological Sciences 15.1 (2019): 58.

4) Harris, Calliandra B., et al. "Dietary pyrroloquinoline quinone (PQQ) alters indicators of inflammation and mitochondrial-related metabolism in human subjects." The Journal of nutritional biochemistry 24.12 (2013): 2076-2084.

5) He, Kai, et al. "Antioxidant and pro-oxidant properties of pyrroloquinoline quinone (PQQ): implications for its function in biological systems." Biochemical pharmacology 65.1 (2003): 67-74.

6) Huang, Caiyun, et al. "Pyrroloquinoline quinone alleviates jejunal mucosal barrier function damage and regulates colonic microbiota in piglets challenged with enterotoxigenic Escherichia coli." Frontiers in microbiology 11 (2020): 1754.

7) Huang, Yuanqing, Dengshun Miao, and Ning Chen. "The treatment effects and mechanisms of pyrroloquinoline quinone on defective teeth and mandible in Bmi-1 knockout mice." Zhonghua kou qiang yi xue za zhi= Zhonghua kouqiang yixue zazhi= Chinese journal of stomatology 50.8 (2015): 496-502.

8) Huang, Yuanqing, Ning Chen, and Dengshun Miao. "Effect and mechanism of pyrroloquinoline quinone on anti-osteoporosis in Bmi-1 knockout mice-Anti-oxidant effect of pyrroloquinoline quinone." American Journal of Translational Research 9.10 (2017): 4361.

9) Hwang, Paul S., et al. "Effects of pyrroloquinoline quinone (PQQ) supplementation on aerobic exercise performance and indices of mitochondrial biogenesis in untrained men." Journal of the American College of Nutrition 39.6 (2020): 547-556.

10) Hwang, Paul, and Darryn S. Willoughby. "Mechanisms behind Pyrroloquinoline Quinone supplementation on skeletal muscle mitochondrial biogenesis: possible synergistic effects with exercise." Journal of the American College of Nutrition 37.8 (2018): 738-748.

11) Itoh, Yuji, et al. "Effect of the antioxidant supplement pyrroloquinoline quinone disodium salt (BioPQQ™) on cognitive functions." Oxygen Transport to Tissue XXXVII. Springer, New York, NY, 2016. 319-325.

12) Jiang, Xiaoxia, et al. "Hepatoprotective effect of pyrroloquinoline quinone against alcoholic liver injury through activating Nrf2-mediated antioxidant and inhibiting TLR4-mediated inflammation responses." Process Biochemistry (2020).

13) Kumar, Narendra, and Anand Kar. "Pyrroloquinoline quinone (PQQ) has potential to ameliorate streptozotocin-induced diabetes mellitus and oxidative stress in mice: a histopathological and biochemical study." Chemico-biological interactions 240 (2015): 278-290.

14) Kumar, Suresh, et al. "A novel quinoline based second-generation mTOR inhibitor that induces apoptosis and disrupts PI3K-Akt-mTOR signaling in human leukemia HL-60 cells." Anti- Cancer Agents in Medicinal Chemistry (Formerly Current Medicinal Chemistry-Anti-Cancer Agents) 15.10 (2015): 1297-1304.

15) Ma, Wenjing, et al. "PQQ ameliorates skeletal muscle atrophy, mitophagy and fiber type transition induced by denervation via inhibition of the inflammatory signaling pathways." Annals of translational medicine 7.18 (2019).

16) Nakano, Masahiko, et al. "Effects of pyrroloquinoline quinone disodium salt intake on the serum cholesterol levels of healthy Japanese adults." Journal of nutritional science and vitaminology 61.3 (2015): 233-240.

17) Naveed, M., et al. "The life history of pyrroloquinoline quinone (PQQ): a versatile molecule with novel impacts on living systems." Mol Biol 1 (2016): 29-46.

18) Saihara, Kazuhiro, et al. "Pyrroloquinoline quinone, a redox-active

*o-quinone, stimulates mitochondrial biogenesis by activating the SIRT1/PGC-1α signaling pathway." Biochemistry 56.50 (2017): 6615-6625.*

*19) Wu, J. Z., et al. "Pyrroloquinoline quinone enhances the resistance to oxidative stress and extends lifespan upon DAF-16 and SKN-1 activities in C. elegans." Experimental gerontology 80 (2016): 43-50.*

*20) Wu, Xuan, et al. "Pyrroloquinoline quinone prevents testosterone deficiency-induced osteoporosis by stimulating osteoblastic bone formation and inhibiting osteoclastic bone resorption." American journal of translational research 9.3 (2017): 1230.*

*21) Yamada, Yasue, et al. "Effects of pyrroloquinoline quinone and imidazole pyrroloquinoline on biological activities and neural functions." Heliyon 6.1 (2020): e03240.*

*22) Yang, Chongfei, et al. "Pyrroloquinoline quinone (PQQ) inhibits lipopolysaccharide induced inflammation in part via downregulated NF-κB and p38/JNK activation in microglial and attenuates microglia activation in lipopolysaccharide treatment mice." PLoS One 9.10 (2014): e109502.*

*23) Zhang, Hongyun, et al. "Pyrroloquinoline quinone inhibits the production of inflammatory cytokines via the SIRT1/NF-κB signal pathway in weaned piglet jejunum." Food & Function 11.3 (2020): 2137-2153.*

*24) Zhang, Jian, et al. "Pyrroloquinoline quinone increases the expression and activity of Sirt1 and-3 genes in HepG2 cells." Nutrition Research 35.9 (2015): 844-849.*

**CHAPTER 26** - SALIDROSIDE

*1) Abidov, M., et al. "Effect of extracts from Rhodiola rosea and Rhodiola crenulata (Crassulaceae) roots on ATP content in mitochondria of skeletal muscles." Bulletin of experimental biology and medicine 136.6 (2003): 585-587.*

*2) Ballmann, Christopher G., et al. "Effects of short-term Rhodiola Rosea (Golden Root Extract) supplementation on anaerobic exercise performance." Journal of sports sciences 37.9 (2019): 998-1003.*

*3) Bai, Hai, et al. "Effects of salidroside on proliferation of bone marrow mesenchymal stem cells." Zhongguo shi yan xue ye xue za zhi 22.4 (2014): 1072-1077.*

4) Cai, Linlin, et al. "Salidroside protects rat liver against ischemia/reperfusion injury by regulating the GSK-3β/Nrf2-dependent antioxidant response and mitochondrial permeability transition." European journal of pharmacology 806 (2017): 32-42.

5) Chen, Jin-Jing, et al. "Salidroside stimulates osteoblast differentiation through BMP signaling pathway." Food and chemical toxicology 62 (2013): 499-505.

6) Chen, Tong, et al. "Suppressing receptor-interacting protein 140: a new sight for salidroside to treat cerebral ischemia." Molecular neurobiology 53.9 (2016): 6240-6250.

7) Chen, Ya-Nan, et al. "Salidroside via ERK1/2 and PI3K/AKT/mTOR signal pathway induces mouse bone marrow mesenchymal stem cells differentiation into neural cells." Yao xue xue bao= Acta pharmaceutica Sinica 48.8 (2013): 1247-1252.

8) Cropley, Mark, Adrian P. Banks, and Julia Boyle. "The effects of Rhodiola rosea L. extract on anxiety, stress, cognition and other mood symptoms." Phytotherapy research 29.12 (2015): 1934-1939.

9) Feng, Tian, et al. "Salidroside, a scavenger of ROS, enhances the radioprotective effect of Ex-RAD® via a p53-dependent apoptotic pathway." Oncology Reports 38.5 (2017): 3094-3102.

10) Gao, Jin, et al. "Salidroside suppresses inflammation in a D-galactose-induced rat model of Alzheimer's disease via SIRT1/NF-κB pathway." Metabolic brain disease 31.4 (2016): 771-778.

11) Gen-Xiang, M. A. O., et al. "Protective role of salidroside against aging in a mouse model induced by D-galactose." Biomedical and Environmental Sciences 23.2 (2010): 161-166

12) Guo, Qiaoyun, et al. "Salidroside improves angiogenesis-osteogenesis coupling by regulating the HIF-1α/VEGF signalling pathway in the bone environment." European Journal of Pharmacology 884 (2020): 173394.

13) Guo, Xiao Qin, et al. "Salidroside accelerates fracture healing through cell-autonomous and non-autonomous effects on osteoblasts." Cell and tissue research 367.2 (2017): 197-211.

14) Hu, Rong, et al. "Salidroside ameliorates endothelial inflammation and oxidative stress by regulating the AMPK/NF-κB/NLRP3 signaling pathway in AGEs-induced HUVECs." European journal of pharmacology 867 (2020): 172797.

15) Jiang, Ya-Ping, et al. "Protective effects of salidroside on spermatogenesis in streptozotocin induced type-1 diabetic male mice by inhibiting oxidative stress mediated blood-testis barrier damage." Chemico-Biological Interactions 315 (2020): 108869.

16) Jin, Huijuan, et al. "Therapeutic intervention of learning and memory decays by salidroside stimulation of neurogenesis in aging." Molecular neurobiology 53.2 (2016): 851-866.

17) Ju, Linjie, et al. "Salidroside, a natural antioxidant, improves β-cell survival and function via activating AMPK pathway." Frontiers in pharmacology 8 (2017): 749.

18) Kasper, Siegfried, and Angelika Dienel. "Multicenter, open-label, exploratory clinical trial with Rhodiola rosea extract in patients suffering from burnout symptoms." Neuropsychiatric disease and treatment 13 (2017): 889.

19) Kumar, Hemant, and Dong-Kug Choi. "Hypoxia inducible factor pathway and physiological adaptation: a cell survival pathway?." Mediators of inflammation 2015 (2015).

20) Li, Fenglin, et al. "Protective effect of salidroside from Rhodiolae Radix on diabetes-induced oxidative stress in mice." Molecules 16.12 (2011): 9912-9924.

21) Li, Ma, et al. "Anti-fatigue effects of salidroside in mice." Journal of medical colleges of PLA 23.2 (2008): 88-93.

22) Li, Ruru, et al. "Salidroside protects dopaminergic neurons by preserving complex I activity via DJ-1/Nrf2-mediated antioxidant pathway." Parkinson's Disease 2019 (2019).

23) Li, Xue, et al. "Salidroside stimulates DNA repair enzyme Parp-1 activity in mouse HSC maintenance." Blood, The Journal of the American Society of Hematology 119.18 (2012): 4162-4173.

24) Li, Xue, et al. "Binding to WGR domain by salidroside activates PARP1 and protects hematopoietic stem cells from oxidative stress." Antioxidants & redox signaling 20.12 (2014): 1853-1865.

25) Li, Yonghong, et al. "Rhodiola rosea L.: an herb with anti-stress, anti-aging, and immunostimulating properties for cancer chemoprevention." Current pharmacology reports 3.6 (2017): 384-395.

26) Ling, Zhao, et al. "Salidroside induces the differentiation of mouse bone marrow mesenchymal stem cells into neuron-like cells mediated by calcium/ calmodulin signaling pathway." Journal of Clinical Rehabilitative Tissue Engineering Research 18.37 (2014): 6019-6023.

27) Luo, Fen, et al. "Salidroside alleviates cigarette smoke-induced COPD in mice." Biomedicine & Pharmacotherapy 86 (2017): 155-161.

28) Ma, Gou-ping, et al. "Rhodiola rosea L. Improves learning and memory function: preclinical evidence and possible mechanisms." Frontiers in pharmacology 9 (2018): 1415.

29) Mao, Gen-Xiang, et al. "Salidroside delays cellular senescence by stimulating mitochondrial biogenesis partly through a miR-22/SIRT-1 pathway." Oxidative medicine and cellular longevity 2019 (2019).

30) Ni, Gui-Lian, et al. "Salidroside ameliorates diabetic neuropathic pain in rats by inhibiting neuroinflammation." Journal of Molecular Neuroscience 63.1 (2017): 9-16.

31) Panossian, Alexander, and Georg Wikman. "Evidence-based efficacy of adaptogens in fatigue, and molecular mechanisms related to their stress-protective activity." Current clinical pharmacology 4.3 (2009): 198-219.

32) Perfumi, Marina, and Laura Mattioli. "Adaptogenic and central nervous system effects of single doses of 3% rosavin and 1% salidroside Rhodiola rosea L. extract in mice." Phytotherapy Research: An International Journal Devoted to Pharmacological and Toxicological Evaluation of Natural Product Derivatives 21.1 (2007): 37-43.

33) Qu, Ze-qiang, et al. "Protective effects of a Rhodiola crenulata extract and salidroside on hippocampal neurogenesis against streptozotocin-induced neural injury in the rat." PloS one 7.1 (2012): e29641.

34) Sun, Siyu, et al. "Antioxidant Effects of Salidroside in the Cardiovascular System." Evidence-Based Complementary and Alternative Medicine 2020 (2020).

35) Wu, Dan, et al. "Salidroside suppresses solar ultraviolet-induced skin inflammation by targeting cyclooxygenase-2." Oncotarget 7.18 (2016): 25971.

36) Xie, Zeping, et al. "Salidroside Attenuates Cognitive Dysfunction in Senescence-Accelerated Mouse Prone 8 (SAMP8) Mice and Modulates

*Inflammation of the Gut-Brain Axis." Frontiers in pharmacology 11 (2020).*

37) Xing, Shasha, et al. "Salidroside stimulates mitochondrial biogenesis and protects against H2O2-induced endothelial dysfunction." Oxidative medicine and cellular longevity 2014 (2014).

38) Xing, Sha-Sha, et al. "Salidroside attenuates endothelial cellular senescence via decreasing the expression of inflammatory cytokines and increasing the expression of SIRT3." Mechanisms of Ageing and Development 175 (2018): 1-6.

39) Xu, Jiansheng, and Ying Li. "Effects of salidroside on exhaustive exercise-induced oxidative stress in rats." Molecular medicine reports 6.5 (2012): 1195-1198.

40) Xu, Mao-Chun, et al. "Salidroside protects against hydrogen peroxide-induced injury in HUVECs via the regulation of REDD1 and mTOR activation." Molecular medicine reports 8.1 (2013): 147-153.

41) You, Baiyang, et al. "Anti-insulin resistance effects of salidroside through mitochondrial quality control." Journal of Endocrinology 244.2 (2020): 383-393.

42) Zhang, Jin-Kang, et al. "Protection by salidroside against bone loss via inhibition of oxidative stress and bone-resorbing mediators." PloS one 8.2 (2013): e57251.

43) Zhang, Jinping, et al. "Salidroside protects cardiomyocyte against hypoxia-induced death: A HIF-1α-activated and VEGF-mediated pathway." European Journal of Pharmacology 607.1-3 (2009): 6-14.

44) Zhao, Dongming, et al. "Salidroside attenuates oxidized low-density lipoprotein-induced endothelial cell injury via promotion of the AMPK/SIRT1 pathway." International Journal of Molecular Medicine 43.6 (2019): 2279-2290.

45) Zhao, Hong-Bin, et al. "Salidroside induces neuronal differentiation of mouse mesenchymal stem cells through Notch and BMP signaling pathways." Food and chemical toxicology 71 (2014): 60-67.

46) Zheng, Tao, et al. "Salidroside ameliorates insulin resistance through activation of a mitochondria-associated AMPK/PI3K/A kt/GSK 3β pathway." British journal of pharmacology 172.13 (2015): 3284-3301

47) Zheng, Xiang-Tao, et al. "Induction of autophagy by salidroside through

the AMPK-mTOR pathway protects vascular endothelial cells from oxidative stress-induced apoptosis." Molecular and cellular biochemistry 425.1-2 (2017): 125-138.

48) Zhong, Zhifeng, et al. "Pharmacological activities, mechanisms of action, and safety of salidroside in the central nervous system." Drug design, development and therapy 12 (2018):1479.

49) Zhu, Lingpeng, et al. "Salidroside ameliorates arthritis-induced brain cognition deficits by regulating Rho/ROCK/NF-κB pathway." Neuropharmacology 103 (2016): 134-142.

50) Zhu, Yao, et al. "Salidroside suppresses HUVECs cell injury induced by oxidative stress through activating the Nrf2 signaling pathway." Molecules 21.8 (2016): 1033.

## CHAPTER 27 - SHILAJIT

1) Bhardwaj, Payal, Mehak Goel, and Durg Vijay Rai. "Biophysical studies of administered Shilajit on rat bone tissue." Indian J. Applied & Pure Bio. Vol 31.1 (2016): 27-34.

2) Bhattacharya, Salil K., Ananda P. Sen, and Shibnath Ghosal. "Effects of Shilajit on biogenic free radicals." Phytotherapy Research 9.1 (1995): 56-59.

3) Bhattacharya, Salil K. "Shilajit attenuates streptozotocin induced diabetes mellitus and decrease in pancreatic islet superoxide dismutase activity in rats." Phytotherapy research 9.1 (1995): 41-44.

4) Bhavsar, Shailesh K., Aswin M. Thaker, and Jitendra K. Malik. "Shilajit." Nutraceuticals. Academic Press, 2016. 707-716.

5) Biswas, Tuhin Kanti, et al. "Clinical evaluation of spermatogenic activity of processed Shilajit in oligospermia." Andrologia 42.1 (2010): 48-56.

6) Carrasco-Gallardo, Carlos, Leonardo Guzmán, and Ricardo B. Maccioni. "Shilajit: a natural phytocomplex with potential procognitive activity." International Journal of Alzheimer's disease 2012 (2012).

7) Das, Amitava, et al. "Skin Transcriptome of Middle-Aged Women Supplemented With Natural Herbo-mineral Shilajit Shows Induction of Microvascular and Extracellular Matrix Mechanisms." Journal of the American College of Nutrition 38.6 (2019): 526-536.

8) Ghosal, S., et al. "Effects of Shilajit and its active constituents on learning and memory in rats." Phytotherapy research 7.1 (1993): 29-34.

9) Hemant, Dhongade, et al. "Free radical scavenging effects of the hydro-alcoholic extract of Shilajit." Hamdard Medicus 53.1 (2010): 87-90.

10) Ikram-ul-Haq, Muhammad, et al. "Effects of asphaltum (Shilajit) on scrotal circumference and semen quality parameters of adult Lohi rams." (2016).

11) Jaiswal, A. K., and S. K. Bhattacharya. "Effects of Shilajit on memory, anxiety and brain monoamines in rats." Indian J Pharmacol 24.1 (1992): 12-7.

12) Khaksari, Mohammad, et al. "The effects of shilajit on brain edema, intracranial pressure and neurologic outcomes following the traumatic brain injury in rat." Iranian journal of basic medical sciences 16.7 (2013): 858.

13) Meena, Harsahay, et al. "Shilajit: A panacea for high-altitude problems." International journal of Ayurveda research 1.1 (2010): 37.

14) Mirza, Mohd Aamir, et al. "Shilajit: An ancient panacea." Int J Curr Pharmaceut Rev Res 1 (2010): 2-11.

15) Mishra, Raghav Kumar, Ashish Jain, and Shio Kumar Singh. "Profertility effects of Shilajit on cadmium-induced infertility in male mice." Andrologia 50.8 (2018): e13064.

16) Pandit, S., et al. "Clinical evaluation of purified Shilajit on testosterone levels in healthy volunteers." Andrologia 48.5 (2016): 570-575.

17) Pant, Kishor, Bimala Singh, and Nagendra Thakur. "Shilajit: a humic matter panacea for cancer." (2012).

18) Park, Jeong-Sook, Gee-Young Kim, and Kun Han. "The spermatogenic and ovogenic effects of chronically administered Shilajit to rats." Journal of ethnopharmacology 107.3 (2006): 349-353.

19) Rege, Anuya, Parikshit Juvekar, and Archana Juvekar. "In vitro antioxidant and anti-arthritic activities of Shilajit." Int J Pharm Pharm Sci 4.2 (2012): 650-3.

20) Sharma, Praveen, et al. "Shilajit: evalution of its effects on blood chemistry of normal human subjects." Ancient science of life 23.2 (2003): 114.

21) Shilajiyin Hizli, Maksiller Genisletme Tedavisinde, et al. "Does shillajit

have an effect on new bone remodeling in the rapid maxillary expansion treatment? TA Biomedical Histopathological and Immunohistochemical study.." Med J SDU/SDÜ Tıp Fak Derg 26.1 (2019): 96-103.

22) Stohs, Sidney J. "Safety and efficacy of shilajit (mumie, moomiyo)." Phytotherapy research28.4 (2014): 475-479.

23) Stohs, Sidney J., et al. "Energy and Health Benefits of Shilajit." Sustained Energy for Enhanced Human Functions and Activity. Academic Press, 2017. 187-204. 24) Surapaneni, Dinesh Kumar, et al. "Shilajit attenuates behavioral symptoms of chronic fatigue syndrome by modulating the hypothalamic–pituitary–adrenal axis and mitochondrial bioenergetics in rats." Journal of ethnopharmacology 143.1 (2012): 91-99.

25) Talbert, Robert, et al. "Chemistry of Shilajit." Complement (2014).

26) Talbert, Robert, et al. "Therapeutic Potentials of "Shilajit Rasayana." Complement (2014).

27) Velmurugan, C., et al. "Evaluation of safety profile of black shilajit after 91 days repeated administration in rats." Asian Pacific journal of tropical biomedicine 2.3 (2012): 210-214.

28) Zubair, Muhammad. "Effects of asphaltum (Shilajit) on male reproductive system (mini review)." sexual dysfunction 3: 1.

## CHAPTER 28 - SPERMIDINE

1) Ali, Mohamed Atiya, et al. "Polyamines: total daily intake in adolescents compared to the intake estimated from the Swedish Nutrition Recommendations Objectified (SNO)." Food & Nutrition Research 55.1 (2011): 5455.

2) Alsaleh, Ghada, et al. "Autophagy in T cells from aged donors is maintained by spermidine, and correlates with function and vaccine responses." BioRxiv (2020)

3) Brooks, Wesley H. "Systemic lupus erythematosus and related autoimmune diseases are antigen-driven, epigenetic diseases." Medical hypotheses 59.6 (2002): 736-741.

4) Brysont, K., and R. J. Greenall. "Binding sites of the polyamines putrescine,

cadaverine, spermidine and spermine on A-and B-DNA located by simulated annealing." *Journal of Biomolecular Structure and Dynamics* 18.3 (2000): 393-412.

5) Carriche, Guilhermina M., et al. "Regulating T cell differentiation through the polyamine spermidine." *Journal of Allergy and Clinical Immunology* (2020).

6) Chiu, Jen-Fu, and S. C. Sung. "Effect of spermidine on the activity of DNA polymerases." *Biochimica et Biophysica Acta (BBA)-Nucleic Acids and Protein Synthesis* 281.4 (1972): 535-542.

7) Deng, Hong, et al. "Structural basis of polyamine–DNA recognition: spermidine and spermine interactions with genomic B-DNAs of different GC content probed by Raman spectroscopy." *Nucleic acids research* 28.17 (2000): 3379-3385.\

8) Eisenberg, Tobias, et al. "Cardioprotection and lifespan extension by the natural polyamine spermidine." *Nature medicine* 22.12 (2016): 1428-1438.

9) Fischer, Maximilian, et al. "Spermine and spermidine modulate T-cell function in older adults with and without cognitive decline ex vivo." *Aging (Albany NY)* 12.13 (2020): 13716.

10) Gao, Mingyue, et al. "Spermidine ameliorates non-alcoholic fatty liver disease through regulating lipid metabolism via AMPK." *Biochemical and biophysical research communications* 505.1 (2018): 93-98.

11) Ghosh, Indrani, et al. "Spermidine, an autophagy inducer, as a therapeutic strategy in neurological disorders." *Neuropeptides* (2020): 102083.

12) Gugliucci, A., and T. Menini. "The polyamines spermine and spermidine protect proteins from structural and functional damage by AGE precursors: a new role for old molecules?." *Life sciences* 72.23 (2003): 2603-2616.

13) Khan, Ahsan U., et al. "Spermine and spermidine protection of plasmid DNA against single-strand breaks induced by singlet oxygen." *Proceedings of the National Academy of Sciences* 89.23 (1992): 11428-11430.

14) Kiechl, Stefan, et al. "Higher spermidine intake is linked to lower mortality: a prospective population-based study." *The American journal of clinical nutrition* 108.2 (2018): 371-380.

15) LaRocca, Thomas J., et al. "The autophagy enhancer spermidine reverses

*arterial aging." Mechanisms of ageing and development 134.7-8 (2013): 314-320.*

16) Liu, Shuai, et al. "Spermidine Suppresses Development of Experimental Abdominal Aortic Aneurysms." *Journal of the American Heart Association 9.8 (2020): e014757.*

17) Liquori, A. M., et al. "Complexes between DNA and polyamines: a molecular model." *Journal of Molecular Biology 24.1 (1967): 113-122.*

18) Madeo, Frank, et al. "Spermidine in health and disease." *Science 359.6374 (2018).*

19) Madeo, Frank, et al. "Spermidine delays aging in humans." *Aging (Albany NY) 10.8 (2018): 2209.*

20) Madeo, Frank, et al. "Spermidine: a physiological autophagy inducer acting as an anti-aging vitamin in humans?." *Autophagy 15.1 (2019): 165-168.*

21) Maglione, Marta, et al. "Spermidine protects from age-related synaptic alterations at hippocampal mossy fiber-CA3 synapses." *Scientific reports 9.1 (2019): 1-12.*

22) Méndez, José D. "Polyamine Metabolism in Sperm Cell." *Diabetes Complications 2.2 (2018): 1-3.*

23) Metur, Shree Padma, and Daniel J. Klionsky. "The curious case of polyamines: spermidine drives reversal of B cell senescence." *(2020): 389-390.*

24) Minois, Nadège. "Molecular Basis of the 'Anti-Aging'Effect of Spermidine and Other Natural Polyamines-A Mini-Review." *Gerontology 60.4 (2014): 319-326.*

25) Morgan, James E., James W. Blankenship, and Harry R. Matthews. "Association constants for the interaction of double-stranded and single-stranded DNA with spermine, spermidine, putrescine, diaminopropane, N1- and N8-acetylspermidine, and magnesium: determination from analysis of the broadening of thermal denaturation curves." *Archives of biochemistry and biophysics 246.1 (1986): 225-232.*

26) Muñoz-Esparza, Nelly C., et al. "Polyamines in food." *Frontiers in nutrition 6 (2019).*

27) Park, In-Hwan, and Moon-Moo Kim. "Spermidine inhibits MMP-2 via

*modulation of histone acetyltransferase and histone deacetylase in HDFs." International journal of biological macromolecules 51.5 (2012): 1003-1007.*

*28) Ramos-Molina, Bruno, et al. "Dietary and gut microbiota polyamines in obesity-and age-related diseases." Frontiers in Nutrition 6 (2019): 24.*

*29) Sacitharan, Pradeep K., et al. "Spermidine restores dysregulated autophagy and polyamine synthesis in aged and osteoarthritic chondrocytes via EP300." Experimental & molecular medicine 50.9 (2018): 1-10.*

*30) Spotheim-Maurizot, M., et al. "Radioprotection of DNA by polyamines." International journal of radiation biology 68.5 (1995): 571-577.*

*31) Wang, Junying, et al. "Spermidine alleviates cardiac aging by improving mitochondrial biogenesis and function." Aging (Albany NY) 12.1 (2020): 650.*

*32) Wirth, Miranka, et al. "The effect of spermidine on memory performance in older adults at risk for dementia: A randomized controlled trial." Cortex 109 (2018): 181-188.*

*33) Xu, Ting-Ting, et al. "Spermidine and spermine delay brain aging by inducing autophagy in SAMP8 mice." Aging (Albany NY) 12.7 (2020): 6401.*

*34) Yan, Jing, et al. "Spermidine-enhanced autophagic flux improves cardiac dysfunction following myocardial infarction by targeting the AMPK/mTOR signalling pathway." British journal of pharmacology 176.17 (2019): 3126-3142.*

*35) Zhang, Hanlin, and Anna Katharina Simon. "Polyamines reverse immune senescence via the translational control of autophagy." Autophagy 16.1 (2020): 181-182.*

**CHAPTER 29** - URSOLIC ACID

*1) Bahrami, Soroush Alaghehband, and Nuredin Bakhtiari. "Ursolic acid regulates aging process through enhancing of metabolic sensor proteins level." Biomedicine & Pharmacotherapy 82 (2016): 8-14.*

*2) Bakhtiari, Nuredin, et al. "Ursolic acid ameliorates aging-metabolic phenotype through promoting of skeletal muscle rejuvenation." Medical Hypotheses 85.1 (2015): 1-6.*

*3) Bakhtiari, Nuredin, Elham Moslemee-Jalalvand, and Jahanfard Kazemi.*

*"Ursolic acid: a versatile triterpenoid compound in regulating the aging." Physiology and Pharmacology 21.1 (2017): 15-24.*

4) Bhat, Rafia A., et al. "Effect of ursolic acid in attenuating chronic constriction injury-induced neuropathic pain in rats." *Fundamental & clinical pharmacology* 30.6 (2016): 517-528.

5) Chen, Jihang, et al. "Ursolic acid induces mitochondrial biogenesis through the activation of AMPK and PGC-1 in C2C12 myotubes: A possible mechanism underlying its beneficial effect on exercise endurance." *Food & Function* (2017).

6) Chu, Xia, et al. "Ursolic acid increases energy expenditure through enhancing free fatty acid uptake and β-oxidation via an UCP3/AMPK-dependent pathway in skeletal muscle." *Molecular nutrition & food research* 59.8 (2015): 1491-1503.

7) Colla, André RS, et al. "Serotonergic and noradrenergic systems are implicated in the antidepressant-like effect of ursolic acid in mice." *Pharmacology Biochemistry and Behavior* 124 (2014): 108-116.

8) Gharibi, Shadi, Nuredin Bakhtiari, and Elham-Moslemee Jalalvand. "Ursolic Acid Mediates Hepatic Protection through Enhancing of anti-aging Biomarkers." *Current aging science* (2017).

9) Habtemariam, Solomon. "Antioxidant and anti-inflammatory mechanisms of neuroprotection by ursolic acid: addressing brain injury, cerebral ischemia, cognition deficit, anxiety, and depression." *Oxidative medicine and cellular longevity* 2019 (2019).

10) Hao, Wangjun, et al. "Ursolic acid alleviates hypercholesterolemia and modulates the gut microbiota in hamsters." *Food & Function* 11.7 (2020): 6091-6103.

11) He, Yonghan, et al. "Ursolic acid inhibits adipogenesis in 3T3-L1 adipocytes through LKB1/ AMPK pathway." *PloS one* 8.7 (2013): e70135.

12) Jia, Yaoyao, et al. "Ursolic acid improves lipid and glucose metabolism in high-fat-fed C57BL/6J mice by activating peroxisome proliferator-activated receptor alpha and hepatic autophagy." *Molecular nutrition & food research* 59.2 (2015): 344-354.

13) Jiang, Qixiao, et al. "SIRT1/Atg5/autophagy are involved in the antiatherosclerosis effects of ursolic acid." *Molecular and cellular biochemistry*

420.1-2 (2016): 171-184.

14) Katashima, Carlos K., et al. "Ursolic acid and mechanisms of actions on adipose and muscle tissue: a systematic review." Obesity Reviews 18.6 (2017): 700-711.

15) Kim, Hyuck, et al. "Epigenetic modifications of triterpenoid ursolic acid in activating Nrf2 and blocking cellular transformation of mouse epidermal cells." The Journal of nutritional biochemistry 33 (2016): 54-62.

16) Kim, Minjung, et al. "The combination of ursolic acid and leucine potentiates the differentiation of C2C12 murine myoblasts through the mTOR signaling pathway." International journal of molecular medicine 35.3 (2015): 755-762.

17) Kunkel, Steven D., et al. "Ursolic acid increases skeletal muscle and brown fat and decreases diet-induced obesity, glucose intolerance and fatty liver disease." PloS one 7.6 (2012): e39332.

18) Leng, Shuilong, et al. "Ursolic acid enhances macrophage autophagy and attenuates atherogenesis." Journal of lipid research 57.6 (2016): 1006-1016.

19) Li, Litao, et al. "Ursolic acid promotes the neuroprotection by activating Nrf2 pathway after cerebral ischemia in mice." Brain research 1497 (2013): 32-39.

20) Lu, Jun, et al. "Ursolic acid attenuates D-galactose-induced inflammatory response in mouse prefrontal cortex through inhibiting AGEs/RAGE/NF-κB pathway activation." Cerebral Cortex 20.11 (2010): 2540-2548.

21) Navina, Ray, Yulia G Lee, and Soo M Kim. "Molecular Biological Roles of Ursolic Acid in the Treatment of Human Diseases." Current Bioactive Compounds 13.3 (2017): 177-185.

22) Ou, Xiang, et al. "Ursolic acid inhibits leucine-stimulated mTORC1 signaling by suppressing mTOR localization to lysosome." PloS one 9.4 (2014): e95393.

23) Park, Hyun Jun, et al. "Ursolic acid inhibits pigmentation by increasing melanosomal autophagy in B16F1 cells." Biochemical and Biophysical Research Communications 531.2 (2020): 209-214.

24) Rai, Sachchida Nand, et al. "Ursolic acid attenuates oxidative stress in nigrostriatal tissue and improves neurobehavioral activity in MPTP-induced

*Parkinsonian mouse model." Journal of chemical neuroanatomy 71 (2016): 41-49.*

25) Ramírez-Rodríguez, Alejandra M., et al. "Effect of ursolic acid on metabolic syndrome, insulin sensitivity, and inflammation." Journal of medicinal food 20.9 (2017): 882-886.

26) Ramos, Alice A., Cristina Pereira-Wilson, and Andrew R. Collins. "Protective effects of ursolic acid and luteolin against oxidative DNA damage include enhancement of DNA repair in Caco-2 cells." Mutation Research/Fundamental and Molecular Mechanisms of Mutagenesis 692.1-2 (2010): 6-11.

27) Ramos-Hryb, Ana B., et al. "Therapeutic potential of ursolic acid to manage neurodegenerative and psychiatric diseases." CNS drugs 31.12 (2017): 1029-1041.

28) Rao, Ajmeera Rama, Ciddi Veeresham, and Kaleab Asres. "In vitro and in vivo inhibitory activities of four Indian medicinal plant extracts and their major components on rat aldose reductase and generation of advanced glycation endproducts." Phytotherapy Research 27.5 (2013): 753-760.

29) Wang, Hong, et al. "Potential protective effects of ursolic acid against gamma irradiationinduced damage are mediated through the modulation of diverse inflammatory mediators." Frontiers in pharmacology 8 (2017): 352.

30) Wang, Y. L., et al. "Effects of artesunate and ursolic acid on hyperlipidemia and its complications in rabbit." European Journal of Pharmaceutical Sciences 50.3-4 (2013): 366-371.

31) Wang, Yanwen, and Yonghan He. "Ursolic acid, a promising dietary bioactive compound of anti-obesity (1045.40)." The FASEB Journal 28.1 Supplement (2014): 1045-40.

32) Woźniak, Łukasz, Sylwia Skąpska, and Krystian Marszałek. "Ursolic acid—a pentacyclic triterpenoid with a wide spectrum of pharmacological activities." Molecules 20.11 (2015): 20614-20641.

33) Yin, Mei-Chin. "Inhibitory effects and actions of pentacyclic triterpenes upon glycation." BioMedicine 5.3 (2015).

34) Zhang, Chengyue, et al. "Pharmacokinetics and pharmacodynamics of the triterpenoid ursolic acid in regulating the antioxidant, anti-inflammatory, and epigenetic gene responses in rat leukocytes." Molecular pharmaceutics 14.11 (2017): 3709-3717.

# CHAPTER 30 - VITAMIN C

1) Arvinte, Cristian, Maharaj Singh, and Paul E. Marik. "Serum levels of vitamin C and vitamin D in a cohort of critically ill COVID-19 patients of a north American community hospital intensive care unit in May 2020: A pilot study." Medicine in drug discovery 8 (2020): 100064.

2) Blaschke, Kathryn, et al. "Vitamin C induces Tet-dependent DNA demethylation and a blastocyst-like state in ES cells." Nature 500.7461 (2013): 222-226.

3) Carr, Anitra C., and Silvia Maggini. "Vitamin C and immune function." Nutrients 9.11 (2017): 1211.

4) Chen, Yuanyuan, et al. "Vitamin C mitigates oxidative stress and tumor necrosis factor-alpha in severe community-acquired pneumonia and LPS-induced macrophages." Mediators of inflammation 2014 (2014).

5) D'Aniello, Cristina, et al. "Vitamin C in stem cell biology: impact on extracellular matrix homeostasis and epigenetics." Stem cells international 2017 (2017).

6) Granger, Matthew, and Peter Eck. "Dietary vitamin C in human health." Advances in food and nutrition research 83 (2018): 281-310.

7) Hemilä, Harri. "Vitamin C and infections." Nutrients 9.4 (2017): 339.

8) Jafari, Davood, et al. "Vitamin C and the immune system." Nutrition and Immunity. Springer, Cham, 2019. 81-102.

9) Juan I., Stephan Züchner, and Gaofeng Wang. "Regulation of the epigenome by vitamin C." Annual review of nutrition 35 (2015): 545-564.

10) Kaźmierczak-Barańska, Julia, et al. "Two Faces of Vitamin C—Antioxidative and Pro- Oxidative Agent." Nutrients 12.5 (2020): 1501.

11) Padayatty, Sebastian J., and Mark Levine. "Vitamin C: the known and the unknown and Goldilocks." Oral diseases 22.6 (2016): 463-493.

12) Pallauf, K., et al. "Vitamin C and lifespan in model organisms." Food and Chemical Toxicology 58 (2013): 255-263.

13) Pullar, Juliet M., Anitra C. Carr, and Margreet Vissers. "The roles of vitamin C in skin health." Nutrients 9.8 (2017): 866.

*14) Monacelli, Fiammetta, et al. "Vitamin C, aging and Alzheimer's disease." Nutrients 9.7 (2017): 670.*

*15) Mousavi, Soraya, Stefan Bereswill, and Markus M. Heimesaat. "Immunomodulatory and antimicrobial effects of vitamin C." European Journal of Microbiology and Immunology 9.3 (2019): 73-79.*

*16) Wei, Fulan, et al. "Vitamin C treatment promotes mesenchymal stem cell sheet formation and tissue regeneration by elevating telomerase activity." Journal of cellular physiology 227.9 (2012): 3216-3224.*

*17) Young, Juan I., Stephan Züchner, and Gaofeng Wang. "Regulation of the epigenome by vitamin C." Annual review of nutrition 35 (2015): 545-564.*

# ACCESSORY INFORMATION

## TABLE ONE: KAUFMANN RATINGS

| AGENT | TENET 1 | TENET 2 | TENET 3 | TENET 4 | TENET 5 | TENET 6 | TENET 7 | |
|---|---|---|---|---|---|---|---|---|
| ALPHA-KETOGLUTARATE | 3 | 2 | 1 | 1 | 2 | 2 | 0 | 11 |
| ALOE VERA | 1 | 3 | 0 | 0 | 3 | 2 | 2 | 11 |
| ANDROGRAPHOLIDE | 0 | 3 | 1 | 1 | 3 | 1 | 2 | 11 |
| BERBERINE | 0 | 2 | 3 | 0 | 3 | 0 | 3 | 11 |
| BLACK SEED OIL | 1 | 2 | 2 | 0 | 3 | 0 | 3 | 11 |
| CENTELLA ASIATICA | 1 | 2 | 2 | 0 | 2 | 2 | 2 | 11 |
| CHLOROGENIC ACID | 1 | 2 | 2 | 0 | 2 | 0 | 3 | 10 |
| CISTANCHE DESERTICOLA | 2 | 2 | 0 | 0 | 2 | 2 | 2 | 10 |
| COLLAGEN | 0 | 2 | 0 | 0 | 2 | 3 | 1 | 8 |
| COENZYME Q10 | 0 | 3 | 0 | 0 | 3 | 0 | 0 | 6 |
| DELPHINIDIN | 2 | 3 | 2 | 1 | 2 | 1 | 2 | 13 |
| ECKLONIA CAVA | 0 | 2 | 1 | 0 | 2 | 2 | 3 | 10 |
| ELLAGIC ACID | 1 | 2 | 2 | 2 | 2 | 1 | 2 | 12 |
| FISETIN | 3 | 2 | 2 | 2 | 3 | 3 | 2 | 17 |
| GANODERMA LUCIDUM | 0 | 3 | 1 | 1 | 3 | 0 | 3 | 11 |
| HYALURONIC ACID | 0 | 2 | 0 | 0 | 2 | 0 | 0 | 4 |
| KAEMPFERIA PARVIFLORA | 0 | 3 | 2 | 0 | 2 | 1 | 2 | 10 |
| LACTOFERRIN | 1 | 3 | 1 | 2 | 3 | 2 | 2 | 14 |
| LEUCINE | 0 | 2 | 3 | 0 | 2 | 0 | 1 | 8 |
| MAGNESIUM THREONATE | 2 | 2 | 0 | 0 | 3 | 3 | 2 | 12 |
| NARINGENIN | 0 | 3 | 2 | 1 | 2 | 3 | 2 | 13 |
| POLYPODIUM LEUCOTOMOS | 0 | 2 | 0 | 3 | 3 | 0 | 0 | 8 |
| PYRROLOQUINOLINE QUINONE | 0 | 3 | 2 | 0 | 2 | 0 | 0 | 7 |
| SALIDROSIDE | 0 | 3 | 2 | 2 | 2 | 2 | 2 | 13 |
| SHILAJIT | 0 | 2 | 0 | 0 | 2 | 1 | 0 | 5 |
| SPERMIDINE | 3 | 2 | 2 | 3 | 2 | 1 | 2 | 15 |
| URSOLIC ACID | 1 | 2 | 2 | 2 | 2 | 0 | 2 | 11 |
| VITAMIN C | 2 | 3 | 0 | 1 | 3 | 2 | 1 | 12 |

## TABLE TWO: BENEFITS

| AGENT | ▲ | Bone | Brain | Immune | Body | Hair | Heart | Skin | Joint | Lungs | Sexual | Strength | Cellular |
|---|---|---|---|---|---|---|---|---|---|---|---|---|---|
| ALPHA-KETOGLUTARATE | ● | | ● | ● | | | ● | | | | | ● | |
| ALOE VERA | ● | | ● | | | | | | | | | | ● |
| ANDROGRAPHOLIDE | ● | ● | | ● | | | | | | | | | |
| BERBERINE | | | ● | ● | ● | | ● | | | | | | |
| BLACK SEED OIL | | | ● | ● | | | ● | | | | ● | | ● |
| CENTELLA ASIATICA | | | ● | | | ● | | | | | | | ● |
| CHLOROGENIC ACID | | | ● | | | ● | | | | | | | |
| CISTANCHE DESERTICOLA | ● | ● | | | ● | | ● | | | ● | | | |
| COLLAGEN | | | | | | | ● | | ● | | | | ● |
| COENZYMEQ10 | | | | | | | ● | | | | | | |
| DELPHINIDIN | ● | | | | | | | | | | | | ● |
| ECKLONIA CAVA | | | ● | | | ● | | | | | | | ● |
| ELLAGIC ACID | | | | ● | | ● | | | | | | | |
| FISETIN | | ● | ● | ● | | | | | | | ● | | |
| GANODERMA LUCIDUM | ● | | ● | ● | ● | | | ● | | ● | | | |
| HYALURONIC ACID | | | | | | | | | ● | | | | ● |
| KAEMPFERIA PARVIFLORA | | | ● | | | | | | | | ● | ● | ● |
| LACTOFERRIN | | ● | ● | | | | ● | | | | | | |
| LEUCINE | | | ● | | | | ● | | | | | ● | |
| MAGNESIUM THREONATE | | ● | ● | | | | ● | | | | | | |
| NARINGENIN | | ● | ● | ● | | | | | | | | | |
| POLYPODIUM LEUCOTOMOS | | | | | | | | ● | | | | | ● |
| PYRROLOQUINOLINE QUINONE | | ● | ● | | | | | | | | | | ● |
| SALIDROSIDE | ● | ● | ● | | | | | | | | | | |
| SHILAJIT | ● | | ● | | | | | | | | ● | | |
| SPERMIDINE | | | ● | ● | | | ● | | | | | | |
| URSOLIC ACID | | | ● | ● | | | | | | | | ● | |
| VITAMIN C | | | ● | | | | ● | | | | | | ● |

## DOSAGE

| AGENT | DOSE |
|---|---|
| ALPHA KETOGLUTARATE | 500-1,000 MG/DAY |
| ALOE VERA | 300-600 MG/DAY |
| ANDROGRAPHOLIDE | 400-800 MG/DAY |
| BERBERINE | 500-1,000 MG/DAY |
| BLACK SEED OIL | 1-2 GRAMS/DAY |
| CENTELLA ASIATICA | 600 MG/DAY |
| CHLOROGENIC ACID | 500-1,000 MG/DAY |
| CISTANCHE DESERTICOLA | 500-1,000 MG/DAY |
| COLLAGEN | 2.5-10 GRAMS/DAY |
| COENZYME Q10 | 100-400 MG/DAY |
| DELPHINIDIN | 200-300 MG/DAY |
| ECKLONIA CAVA | 300-360 MG/DAY |
| ELLAGIC ACID | 100-200 MG/DAY |
| FISETIN | 1,200-1,400 MG ON TWO CONSECUTIVE DAYS, ONCE A MONTH. ON THE OTHER DAYS, 100 MG ONLY |
| GANODERMA LUCIDUM | 2-5 GRAMS/DAY |
| HYALURONIC ACID | 200 MG/DAY |
| KAEMPFERIA PARVIFLORA | 200 MG/DAY |
| LACTOFERRIN | 250 MG/DAY |
| LEUCINE | 1-1.5 GRAMS/DAY |
| MAGNESIUM THREONATE | 500 MG/DAY |
| NARINGENIN | 300-600 MG/DAY |
| POLYPODIUM LEUCOTOMOS | 400-500 MG/DAY |
| PYRROLOQUINOLINE QUINONE | 20-60 MG/DAY |
| SALIDROSIDE | 200-400 MG/DAY |
| SHILAJIT | 200-300 MG/DAY |
| SPERMIDINE | 1-2 MG/DAY |
| URSOLIC ACID | 150 MG/DAY |
| VITAMIN C | 200 MG/DAY (LONGEVITY) - 5 GRAMS/DAY (TREATING COLDS) |